Málek · Nečas · Rokyta (Eds.)
Advances in Mathematical Fluid Mechanics

Springer
Berlin
Heidelberg
New York
Barcelona
Hong Kong
London
Milan
Paris
Singapore
Tokyo

Josef Málek   Jindřich Nečas
Mirko Rokyta (Eds.)

Advances in Mathematical Fluid Mechanics

Lecture Notes of the Sixth International School
Mathematical Theory in Fluid Mechanics,
Paseky, Czech Republic,
Sept. 19 – 26, 1999

 Springer

Editors

Josef Málek
Mathematical Institute
Charles University
Sokolovská 83
18675 Prague 8
Czech Republic
e-mail: malek@karlin.mff. cuni.cz

Mirko Rokyta
Mathematical Institute
Charles University
Sokolovská 83
18675 Prague 8
Czech Republic
e-mail: rokyta@karlin.mff. cuni.cz

Jindřich Nečas
Mathematical Institute
Charles University
Sokolovská 83
18675 Prague 8
Czech Republic
e-mail: necas@karlin.mff. cuni.cz

and

Department of Mathematical Sciences
Northern Illinois University
DeKalb, IL 60115-2888, USA
e-mail: necas@math.niu.edu

Cataloging-in-Publication Data applied for

Die Deutsche Bibliothek – CIP-Einheitsaufnahme

Advances in mathematical fluid mechanics : lecture notes of the
Sixth International School "Mathematical Theory in Fluid Mechanics",
Paseky, Czech Republic, Sept. 19 - 26, 1999 / Josef Málek ... (ed.). –
Berlin ; Heidelberg ; New York ; Barcelona ; Hong Kong ; London ;
Milan ; Paris ; Singapore ; Tokyo : Springer, 2000
ISBN 3-540-67786-0

Mathematics Subject Classification (2000): 35Qxx, 76Bxx, 76Dxx, 76Nxx, 65Nxx

ISBN 3-540-67786-0 Springer-Verlag Berlin Heidelberg New York

Springer-Verlag Berlin Heidelberg New York
a member of BertelsmannSpringer Science+Business Media GmbH

© Springer-Verlag Berlin Heidelberg 2000
Printed in Germany

Typesetting by the authors using a Springer T$_E$X macro package
Cover design by *design & production* GmbH, Heidelberg

SPIN 10773964   46/3142LK – 5 4 3 2 1 0 – Printed on acid-free paper

Preface

This book consists of six survey contributions that are focused on several open problems of theoretical fluid mechanics both for incompressible and compressible fluids.

The first article "Viscous flows in Besov spaces" by *Marco Cannone* addresses the problem of global existence of a uniquely defined solution to the three-dimensional Navier-Stokes equations for incompressible fluids. Among others the following topics are intensively treated in this contribution: (i) the systematic description of the spaces of initial conditions for which there exists a unique local (in time) solution or a unique global solution for small data, (ii) the existence of forward self-similar solutions, (iii) the relation of these results to Leray's weak solutions and backward self-similar solutions, (iv) the extension of the results to further nonlinear evolutionary problems. Particular attention is paid to the critical spaces that are invariant under the self-similar transform. For sufficiently small Reynolds numbers, the conditional stability in the sense of Lyapunov is also studied. The article is endowed by interesting personal and historical comments and an exhaustive bibliography that gives the reader a complete picture about available literature.

The papers "The dynamical system approach to the Navier-Stokes equations for compressible fluids" by *Eduard Feireisl*, and "Asymptotic problems and compressible-incompressible limits" by *Nader Masmoudi* are devoted to the global (in time) properties of solutions to the Navier-Stokes equations for compressible fluids. The global (in time) analysis of two and three-dimensional motions of compressible fluids were left open for many years. Only recently, the global existence of solutions has been proved by P.-L. Lions in case of isentropic and barotropic processes. These results are discussed in both papers.

Eduard Feireisl deals with large time behavior of solutions of such systems. He, together with Hana Petzeltová, is the first one who have obtained the nontrivial results in more space dimensions. The existence of an absorbing set and the existence of a global attractor (in the sense of short trajectories or in the classical sense with additional regularity) in two and three dimensions are presented. The difficulties that have to be overcome stem from the missing uniqueness and regularity properties. The reader can also benefit from finding here an alternative proof of the compactness for the effective viscous flux.

Nader Masmoudi reflects in the main part of his article a joint work with P.-L. Lions on the limits from compressible equations of motions to incompressible models. Both the Navier-Stokes and Euler equations are treated. For reader's convenience this task is presented on three levels. First, the author gives a formal proof, which is then followed by a proof via global group method, which is convenient for studying the Cauchy and space-periodic problem. Finally, he presents a proof treating the problem locally, which enables to cover also the case of Dirichlet boundary conditions. In the final part of the article, the passage from the compressible Navier-Stokes equations with the pressure-density constitutive law of the form $p = \rho^\gamma$ to the incompressible models for $\gamma \to \infty$ is presented. Another fundamental open problem, the passage from the Navier-Stokes equations with Dirichlet boundary conditions to the Euler equations, is also touched. It is shown that the problem can be solved if two different types of viscosities (horizontal and vertical) are introduced and vertical viscosity vanishes quicker than the horizontal one.

The question of fluid-mechanics limit is also the central theme of the article "On the mathematical theory of fluid dynamic limits to conservation laws" by *Athanasios Tzavaras*. The open problem behind this report is the rigorous justification of the fluid-limit from the Boltzmann equation to macroscopic models for compressible fluids. To understand this task various kinetic (or discrete kinetic) models equipped with one or two conservation laws and their fluid limits have been recently studied. A survey of the current state of the mathematical theory of fluid-dynamic limits to entropy solutions of scalar hyperbolic conservation law is first presented. This is followed by an outline of recent results on fluid-dynamic limits for discrete kinetic models (of relaxation type) leading to systems of two conservation laws. The link between the theory of conservation laws and the theory of transport equations via the kinetic formulation is also underlined.

The contribution "Adaptive wavelet solvers for the unsteady incompressible Navier-Stokes equations" by *Michael Griebel* and *Frank Koster* is aimed on the development of a new method for finding numerically the accurate solution of the evolutionary Navier-Stokes equations for incompressible fluids in problems with (at least) moderate Reynolds number. As presented numerical tests show, this goal can be successfully achieved by wavelet solvers that allow: (i) higher accuracy due to the presence of polynomials of a certain degree in the space generated by the wavelet basis, (ii) the adaptivity both in space and time, and, (iii) the efficient preconditioning of the large linear systems that have to be solved. The equations are discretized by a Petrov-Galerkin scheme. To overcome algorithmical difficulties coming from the complicated wavelet structure the authors developed a new approach based on tensor products of interpolet wavelets of arbitrary levels of refinement generalizing multi-resolution analysis. This approach brings several advantages explained inside the paper in detail. A time step method, the projection method with a consistent discretization of the elliptic equation for the pressure, and an upwind-downwind discretization of the pressure gradient and divergence operator then allow to keep obtained advantages and make this wavelet solver very efficient and robust.

Finally, the article "Weighted spaces with detachted asymptotics in application to the Navier-Stokes equations" by *Sergueï Nazarov* contributes significantly to the problem of the existence of solution of the (incompressible) Navier-Stokes equations in exterior and aperture domains. The analysis of PDEs in unbounded domains is performed essentially in two directions that differ in the functional setting which is employed. On the one hand it is L^q-theory and on the other hand *weighted* or *Kondrati'ev theory*. The basis for treating the nonlinear equations in both approaches is a detailed analysis of the linear part, known as Stokes system and a contraction principle. Although the weighted theory is a very powerful and general tool for treating linear problems, any standard fixed point argument fails to treat the nonlinear Navier-Stokes equations, since the nonlinearity is not a continuous mapping between the corresponding spaces. This is the reason why it has been common believe for a long time that one cannot deduce the corresponding nonlinear theory in the weighted setting. However, a new approach of Sergueï Nazarov based on using *detached asymptotics* to treat the Stokes system, enables to construct the nonlinear theory also in weighted spaces. It turns out that also the *transport equation* can be considered in this frame which is very useful in the solvability theory of other, even non-Newtonian incompressible problems. Here, the approach is applied successfully to the Navier-Stokes equations of steady flows in $3d$ exterior and $2d$ aperture domain, and to the transport equation.

All articles present significant results and provide a better understanding of the problems in areas that enjoy a long-lasting attention of analysts in the area of fluid mechanics PDEs. Although the papers have the character of detailed summaries, their central parts contain the newest results achieved by the authors. This is another feature common to all papers.

All contributions also reflect series of lectures delivered by Marco Cannone, Eduard Feireisl, Michael Griebel, Nader Masmoudi, Sergueï Nazarov and Athanasios Tzavaras at the Sixth International School "Mathematical Theory in Fluid Mechanics", held September 19–26, 1999, at Paseky, Czech Republic. Those who are interested in knowing more on this school can look at

http://www.karlin.mff.cuni.cz/paseky-fluid/

where information on forthcoming Seventh Fluid-Mechanics Paseky School, June 3–10, 2001, can also be found.

We would like to use this opportunity and thank all speakers for their excellent lecturing and for preparing the articles of this volume. At the same time we thank all participants for their interest, stimulating questions and discussions during the course. We also thank Michal Netušil, Milan Pokorný and Gudrun Thäter for helping us in the process of preparation of this volume.

Without enthusiasm of Martin Peters from Springer-Verlag who agreed to realize our project, and unselfish help of Michael Růžička who made the contact possible, this book would hardly occur so soon.

May 2000   Josef Málek
                                                                     Jindřich Nečas
                                                                     Mirko Rokyta

Table of Contents

Viscous Flows in Besov Spaces

Marco Cannone

UMR 7599 du CNRS, UFR Mathématiques, Université Denis Diderot Paris VII,
2 Place Jussieu, 75251 Paris Cedex 05, France
e-mail: cannone@math.jussieu.fr

A tribute to the memory of Tosio Kato

Prologue

The Navier-Stokes equations did not yet exist when J. Fourier gave the explicit solution of the heat equation

$$\begin{cases} \dfrac{\partial u}{\partial t} - \Delta u = f \\ u(0) = u_0 \, . \end{cases} \tag{1}$$

This equation, governing the evolution of temperature $u(x, t)$, in the presence of an exterior source of heat $f(x, t)$, at a point x and time t of a body assumed here to fill the whole space $\mathbb{R}^3$, becomes, when we consider its partial Fourier transform with respect to x, an ordinary differential equation in t, whose solution is given by

$$u(t, x) = S(t)u_0 + \int_0^t S(t - s)f(s)\,ds, \tag{2}$$

$S(t)$ being the convolution operator defined by

$$S(t) = \exp(t\Delta) = \left(\frac{1}{4\pi t}\right)^{\frac{3}{2}} \exp\left(-\frac{|x|^2}{4t}\right) . \tag{3}$$

The Navier-Stokes equations, that describe the motion of a viscous fluid, were introduced in 1822 [109], the same year that, by a curious coincidence, J. Fourier published the treatise "Théorie Analytique de la Chaleur" [54], in which he developed in a systematic way the ideas contained in a famous paper of 1807.

If we follow Fourier's method to solve the Navier-Stokes equations for a viscous incompressible fluid

$$\begin{cases} \dfrac{\partial v}{\partial t} - \nu\Delta v = -(v \cdot \nabla)v - \nabla p + \phi \\ \nabla \cdot v = 0 \\ v(0) = v_0 \end{cases} \tag{4}$$

(here the velocity v and the pressure p are the unknowns of the problem, whereas the viscosity ν and the external force $\phi(t, x)$ are, with the velocity v_0, initially known), we obtain an integral equation satisfied by the velocity field v, say

$$v(t) = S(t)v_0 + B(v, v)(t) + \int_0^t S(t - s)\mathbb{P}\phi(s)\, ds, \tag{5}$$

where

$$B(v, u)(t) = -\int_0^t S(t - s)\mathbb{P}\nabla \cdot (v \otimes u)(s)\, ds, \tag{6}$$

$\mathbb{P}$ is the Leray-Hopf projector onto the divergence free vector field and, without loss of generality $\nu = 1$.

The precise meaning of the integral defined by (6), in different function spaces, is one of the main problems arising from this approach. Let us recall that for a function $u(t, \cdot)$ that takes values in a Banach space E, the integral $\int_0^T u(t, \cdot)\, dt$ exists either because $\int_0^T \|u(t, \cdot)\|_E\, dt < \infty$ (in this case we say that the integral is defined in the sense of Bochner) or because $\int_0^T |\langle u(t, \cdot), y\rangle|\, dt$ converges for any vector y of the dual (or predual) E' of E (the integral is said to be weakly convergent). The weak convergence is ensured by the oscillatory behaviour of $u(t, \cdot)$ in the Banach space E.

Now, the oscillatory property of the bilinear term arising from the Navier-Stokes equations is systematically taken into account in all papers that are based on the energy inequality, in particular $\langle B(v, v), v\rangle = 0$ as long as $\nabla \cdot v = 0$. In the following pages, we will never take advantage of this remarkable property, for we will only consider functional spaces where it is not possible to write $\langle B(v, v), v\rangle$. In fact, $B(v, v)$ will never belong to a space that is a dual of the one to which v belongs. This is the reason why our works ([29,30] excepted) are not based on the innermost structure of the Navier-Stokes equations and will be easily extended to other nonlinear parabolic equations (Section 6).

Following the pioneering papers of T. Kato and his collaborators, we got used to calling *mild solutions* those solutions whose existence is obtained by a fixed point algorithm applied to the integral formulation (5). Of course, we do not expect to use the energy inequality, but we hope to ensure in such a way the uniqueness of the solution. This is in contrast with Leray's construction of *weak solutions*, relying on compactness arguments and *a priori* energy estimates. Moreover, the fixed point algorithm is stable and constructive. Thus the problem of defining mild solutions is closely akin to the question of knowing whether the Cauchy problem for Navier-Stokes equations is well-posed in the sense of Hadamard. This question will be discussed in Section 4 in connection with the theory of stability and Lyapunov functions. Without trying to define the concept of mild solutions by these remarks, we will give in each particular case a definition having a precise mathematical meaning.

Through lack of a physical interpretation of the functional spaces that they involve – global solutions in $L^3(\mathbb{R}^3)$ instead of $L^2(\mathbb{R}^3)$; in $\dot{H}^{\frac{1}{2}}(\mathbb{R}^3)$ instead of $\dot{H}^1(\mathbb{R}^3)$, but we will come back to this particular case in the following pages – the notion of mild solutions appeared only in the beginning of the 1960s (T. Kato and H. Fujita, [73,55]), whereas the existence of weak solutions was discovered much earlier (J. Leray, [94,95]).

Whatever solution one obtains, the main difficulty in the study of the Navier-Stokes equations is related to the nonlinear kinematic term $-(v \cdot \nabla)v$, that does not appear in the aforementioned heat equation. To overcome this difficulty, we have suggested since 1994, following J.-M. Bony's example [17], the introduction of new techniques, directly issued from Fourier analysis, but better suited to nonlinear problems: the Littlewood-Paley decomposition and paraproducts from one side, wavelets from the other. Our purpose was to compare these two approaches in the case of the Navier-Stokes equations.

Now, when we think of the Littlewood-Paley analysis and the paraproduct or wavelet decompositions, we also think of the spaces introduced by the Russian mathematician Oleg Vladimirovitch Besov [12]. These spaces, that without hesitating we may qualify as "diabolical", contain several indices, $(\dot{B}_q^{s,p}(\mathbb{R}^3))$, making them appalling to the eyes of any neophyte. But I write "diabolical" because in Russian, a language with declensions, their name "bes-ov" represents the genitive of "bes", whose translation in English is the word "devil"! In the following pages, the Besov spaces, like the thermodynamic demon imagined by J. C. Maxwell, will appear, disappear and finally come up again – while swinging and oscillating – in a mischievous way, definitely worthy of their name.

Does all this mean that the Littlewood-Paley analysis and the Besov spaces play a crucial role for the Navier-Stokes equations, analogous to the one played by the Fourier transform and the Sobolev spaces in the resolution of the heat equation? Such hopes motivated our program of 1995, "Wavelets, paraproducts and Navier-Stokes" [22].

The interest shown in our work, expressed by the multitude and the importance of the results that have followed ever since, must now be examined in a historical perspective.

Benefiting from this supplement of intelligibility provided by time, we reach, five years later, the following conclusion: although the Besov spaces seemed essential in the proof of some fundamental results (the existence of self-similar solutions, the uniqueness of the solutions in $L^3(\mathbb{R}^3)$), their role turns out to be fully anecdotal, so that these spaces can now be replaced by less complicated ones (Le Jan-Sznitman space, weak Lebesgue space).

But there are other places where Besov spaces appear in an unpredictable and somehow diabolical way. We already pointed out in [22] that, for the existence of Kato's solution in the space $\mathcal{C}([0, \infty); L^3(\mathbb{R}^3))$, the smallness condition $\|v_0\|_3 < \delta$ could be replaced by the smallness of its norm in the space $\dot{B}_q^{-(1-\frac{3}{q}),\infty}(\mathbb{R}^3)$ (for a certain $3 < q < \infty$), and this by means of the inequality

$$\sup_{t>0} t^{\frac{1}{2}(1-\frac{3}{q})} \|S(t)v_0\|_q \leq C\|v_0\|_3 \,. \tag{7}$$

Moreover, we remarked that the norm of v_0 in a Besov space of negative index is small provided v_0 is sufficiently oscillating.

Recently, R. Temam [135] informed us that this result was implicitly contained in the pioneering papers of T. Kato and H. Fujita of 1962 [73,55]. In fact, as we are going to see in Section 4, it is possible, by a simple "bootstrap" argument, to obtain a global mild solution in the space $\mathcal{C}([0,\infty); \dot{H}^1(\mathbb{R}^3))$ as long as the initial data $v_0 \in \dot{H}^1(\mathbb{R}^3)$ has a sufficiently small norm in the space $\dot{H}^{\frac{1}{2}}(\mathbb{R}^3)$. Now, if $v_0 \in \mathcal{S}'(\mathbb{R}^3)$ is such that $\hat{v}_0(\xi) = 0$ if $|\xi| \leq K$, then

$$\|v_0\|_{\dot{H}^{\frac{1}{2}}} \leq K^{-\frac{1}{2}} \|v_0\|_{\dot{H}^1} \tag{8}$$

and thus, as suggested by R. Temam, we get the existence of a global mild solution in $\mathcal{C}([0,\infty); \dot{H}^1(\mathbb{R}^3))$ provided that the initial data is concentrated at high frequencies ($K \gg 1$), say highly oscillating.

In his Ph.D. thesis [116,117] F. Planchon gave the precise interpretation of this fact, by replacing in the theorem of Kato and Fujita the smallness of the $\dot{H}^{\frac{1}{2}}(\mathbb{R}^3)$ norm of the initial data, with its smallness (or oscillation) in a Besov space. Everything takes place as in [22] for $L^3(\mathbb{R}^3)$: there exists an *absolute* constant $\beta > 0$ such that if $\|v_0\|_{\dot{B}_4^{-\frac{1}{4}},\infty} < \beta$ and $v_0 \in \dot{H}^1(\mathbb{R}^3)$, then there exists a global solution in $\mathcal{C}([0,\infty); \dot{H}^1(\mathbb{R}^3))$. The importance of such a result is that it allows to obtain *global* and *regular* solutions in an energy space, under the only hypothesis of oscillation of the initial data. In other words, at variance with the $L^3(\mathbb{R}^3)$ setting, we can establish a link between Leray's weak solutions and Kato's mild ones.

Are there other surprises in store arising from Besov spaces?

1   A Refractory Question

In 1994 Jean Leray summarized the state of the art for the Navier-Stokes equations in the following way [96]: "*A fluid flow initially regular remains so over a certain interval of time; then it goes on indefinitely; but does it remain regular and well-determined? We ignore the answer to this double question. It was addressed sixty years ago in an extremely particular case [95]. At that time H. Lebesgue, questioned, declared: "Don't spend too much time for such a refractory question. Do something different!*" "

At the beginning of 2000, the problem of the uniqueness and the possible blow up in finite time of Leray's weak solutions is still open. "*This is perhaps the most celebrated problem in partial differential equations. [...]. The solution of this problem might well be a fundamental step toward the very big problem of understanding turbulence"*, writes S. Smale, who, inspired by Hilbert's well-known list of 1900, includes the possible blow up of Leray's weak solution as one of the 18 open problems for the next century [133]. Even if, as pointed out by P. Constantin [45] there is *"no evidence that this problem has physical significance; blow up of solutions of the Navier-Stokes requires infinite momentum, that is breakdown of the model. Rather, the physically important problem is posed by Kolmogorov theory."*

The situation is quite different for T. Kato's mild (regular) solutions for which a general uniqueness theorem, that is the subject of this section, has been available since 1997. But the existence and regularity of these global solutions can be assured only if the initial data is sufficiently small, say highly oscillating, and it seems difficult to obtain a better result in this direction (see Section 4).

Since the introduction of the mild formulation of the Navier-Stokes equations by T. Kato and H. Fujita, at the beginning of the 1960s [73,55], the uniqueness of the corresponding solution seemed to be guaranteed only under some strong regularity hypotheses near $t = 0$. In the simplest case, when the solutions belong to $\mathcal{C}([0,T]; L^3(\mathbb{R}^3))$, these additional conditions are written [70] $\lim_{t \to 0} t^{\frac{\alpha}{2}} \|v(t)\|_q = 0$, $\alpha = 1 - \frac{3}{q}$, $3 < q < \infty$, or [60], for the same values of α and q, $v \in L^{\frac{2}{\alpha}}((0,T), L^q)$. In fact, by using one of these two auxiliary norms, it is possible to apply the Picard fixed point algorithm to obtain the existence of mild solutions in $\mathcal{C}([0,T]; L^3(\mathbb{R}^3))$.

H. Brezis is credited with the first uniqueness theorem [18] which avoids imposing one of the two previous regularity conditions. But a third much weaker condition will appear, so that the uniqueness in the whole natural space $\mathcal{C}([0,T]; L^3(\mathbb{R}^3))$ is not yet obtained.

We will follow here the presentation of Brezis' theorem given by Y. Giga [61], showing that this uniqueness result applies to a wider class of parabolic equations. To this end, let X and Y be two Banach spaces and let $S(t)$ be a C_0 (nonlinear) semi-group acting on X. We denote by K a set contained in the intersection $Z = X \cap Y$. If $\alpha > 0$, we say that $S(t)$ is α–regularising if $\sup\{t^\alpha \|S(t)a\|_Y; \ a \in K\}$ tends to zero when t tends to zero. Here K is a pre-compact set in X. A solution v in $\mathcal{C}([0,T); X)$ of an integral equation often satisfies the following uniqueness condition:

$$\text{if } s \geq 0: \ \lim_{t \to 0} t^\alpha \|v(t+s)\|_Y = 0 \quad \text{implies} \quad v(t+s) = S(t)v(s)$$
$$\text{for some } \alpha > 0. \tag{9}$$

Then, Brezis' result reads as follows. Let v in $\mathcal{C}([0,T); X)$ and suppose that (9) is verified for some α (independent of s), $0 \leq s \leq 1$ and that $S(t)$ is regularising for the same value of α. If v is locally bounded and takes values in Y for $0 < t < T$, then $v(t) = S(t)v(0)$ for $0 \leq t < T$. In fact, as v is locally bounded with values in Y, (9) for $0 < s \le 1$ implies $v(t + s) = S(t)v(s)$, $0 < s \le 1$. Now, as $S(t)$ is regularising and $K = v((0, 1])$ is a precompact set of X, we get $\sup\{t^\alpha \|v(t+s)\|_Y; 0 < s \le 1\} \to 0$ when $t \to 0$. This implies the condition

$$\lim_{t \to 0} t^\alpha \|v(t)\|_Y = 0. \tag{10}$$

Finally, by using (9) for $s = 0$, we get $v(t) = S(t)v(0)$.

In [18] H. Brezis proves the uniqueness of the solution to the Navier-Stokes equations in two dimensions, having the vorticity as unknown, and choosing $X = L^1(\mathbb{R}^2)$, $Y = L^\infty(\mathbb{R}^2)$ and $S(t)$ the operator that gives the solution of the vorticity equation. By using the results of a paper by M. Ben-Artzi [11], Brezis applies the abstract theorem above, to prove the uniqueness of the solution v, continuous in time and taking values in $L^1(\mathbb{R}^2)$, of the vorticity equation, under the hypothesis that v is locally bounded and belongs to $L^\infty(\mathbb{R}^2)$ for $t > 0$. In other words, for the uniqueness we can get rid of (10), provided $v \in L^\infty_{loc}((0, T); Y)$.

An application of the previous method is presented in a paper by H. Brezis and T. Cazenave [19]. Here the authors consider the equation $v_t - \Delta v = |v|^{p-1}v$ on $(0, T) \times \Omega$ with a homogeneous boundary data of the Dirichlet type and initial value v_0. The domain $\Omega \subset \mathbb{R}^n$ is bounded and regular, and $p > 1$. It is well-known that if $v_0 \in L^\infty(\Omega)$, then the problem yields a unique solution defined on a maximal interval $[0, T_{max})$ and that this solution is a classic one on $(0, T_{max}) \times \overline{\Omega}$. H. Brezis and T. Cazenave prove an analogous result for $v_0 \in L^q(\Omega), 1 \le q < \infty$, provided either $q > n(p-1)/2$, or $q = n(p-1)/2 > 1$. The importance of their work is that here the uniqueness, at variance with the previous results, does not impose $\lim_{t \downarrow 0} t^\alpha \|v(t)\|_{L^\infty} = 0$ but only that $v \in L^\infty_{loc}((0, T); L^\infty(\Omega))$. When $p = q = n(p - 1)/2$, $\Omega = \mathbb{R}^n$ and $n \ge 3$, E. Terraneo [137] proves, three years later, the nonuniqueness in the whole space $\mathcal{C}([0, T); L^p(\mathbb{R}^n))$ of the solution of the same equation $v_t - \Delta v = |v|^{p-1}v$ studied by Brezis and Cazenave, while Y. Giga [60] had proved the uniqueness of the solution in a subspace of $\mathcal{C}([0, T); L^p(\mathbb{R}^n))$.

The application of Brezis' abstract method for the three dimensional Navier-Stokes equations is contained in the Ph.D. thesis of F. Planchon [116] where the following result is stated by way of a corollary: there exists a unique solution $v \in \mathcal{C}([0, T]; L^3(\mathbb{R}^3)) \cap L^\infty_{loc}((0, T); L^\infty(\mathbb{R}^3))$ of the Navier-Stokes equations.

Another way to ensure the uniqueness and to weaken the regularity condition at zero of $v \in L^{\frac{2}{\alpha}}((0, T), L^q(\mathbb{R}^3))$ was given by T. Kawanago [76]. More precisely, if $\alpha = 1 - \frac{3}{q}$, $3 < q < \infty$, then the mild solution is unique in the space $\mathcal{C}([0, T), L^3(\mathbb{R}^3)) \cap L^{\frac{2}{\alpha}}_{loc}((0, T), L^q(\mathbb{R}^3))$. The proof of this result is based on some estimates arising from the stability theory (Section 4) and does not make use of the techniques introduced by H. Brezis. But, like Brezis' result, Kawanago's one is not completely satisfactory, for the uniqueness in $\mathcal{C}([0, T]; L^3(\mathbb{R}^3))$ is not yet guaranteed.

At that moment, Tosio Kato (1917-1999), in a survey paper on Navier-Stokes and before presenting Brezis' and Kawanago's theorems, described the results obtained in [22] by quoting a saying (*kakugen*) in Japanese [72] "*If you are not able to prove the assertion, you should generalise the conjecture*". The "assertion" represents here the uniqueness of mild solutions in $L^3(\mathbb{R}^3)$, whereas the "generalisation of the conjecture" concerns the smallness condition on the initial data, measured in [22,23] by the quickness of its oscillations.

This was the situation in November 1996 when I was invited by M. Struwe at the ETH in Zurich to give a series of lectures on the Navier-Stokes equations. Among the audience were J.-Y. Chemin and W. von Wahl. Following a question addressed the first day by von Wahl, I was led to modify my program and to devote a lecture to the uniqueness problem for mild solutions, and in particular to the results of H. Brezis and T. Kawanago. Shortly after the meeting in Zurich, J.-Y. Chemin obtained a uniqueness result for Navier-Stokes [42], announced in the "Séminaire Equations aux Dérivées Partielles de l'Ecole Polytechnique" [41]. J.-Y. Chemin's theorem applies to the solutions of Navier-Stokes satisfying the energy inequality and is closely related to a previous work published in 1985 by W. von Wahl [143]. More exactly, if we denote by $\tilde{B}_p^{q,\infty}(\mathbb{R}^3)$ the closure of the compactly supported infinitely differentiable functions in the space of tempered distributions for the Besov norm $B_p^{q,\infty}(\mathbb{R}^3)$ and by $\dot{H}^s(\mathbb{R}^3)$ the homogeneous Sobolev space, then J.-Y. Chemin's uniqueness result is the following [42]: let $3 \leq p < \infty$ and $v_0 \in \tilde{B}_p^{\frac{3}{p}-1,\infty}(\mathbb{R}^3) \cap L^2(\mathbb{R}^3)$, there is at most one solution to Navier-Stokes in the space $\mathcal{C}([0,T); \tilde{B}_\infty^{-1,\infty}(\mathbb{R}^3)) \cap L^2([0,T); \dot{H}^1(\mathbb{R}^3))$. This theorem reaches the largest critical Besov space, say $\tilde{B}_\infty^{-1,\infty}(\mathbb{R}^3)$ (this will be dealt with in Section 7), even if the additional energy hypothesis, say $v \in L^2([0,T); \dot{H}^1(\mathbb{R}^3))$, does not arise directly in Kato's theory.

During the conference in Zurich I met Alain-Sol Sznitman, director of the Mathematical Center at the ETH, who was also interested in the use of Besov spaces and harmonical analysis. He informed me of a result he had obtained in collaboration with Y. Le Jan [88], showing the existence and the uniqueness of a solution to Navier-Stokes in an uncommon functional space, defined by means of the Fourier transform. First of all, he questioned me if, in the huge literature devoted to the subject, such a space had been used before, then asked whether it was possible to compare their space with the Besov ones. Le Jan and Sznitman's space is made up by the tempered distributions $f \in \mathcal{S}'(\mathbb{R}^3)$ such that $\sup_{\xi \in \mathbb{R}^3} |\xi|^2 |\hat{f}(\xi)| < \infty$. Now, if in the latter expression we consider $\int_{\xi \in \mathbb{R}^3}$ instead of $\sup_{\xi \in \mathbb{R}^3}$, we obtain a homogeneous Sobolev space. This is not the case: the functions whose Fourier transform is bounded define the pseudo-measure space $PM(\mathbb{R}^3)$ of J.-P. Kahane. In other words, a function f belongs to the space introduced by Le Jan and Sznitman if and only if $\Delta f \in PM(\mathbb{R}^3)$, Δ being here the Laplacian (in 3D). A simple calculation shows that the latter condition is written, in the dyadic decomposition Δ_j of Littlewood and Paley in the form $4^j\|\Delta_j f\|_{PM} \in \ell^\infty(\mathbb{Z})$, and defines in this way "the Besov space" $\dot{B}^{2,\infty}_{PM}(\mathbb{R}^3)$. Y. Le Jan and A.-S. Sznitman's viewpoint [88, page 825], [89, page 345]: *"These statements seem to be new [...]. Although not directly comparable, some of them share a common flavor with Cannone's existence and uniqueness results in Besov space, see [22]"* was totally justified.

The space of Le Jan and Sznitman was a new one and deserved a more attentive analysis. The main idea contained in the paper [89] was to study the nonlinear integral equation verified by the Fourier transform of the Laplacian of the velocity vector field associated with the "deterministic" equations of Navier-Stokes. This integral representation involves a Markovian kernel K_ξ, associated to the branching process, called stochastic cascades, in which each particle located at $\xi \neq 0$, after an exponential holding time of parameter $\nu|\xi|^2$, with equal probability either dies out or gives birth to two descendants, distributed according to K_ξ. By taking the inverse Fourier transform one can thus obtain a solution to the Navier-Stokes equations ... arising from a sequence of cascades!

The day when I thought to have understood, thanks to many fruitful discussions with G. Kerkyacharian, the mechanism of the stochastic cascades that are involved in the paper by Y. Le Jan and A.-S. Sznitman, I found out that their proof was much easier using the Picard fixed point theorem. More precisely, it is sufficient to remark that the convolution product of two functions f and g verifying $\sup_{\xi \in \mathbb{R}^3} |\xi|^2|\hat{f}(\xi)| < \infty$ and $\sup_{\xi \in \mathbb{R}^3} |\xi|^2|\hat{g}(\xi)| < \infty$ is such that $\sup_{\xi \in \mathbb{R}^3} |\xi||\hat{f} \star \hat{g}(\xi)| < \infty$, to be able to deduce the continuity of the bilinear term appearing in the Navier-Stokes equations and, *a fortiori*, the existence of the solution of Le Jan and Sznitman (let us note, in passing, that such an argument also applies to the space, $L^1_\xi(L^2_t)$). Moreover, the homogeneous functions of degree -1 being contained in such a space, the existence of self-similar solutions is also trivially guaranteed (Section 3).

Faced with the perseverance of Giulia Furioli and Elide Terraneo, who invited me to give a talk at the University of Evry, I accepted, proposing the following title: "Fluids in cascades". Even if Le Jan and Sznitman's theorem seemed to be a straightforward application of Kato's approach, the effort to understand the stochastic cascades deserved to be shared. The day after the talk, G. Furioli, P.-G. Lemarié and E. Terraneo obtained the first uniqueness result for mild solutions in the natural space $\mathcal{C}([0,T]; L^3(\mathbb{R}^3))$, without any additional hypotheses. Of course, it was a refractory question, left unsolved since the pioneering papers of T. Kato and H. Fujita at the beginning of the 1960s. What did not escape Lemarié's notice in the simplified proof of the theorem of Le Jan and Sznitman, was the presence of a Besov space with positive regularity index, say $\dot{B}^{2,\infty}_{PM}(\mathbb{R}^3)$, prompting him to introduce the space $\dot{B}^{\frac{1}{2},\infty}_2(\mathbb{R}^3)$ [58, page 1256] at variance with the Besov spaces $\dot{B}^{-1+\frac{3}{q},\infty}_q(\mathbb{R}^3)$ with $q \geq 3$ used in [22].

The first proof of uniqueness in $\mathcal{C}([0,T]; L^3(\mathbb{R}^3))$ (that was followed by at least five different other proofs [93]) was based on two well-known ideas. The first one is that it is more simple to study the bilinear operator $B(v,u)(t)$ in a Besov frame [22]; the second is that it is helpful to distinguish in the solution v the contribution from the tendency $\exp(t\Delta)v_0$ and from the fluctuation $B(v,v)(t)$, the latter function always being more regular than the former [22]. What is remarkable is that, contrary to what one would expect, the spaces $L^3(\mathbb{R}^3)$ and $\dot{B}_2^{\frac{1}{2},\infty}(\mathbb{R}^3)$ are not comparable. The fact that the Besov space of the positive regularity index played only a minor role in the paper [59] led naturally to the question whether one could do without it. Some months later Y. Meyer showed how to improve the uniqueness theorem of P.-G. Lemarié and his students. The distinction between the fluctuation and the tendency was not used, the time-frequency approach was unnecessary and the Besov spaces did not play a role. Y. Meyer's proof shortened the problem to the bicontinuity of the bilinear term $B(v,u)(t)$ in the weak Lebesgue space $L^{3,\infty}(\mathbb{R}^3)$ and more precisely in $\mathcal{C}([0,T); L^{3,\infty}(\mathbb{R}^3))$ [103]. This result by itself is even more surprising because F. Oru has otherwise proven that, in spite of all the cancellations that it contains, the bilinear term is not continuous in $\mathcal{C}([0,T]; L^3(\mathbb{R}^3))$ [114].

As we have already announced, a plethora of proofs and generalisations of the same uniqueness theorem followed. Here, we will limit ourselves to review some of them; a complete list is contained in [93].

Let us start with the results obtained by N. Depauw in his Ph.D. thesis, that contains a generalisation of the proof of uniqueness to the exterior domains, Ω_e of $\mathbb{R}^n$, or in other words, for a solution $v \in \mathcal{C}([0,T); L^n(\Omega_e))$, with $n \geq 4$ [47] as well as for $n = 3$ [48].

On the other hand, if Ω_b is a bounded domain, by means of the embedding $L^n(\Omega_b) \hookrightarrow L^2(\Omega_b)$ and W. von Wahl's uniqueness theorem, it is possible to prove that Leray's weak solutions coincide with Kato's mild ones, so that their uniqueness follows in a straightforward manner [59].

A completely different proof of the uniqueness in $\mathcal{C}([0,T); L^3(\Omega))$, Ω being the whole space $\mathbb{R}^3$, or a bounded regular subset of $\mathbb{R}^3$ or the periodic torus $\mathbb{T}^3$, was announced recently by P.-L. Lions and N. Masmoudi [97] and applies as well to $L^\infty([0,T); L^n(\mathbb{R}^n))$, for any $n \geq 4$. The main ingredients of the proof are: the introduction of a dual problem, an *ad hoc* decomposition of the solutions and a "bootstrap" argument.

Another proof of the uniqueness in $\mathcal{C}([0,T]; L^3(\mathbb{R}^3))$ was given by S. Monniaux [107] and makes use of the maximal L^p regularity of the Laplacian in $\mathbb{R}^3$. Later, H. Amman gave a systematic approach of the (existence and) uniqueness problem for mild solutions in [1,2].

Finally, P.-G. Lemarié's student R. May seems to dispose of a different idea [100] to solve the same refractory question!

2   Critical Spaces

By means of the Lebesgue space $L^3(\mathbb{R}^3)$ in the previous section, we have already remarked that the functional spaces for which the norm is invariant under the action of the transformation $\|f\| = \|\lambda f(\lambda x)\|$, $\lambda > 0$, seem to be well-suited to the study of the Navier-Stokes equations. In fact, they ensure the existence and uniqueness of local solutions (for arbitrary data) as well as of global ones (for small or highly oscillating data). In the following pages, any Banach space that is continuously embedded into S' and such that the norm is invariant under the scaling $f \longrightarrow \lambda f(\lambda x)$ is called a *critical space*.

Following the names of the founders, T. Kato and H. Fujita, we have also mentioned those of Y. Giga and T. Miyakawa, whose contribution to the study of mild solutions in critical spaces is very important (see Section 3). In passing, we would like to quote here one of their papers [63] that, in spite of its publication year, was written after F.B. Weissler's one of 1981 [148] and before T. Kato's famous article of 1984 [70]. In fact, F.B. Weissler gave the first existence result of mild solutions in the half-space $L^3(\mathbb{R}^3_+)$, then Y. Giga and T. Miyakawa generalized the proof to $L^3(\Omega_b)$, Ω_b an open bounded domain in $\mathbb{R}^3$. Finally T. Kato obtained, by means of a purely analytical proof (involving only Hölder and Young inequalities and without using any estimate of fractional powers of the Stokes operator) an existence theorem in the whole space $L^3(\mathbb{R}^3)$. Y. Giga and T. Miyakawa's paper, accepted by the editors in December 1982, was only published in 1985.

A systematic study of critical and super-critical spaces was given in [22], accompanied by the introduction of some *ad hoc* Besov spaces, useful in obtaining new existence results for Navier-Stokes and in particular for self-similar solutions (Section 3). Two years later F. Planchon followed the same formalism provided by the Besov spaces in his Ph.D. thesis [116], and subsequently we published a series of joint papers on Navier-Stokes [27–31]. Furthermore, some improvements of the results we obtained in the Besov spaces $\dot{B}_q^{-1+\frac{3}{q},\infty}(\mathbb{R}^3)$, $3 < q < \infty$, have been recently announced by Z.-M. Chen and Z. Xin, from Hong Kong University [44]. The most impressive result in this direction is certainly the one obtained by H. Koch and D. Tataru [77] and concerns the Triebel-Lizorkin space $\dot{F}_\infty^{-1,2}(\mathbb{R}^3)$, in other words $BMO^{-1}(\mathbb{R}^3)$. More exactly, if $3 < q < \infty$, we have the following embedding: $\dot{B}_q^{-1+\frac{3}{q},\infty}(\mathbb{R}^3) \hookrightarrow \dot{F}_\infty^{-1,2}(\mathbb{R}^3) \hookrightarrow \dot{B}_\infty^{-1,\infty}(\mathbb{R}^3)$. We will come back to the largest critical space $\dot{B}_\infty^{-1,\infty}(\mathbb{R}^3)$ in Section 7.

New mild solutions for the Navier-Stokes equations were obtained by H. Kozono, M. Yamazaki and Y. Taniuchi in Besov spaces [78–81]. In particular H. Kozono and M. Yamazaki introduced some critical "Besov spaces" built up on some Morrey-Campanato spaces to get existence, uniqueness and self-similar solutions in these new frames. Their construction is somehow reminiscent of the "Besov space" of Y. Le Jan and A.-S. Sznitman, built up on the pseudo-measures space of J.-P. Kahane, as we described in Section 1.

Later, H. Kozono and M. Yamazaki studied the Lorentz space $L^{3,\infty}(\Omega_e)$, Ω_e being an exterior domain of $\mathbb{R}^3$ [82,83]. We have already encountered this space when dealing with the uniqueness problem for mild solutions in Section 1, and we will see in Section 3 how to construct self-similar solutions in it. Rather, H. Kozono and M. Yamazaki are interested in the exterior problem for the stationary Navier-Stokes equations. More precisely, they consider the system $-\Delta w + w \cdot \nabla w + \nabla \pi = \operatorname{div} F$, $\operatorname{div} w = 0$ where the space variable belongs to the exterior domain $\Omega_e \subset \mathbb{R}^3$ with homogeneous boundary conditions and the condition at infinity $w(x) \to 0$ if $|x| \to \infty$. Here F is a given tensor, from which the external force $\operatorname{div} F$ arises, and that, in the literature devoted to the subject, is always considered to be rapidly decreasing at infinity. This is not the case in H. Kozono and M. Yamazaki's paper, who prove that if $F \in L^{3/2,\infty}(\Omega_e)$ and if $\|F\|_{3/2,\infty} < \delta$ for a sufficiently small δ, then the existence, the uniqueness and some regularity properties are guaranteed for the corresponding solutions to the problem.

As far as the Hardy spaces (even if not critical), they have been systematically treated by T. Miyakawa at Kobe University. The advantage in using these spaces is that one can obtain new results on the asymptotic behaviour for the solutions to the incompressible Navier-Stokes system [104–106].

During a visit to Kobe, T. Miyakawa drew my attention to some papers of Calixto P. Calderón [20,21], that surprisingly seemed to have been ignored in the "mild solutions community".

The first paper is devoted to the critical space $L^3(\mathbb{R}^3)$. We have already remarked that the existence of a mild solution in $v \in \mathcal{C}([0,T]; L^3(\mathbb{R}^3))$ is not easy to ensure, because the bilinear term is not continuous in this space [114]. The starting point of C.P. Calderón's work is to change the norm in order to restore the continuity, in the same way as Kato and Fujita for $H^{\frac{1}{2}}$ and F.B. Weissler for $L^3(\mathbb{R}^3)$ had suggested before. Instead of using the norms described in Section 1, Calderón's idea is simply to exchange the L^3 norm with the L^∞ one: in other words, he considers the space $L^3(\mathbb{R}^3; L^\infty[0,T]))$. Without knowing of Calderón's works, we proposed in [22] the same algorithm and showed in [29,30] how to take an advantage of this method in order to improve a result of J.-Y. Chemin [39] for the Navier-Stokes with an external force.

The aim of the second paper by C.P. Calderón is well-explained in its abstract: *"The existence of weak solutions to the Navier-Stokes equations for an infinite cylinder with initial data in L^p is considered in this paper. We study the case of initial data in $L^p(\mathbb{R}^n)$, $2 < p < n$, and $n = 3, 4$. An existence theorem is proven covering these important cases and, therefore, the 'gap' between the Hopf-Leray theory ($p = 2$) and that of Fabes-Jones-Rivière* *(p > n) is bridged. The existence theorem gives a new method of constructing global solutions. The cases p = n are treated at the end of the paper."*

We find an analogous attempt to unify Leray's theory $(p = 2)$, Kato's $(p = n)$ and Fabes-Jones-Rivière's ones $(p > n)$ in a recent work of P.-G. Lemarié [91,92] as well as in the papers of H. Amman [1,2].

Another point of view and a systematic study of critical spaces was given by Y. Meyer in [103]. Not only is T. Kato's algorithm applied to abstract functional spaces and the uniqueness result of G. Furioli, P.-G. Lemarié-Rieusset and E. Terraneo considerably simplified, but one can find in [103] further new results, as well as a discussion on the relevance of wavelets in the theory of turbulence.

Before concluding, let us mention a very precise analysis of fixed point methods for mild solutions to Navier-Stokes in critical spaces given by P. Auscher and P. Tchamitchian in [3] who listed an exhaustive series of examples and counter-examples of "good" and "bad" spaces for Navier-Stokes!

The critical spaces we have dealt with in the previous pages will help us in the following section entirely devoted to the construction of self-similar solutions.

3   Self-Similar Solutions

The viscous flows for which the profiles of the velocity field at different times are invariant under a scaling of variables are called self-similar.

More precisely, we are talking about solutions to the Navier-Stokes equations

$$\begin{cases} \dfrac{\partial v}{\partial t} - \nu \Delta v = -(v \cdot \nabla)v - \nabla p \\ \nabla \cdot v = 0 \\ v(0) = v_0 \end{cases} \tag{11}$$

such that

$$v(t, x) = \lambda(t)V(\lambda(t)x), \qquad p(t, x) = \lambda^2(t)P(\lambda(t)x) \tag{12}$$

$\lambda(t)$ being a function of time, $P(x)$ a function of x and $V(x)$ a divergence free vector field.

Two possibilities arise in what follows. From one side "backward and turbulent" solutions characterized by a function $\lambda(t) = \dfrac{1}{\sqrt{2a(T-t)}}$, $a > 0$ and $T > 0$ being two reals and $t < T$; from the other side, the "forward and regular" solutions such that $\lambda(t) = \frac{1}{\sqrt{2a(T+t)}}$, $a > 0$, $T > 0$ and $t > -T$. In the first case $V(x)$ and $P(x)$ solve the system

$$\begin{cases} -\nu\Delta V + aV + a(x\cdot\nabla)V + (V\cdot\nabla)V + \nabla P = 0 \\ \nabla\cdot V = 0 \end{cases} \tag{13}$$

whereas in the second $V(x)$ and $P(x)$ verify

$$\begin{cases} -\nu\Delta V - aV - a(x\cdot\nabla)V + (V\cdot\nabla)V + \nabla P = 0 \\ \nabla\cdot V = 0 \,. \end{cases} \tag{14}$$

In 1933, J. Leray remarked that if a weak solution v becomes turbulent at a time T, then the quantity $u(t) = \sup_{x\in\mathbb{R}^3}\sqrt{v\cdot v}$ has to blow up like $\frac{1}{\sqrt{2a(T-t)}}$ when t tends to T. Furthermore, he suggested, without proving their existence, to look for *turbulent self-similar solutions*, i.e. solutions of the form (12) with $\lambda = \frac{1}{\sqrt{2a(T-t)}}$. His conclusion was the following [94]:
"[...] unfortunately I was not able to give an example of such a singularity [...]. If I had succeeded in constructing a solution to the Navier equations that becomes irregular, I would have the right to claim that turbulent solutions not simply reducing to regular ones do exist. But if this position were wrong, the notion of turbulent solution, that for the study of viscous fluids will not play a key role any more, would not lose interest : there have to exist some problems of Mathematical Physics such that the physical causes of regularity are not sufficient to justify the hypothesis introduced when the equations are derived; to these problems we can apply similar considerations of the ones advocated so far".

The first proof of the nonexistence of backward, self-similar turbulent solutions sufficiently decreasing at infinity seems to have been given by a physicist at the beginning of the 1970s in a somewhat esoteric paper. The paper in question, written by G. Rosen [132], and that deserves to be better understood, was drawn to my attention by H. Okamoto during a talk I gave at the Department of Aeronautics and Astronomy in Kyōto. Another argument for the nonexistence of nontrivial solutions to the system (13) was given by C. Foias and R. Temam in [52].

But the "true" proof for the nonexistence of self-similar solutions as imagined by J. Leray was available in "true" functional spaces only later, in 1996, thanks to the works of the Czech school of J. Nečas.

In a paper published in the French Academy "Comptes Rendus" [110] – the last one to be presented by J. Leray (1906-1998) – J. Nečas, M. Růžička and V. Šverák announced that any weak solution to the Navier-Stokes equations belonging to the space $L^3(\mathbb{R}^3) \cap W^{1,2}_{\text{loc}}(\mathbb{R}^3)$ reduces to the zero solution. The proof of this remarkable statement [111] is based on asymptotic estimates at infinity (in the Caffarelli-Kohn-Nirenberg sense) for the functions

V and P as well as for their derivatives, and on the maximum principle for the function $\Pi(x) = \frac{1}{2}|V(x)|^2 + P(x) + ax \cdot V(x)$ on a bounded domain of $\mathbb{R}^3$. A different approach to obtaining the same result, without using the Caffarelli-Kohn-Nirenberg theory, but under the more restrictive condition $V \in W^{1,2}(\mathbb{R}^3)$ was proposed afterwards by J. Málek, J. Nečas, M. Pokorný and M.E. Schonbek [99].

Now, if we impose that the norms of v that appear naturally in the energy equality derived from (11) are finite, we get the estimates $\int_{\mathbb{R}^3} |V|^2 < \infty$ and $\int_{\mathbb{R}^3} |\nabla V|^2 < \infty$, or $V \in W^{1,2}(\mathbb{R}^3)$ which implies $V \in L^3(\mathbb{R}^3)$, by Sobolev embedding. But if, on the contrary, we only impose that the local version of the energy equality is finite, in other words $V \in W^{1,2}_{\text{loc}}(\mathbb{R}^3)$, we get some conditions that do not imply $V \in L^3(\mathbb{R}^3)$. This case, left open in [111,99], was solved by T.P. Tsai and gave origin to the following theorem: any weak backward self-similar solution to the Navier-Stokes equations belonging either to the space $L^q(\mathbb{R}^3)$, $3 < q < \infty$ or to $W^{1,2}_{\text{loc}}(\mathbb{R}^3)$ reduces to the zero solution [139].

After these works showing the nonexistence of singular self-similar solutions, the present discussion on the Navier-Stokes equations seems more favorable to the regular (and unique) solutions than to the turbulent (and probably not unique) ones. But the debate is of course still open.

Moreover, since the pioneering paper of Y. Giga and T. Miyakawa [64], we know of the existence of many self-similar forward regular solutions of the type (12) with $\lambda(t) = \frac{1}{\sqrt{t}}$. These solutions cannot be of finite energy. In fact, if we consider the inner product between V and the equation (14) and integrate by parts in the whole space, we get, if V is sufficiently decreasing at infinity

$$\int_{\mathbb{R}^3} |\nabla V|^2 + a \int_{\mathbb{R}^3} |V|^2 = 0. \tag{15}$$

Finally, this equality results in the conclusion that $V = 0$, in particular when $V \in W^{1,2}(\mathbb{R}^3)$. (It is important to stress here that such a conclusion is not true for backward self-similar solutions because of the difference of signs in (13) and (14)).

This is why Y. Giga and T. Miyakawa suggested, as an alternative to Sobolev spaces, to consider the Morrey-Campanato ones. They succeeded in proving the existence and the uniqueness of self-similar solutions to the Navier-Stokes equations written in terms of the vorticity as unknown, without applying their method to the Navier-Stokes equations in terms of the velocity. Four years later, P. Federbush [50,51], considered the super-critical Morrey-Campanato spaces $\dot{M}_2^q(\mathbb{R}^3)$, $3 < q < \infty$ for these equations. The critical space $\dot{M}_2^3(\mathbb{R}^3)$ was considered shortly after by M. Taylor [134] who, surprisingly, did not take advantage of this space which contains homogeneous functions of degree -1, to get the existence of self-similar solutions.

In [22,26,27] we gave the first construction of self-similar solutions for the Navier-Stokes equations (11), by using Besov spaces. In particular, the existence of forward self-similar solutions of the form $\frac{1}{\sqrt{t}}V(\frac{x}{\sqrt{t}})$ with $V \in L^q(\mathbb{R}^3)$ and $3 < q < \infty$ is contained as a corollary in [22]. The main idea in the aforementioned papers is to study the Navier-Stokes equations by the fixed point algorithm in a critical space containing homogeneous functions of degree -1. Furthermore, as noted by F. Planchon [116], the equivalence between the integral mild equation and the elliptic problem (14) is totally justified.

Since 1995, O. Barraza has suggested replacing the Besov spaces with the Lorentz ones $L^{3,\infty}(\mathbb{R}^3)$, always with the existence of self-similar solutions in mind [9], but he did not achieve the bicontinuity of the bilinear operator in this space. This result was proven later by Y. Meyer [103], and was applied not only to obtain the uniqueness of Kato's mild solutions, but also to the existence of self-similar solutions (forward and regular). As we have already pointed out, the space introduced by Y. Le Jan and A.-S. Sznitman gives an even simpler *ad hoc* setting to prove such a result.

Finding self-similar solutions is important because of their possible connection with attractor sets. In other words, they are related to the asymptotic behaviour of global solutions of the Navier-Stokes equations. A heuristic argument is the following: let $v(t, x)$ be a global solution to the Navier-Stokes system, then, for any $\lambda > 0$, the function $v_\lambda(t, x) =: \lambda v(\lambda^2 t, \lambda x)$ is also a solution to the same system. Now, if in a "certain sense" the limit $\lim_{\lambda\to\infty} v_\lambda(t, x) =: u(t, x)$ exists, then it is easy to see that $u(t, x)$ is a self-similar solution and that $\lim_{t\to\infty} \sqrt{t}v(t, \sqrt{t}x) = u(1, x)$. In [116,118,119] F. Planchon gave the precise mathematical frame to explain the previous heuristic argument (see also Y. Meyer [101] and O. Barraza [10]).

As we suggested among the open problems in [22], the existence of self-similar solutions evokes the study of exact solutions for Navier-Stokes. In fact, G. Tian and Z. Xin gave an explicit one parameter family of self-similar solutions, singular in a single point, that becomes an axis when the viscosity coefficient tends to zero [138] (see also [62]).

Finally, before ending this section, we would like to mention here the papers of H. Okamoto [112,113] that contain a systematic study of exact solutions of the systems (13) and (14). These results merit attention, especially since the resolution of these elliptic equations seems very difficult. Among the possible applications, one could apply mild solutions in the sub-critical case, for which neither the existence nor the uniqueness are known (Section 7).

More precisely, let us suppose that we can prove the existence of a non-trivial self-similar solution $v(t, x) = \frac{1}{\sqrt{t}}V(\frac{x}{\sqrt{t}})$ – in other words a solution V of (14) – with $V \in L^p(\mathbb{R}^3)$ and $1 \leq p < 3$. Then the Cauchy problem associated to the zero initial data would allow two different solutions, *viz.* v and 0, both belonging to $\mathcal{C}([0, T); L^p(\mathbb{R}^3))$. In fact, $\lim_{t\to 0} \|\frac{1}{\sqrt{t}}V(\frac{x}{\sqrt{t}})\|_p = 0$, provided

$1 \leq p < 3$. And the Cauchy problem would be ill-posed in $\mathcal{C}([0,T); L^p(\mathbb{R}^3))$, $1 \leq p < 3$ in the same way that it is ill-posed for a semi-linear PDE studied in 1985 by A. Haraux and F.B. Weissler [67].

This point of view should confirm the conjecture formulated by T. Kato [72], according to which the Cauchy problem is ill-posed in the sense of Hadamard when $1 \leq p < 3$. In the case $p = 2$, for example, we will not obtain a unique, global and regular solution and the scenario imagined by J. Leray would finally be possible. We will come back to this question at the end of Section 4.

4   Reynolds Numbers, Stability and Lyapunov Functions

In this section we will examine the smallness (or oscillation) condition of the initial data, by making the link between this property, the stability of the corresponding global solution and the existence of Lyapunov functions. The Besov spaces will play a key role here, whereas this was not the case, *a posteriori*, for the uniqueness of mild solutions (Section 1), nor for the existence of self-similar solutions (Section 3).

The complexity of the Navier-Stokes equations is essentially due to the competition between the nonlinear convection term $\rho(v \cdot \nabla)v$, and the linear term of viscous diffusion, $\mu \Delta v$. The order of magnitude of the quotient between these terms (dimension equation)

$$\frac{|\rho(v \cdot \nabla)v|}{|\mu \Delta v|} \equiv \frac{\rho}{\mu} \frac{V^2/L}{V/L^2} = \frac{LV}{\nu} =: R \tag{16}$$

defines a dimensionless quantity R, called Reynolds number, that allows a comparison of the inertial forces and the viscosity ones. In the absence of boundary conditions and external forces, this implies that the limit $R \to \infty$ transforms the Navier-Stokes equations into the Euler equations of perfect fluids. Another property of the Reynolds number is the notion of similarity: two flows with the same Reynolds numbers and same geometry can be deduced one from the other by a simple changing of scale. For example, in order to determine the behaviour of the rarefied atmospheres around the wing of a plane it is enough to construct a reduced model of the wing, by choosing the characteristic parameters (velocity, viscosities, etc.) in such a way that the corresponding Reynolds numbers turn out to be the same.

The Reynolds numbers also play a key role in the resolution of the Navier-Stokes equations from a theoretical point of view of PDE's. We know, since the pioneering papers of J. Leray, that there exists a weak solution to the Navier-Stokes equations. But it seems difficult to prove that such a solution is sufficiently regular (differentiable) to verify the (classical) starting system.

For a viscous flow in two dimensions, the solution is always regular and the Navier-Stokes equations have a unique solution. In three dimensions, unfortunately, such a result is not available. So far, the only result one can prove is that the solution is regular at any time if the initial data is sufficiently small (in a certain topology) or regular in a finite time interval and with an arbitrary yet reasonable choice of initial data. As we have already noted, (Sections 2–3) the most complete theory in this direction is without any doubt the one established by T. Kato and H. Fujita (1962). Concerning the solutions à la Kato, it is extremely unpleasant that we do not know whether for arbitrary large data the corresponding solution is in fact global. The question is of course intimately related to the problem of regularity. The solution is initially regular and unique, but at the instant t when it ceases to be regular (if such an instant exists) the uniqueness could also be lost. The uniqueness question is the most important unsolved problem in fluid mechanics. For the solutions of the Boltzmann equation for rarefied gases or Enskog equation for dense gases, such a result is not available either.

For the Navier-Stokes equations one might consider the entire question irrelevant, for the solution is unique and regular for small initial data and no viscous flow can be considered incompressible if the initial data are too large. Analogously, for the Boltzmann (resp. Enskog) equation, if the probability density is too large, the gas ceases to be rarefied (resp. moderately dense).

The problem here is different: the set $(\delta > 0)$ of initial data for which one can ensure the existence and the uniqueness $(\|v_0\| < \delta)$ is not known precisely and could be too small, and the result meaningless from a physical point of view. In other words, the initial data as well as the unique corresponding solution would "physically" be zero!

More exactly, the smallness condition is not absolute, but relative to the viscosity ν and is written $\|v_0\|/\nu < \delta$. Now, if we interpret $\|v_0\|$ as the characteristic velocity of the problem and we suppose (in the whole space $\mathbb{R}^3$ or $\mathbb{T}^3$) the characteristic length normalised to unity, the quotient $R =: \|v_0\|/\nu$ is nothing more than a Reynolds number associated to the problem. Thus, the condition giving the existence and uniqueness of Kato's (global and regular) solution is nothing but by the smallness of the dimensionless number R.

At this point it would be tempting to prove that for Reynolds numbers that are too large, the solution does not exist, or is not regular or simply not unique. This point of view would be confirmed by the image of developed turbulence formulated in 1944 by Landau [87]: *"Yet not every solution of the equations of motion, even if exact, can actually occur in Nature. The flows that occur in Nature must not only obey the equations of fluid dynamics, but also be stable. For the flow to be stable it is necessary that small perturbations, if they arise, should decrease with time. If, on the contrary, the small perturbations which inevitably occur in the flow tend to increase with time, then the flow is absolutely unstable. Such a flow unstable with respect to infinitely small perturbations cannot exist."*

From a theoretical point of view, a solution of the Navier-Stokes equations can exist for any positive Reynolds number. However, according to the previous analysis, such solutions would not necessarily correspond to real flows. Above a certain critical Reynolds number R_{crit}, they could not be stable under small perturbations. The criteria to find the value R_{crit} for a given problem are a matter for the theory of hydrodynamics stability and we refer the reader to [24] for a more comprehensive discussion and accurate bibliography on the subject.

In the following pages we would like to concentrate only on the results that are closely related to the approach for the Navier-Stokes equations introduced in [22]. Let us start with the $L^3(\mathbb{R}^3)$ valued mild solutions. First of all, we should note that the application that associates to the initial value $v_0 \in L^3(\mathbb{R}^3)$ the corresponding solution $v(t,x) \in \mathcal{C}([0,T]; L^3(\mathbb{R}^3))$ constructed, as in Kato's theory, by the fixed point theory, is analytical in a neighborhood of zero, as a functional acting on $L^3(\mathbb{R}^3)$ with values in $\mathcal{C}([0,T]; L^3(\mathbb{R}^3))$ [3]. Accordingly, the stability of mild solutions follows immediately because, by virtue of the uniqueness theorem (Section 1), any mild solution arises from the fixed point algorithm. As we will see at the end of this section, this is not the case for the sub-critical case $2 \leq p < 3$ [102].

T. Kawanago proceeds in the opposite direction [75,76]. First, he obtains a stability estimate, then makes use of it to establish a uniqueness theorem for mild solutions (Section 1). His result concerns global solutions $v \in \mathcal{C}([0,\infty); L^3(\mathbb{R}^3))$ and reads as follows. For any $v_0 \in L^3(\mathbb{R}^3)$ there exist two constants $\delta(v_0) > 0$ and $C > 0$ such that, if $\|v_0 - \tilde{v}_0\|_3 < \delta$, then $\tilde{v} \in \mathcal{C}([0,\infty); L^3(\mathbb{R}^3))$ and

$$\|v(t) - \tilde{v}(t)\|_3 \leq \|v(0) - \tilde{v}(0)\|_3 \exp\left\{C \int_0^t \|v(s)\|_5^5 \, ds\right\} \qquad (17)$$

for any $t > 0$.

Finally, O. Barraza obtains some stability and uniqueness results for solutions in $L^{3,\infty}(\mathbb{R}^3)$ [10]. But, as we have already remarked, the theorem by Y. Meyer in the same weak Lebesgue space [103] allows a considerable simplification of these results.

As pointed out by Yudovich in [151], the choice of the norm for proving the stability of an infinite-dimensional system (*e.g.* a viscous fluid) is crucial because the Banach norms are not necessarily equivalent therein. To be more explicit, let us recall the simple example of the linear Cauchy problem [151]

$$\begin{cases} \dfrac{\partial v}{\partial t} = x \dfrac{\partial v}{\partial x} \\ v(0, x) = \varphi(x). \end{cases} \qquad (18)$$

whose unique (for an arbitrary smooth initial function φ) explicit solution $v(t,x) = \varphi(x \exp(t))$ is exponentially asymptotically stable in $L^p(\mathbb{R})$ for

$1 \leq p < \infty$, stable but not asymptotically stable in $L^\infty(\mathbb{R})$ or $W^{1,1}(\mathbb{R})$ and exponentially unstable in any $W^{k,p}(\mathbb{R})$ for $k > 1$, $p \geq 1$ or $k = 1$, $p > 1$.

A sufficient condition for a solution to be stable for a given norm is that $\|v(t,x) - \tilde{v}(t,x)\|$, the norm of the difference between the solution v and a perturbation $\tilde{v}$, is a decreasing-in-time function. The concept of the Lyapunov function appears naturally and leads to the following definition: if v is a solution of the Navier-Stokes equations, any decreasing-in-time function $\mathcal{L}(v)(t)$ is called a Lyapunov function associated to v.

The most well-known example is certainly provided by the energy

$$E(v)(t) = \frac{1}{2}\|v(t)\|_2^2 \tag{19}$$

for, a calculation similar to the one performed in (15), gives

$$\frac{d}{dt}E(t) = -\nu\|\nabla v\|_2^2 < 0. \tag{20}$$

This result can be generalised easily in the homogeneous Sobolev spaces $\dot{H}^s(\mathbb{R}^3)$, for $0 \leq s \leq 1$. For simplicity, we will only consider the cases $s = \frac{1}{2}$ and $s = 1$. First of all, by means of Hölder and Sobolev inequalities we get [73, page 258]

$$\|\mathbb{P}(v \cdot \nabla)v\|_2 \leq C\|v\|_6\|\nabla v\|_3 \leq C\|v\|_{\dot{H}^1}\|v\|_{\dot{H}^{\frac{3}{2}}} \tag{21}$$

that allows us to easily deduce the decreasing property for the function $v = v(t)$

$$\frac{d}{dt}\|v\|_{\dot{H}^{\frac{1}{2}}}^2 \leq -2\|v\|_{\dot{H}^{\frac{3}{2}}}^2\left(\nu - C\|v\|_{\dot{H}^{\frac{1}{2}}}\right) \tag{22}$$

and thus, if the Reynolds number $\|v_0\|_{\dot{H}^{\frac{1}{2}}}/\nu$ is sufficiently small, we get a Lyapunov function associated to the norm $\dot{H}^{\frac{1}{2}}(\mathbb{R}^3)$. A similar argument allows us to obtain for the $\dot{H}^1(\mathbb{R}^3)$ norm the information:

$$\frac{d}{dt}\|v\|_{\dot{H}^1}^2 \leq -2\|v\|_{\dot{H}^1}^2\left(\nu - C\|v\|_{\dot{H}^{\frac{1}{2}}}\right). \tag{23}$$

This estimate shows that the smallness of the number $\|v_0\|_{\dot{H}^{\frac{1}{2}}}/\nu$ also implies the decrease in $\|v\|_{\dot{H}^1}$. Now, the Sobolev spaces $\dot{H}^s(\mathbb{R}^3)$, $s > \frac{1}{2}$ are supercritical. In other words, as far as the scaling is concerned, they have the same invariance as the Lebesgue spaces $L^p(\mathbb{R}^3)$ if $p > 3$. This means that one can prove the existence of a local mild solution for arbitrary initial data [22]. In the case of $\dot{H}^1(\mathbb{R}^3)$ this solution turns out to be global, provided the quantity $\|v_0\|_{\dot{H}^{\frac{1}{2}}}/\nu$ is sufficiently small, thanks to the uniform estimate

$$\|v(t)\|_{\dot{H}^1} \leq \|v_0\|_{\dot{H}^1}, \qquad \forall\, t > 0 \tag{24}$$

that is derived directly from (23).

As we have already pointed out in the Prologue, this property establishes a direct link between the Lyapunov functions, the existence of global regular solutions in an energy space and the oscillatory behaviour of the corresponding initial data.

In a paper that seems to have been completely ignored [71], T. Kato, after treating the classical cases $\dot{H}^s(\mathbb{R}^3)$, $0 \leq s \leq 1$, derives new Lyapunov functions for the Navier-Stokes equations not necessarily arising from an energy norm. More precisely: there exists $\delta > 0$ such that if the Reynolds number $R_3(v_0) = \frac{\|v_0\|_3}{\nu} < \delta$, then the quantity $R_3(v)(t) = \frac{\|v(t)\|_3}{\nu}$ is a Lyapunov function associated to v. The importance of this result comes from its relationship with the stability theory. In fact, as explained by D.D. Joseph [68]: *"It is sometimes possible to find positive definite functionals of the disturbance of a basic flow, other than energy, which decrease on the solutions when the viscosity is larger than a critical value. Such functionals, which may be called generalized energy functionals of the Lyapunov type, are of interest because they can lead to a larger interval of viscosities on which global stability of the basic flow can be guaranteed."*

T. Kato's result also applies to other functional norms, in particular the Besov ones, as we proved in [31,32]. Not only do these results show the stability for Navier-Stokes in very general functional frames, but as we have noted above, they could shed some light on the research of global Navier-Stokes solutions in super-critical spaces.

Before leaving this section, we would like to recall a result obtained by Y. Meyer and announced at the Conference in honor of Jacques-Louis Lions held in Paris in 1998. The theorem in question expresses the dependence on the initial data of the solutions to Navier-Stokes (mild or weak) in the subcritical case and could shed some light on the conjecture formulated by T. Kato in [72] at the end of Section 3. The result is the following: there is no application of class $\mathcal{C}^2$ that associates the (mild or weak) solution $v(t,x) \in \mathcal{C}([0,T]; L^p(\mathbb{R}^3))$, $2 \leq p < 3$ to the initial condition $v_0 \in L^p(\mathbb{R}^3)$ [102]. This implies that if a solution exists in the sub-critical case, it does not arise from a fixed point algorithm.

In particular, there is no application of class $\mathcal{C}^2$ that associates Leray's weak solution $v(t,x) \in L^\infty((0,T); L^2(\mathbb{R}^3))$, to the initial condition $v_0 \in L^2(\mathbb{R}^3)$ [102]. On the other hand, the application that associates Kato's mild solution $v(t,x) \in \mathcal{C}([0,T]; L^3(\mathbb{R}^3))$ to the initial data $v_0 \in L^3(\mathbb{R}^3)$ is analytical in a neighborhood of zero as a functional acting on $L^3(\mathbb{R}^3)$ and taking values in $\mathcal{C}([0,T]; L^3(\mathbb{R}^3))$.

Is this another argument in favour of Kato's global, regular, unique and stable solutions?

5   Other Domains

Among the problems left open in [22], we pointed out the possible generalisation of some results obtained in the whole space to the more physical case of an open, bounded or unbounded, three-dimensional set. In fact, in practical applications, one never looks for a solution in $\mathbb{R}^3$ yet solid bodies, the surfaces of a container, limit the region of space where the flow takes place.

The Littlewood-Paley decomposition, (that systematically involves convolution products) as well as the techniques arising from the Fourier transform do not apply to the case of a bounded set $\Omega \subset \mathbb{R}^3$ unless, of course, if the open set has some periodicity properties, say the torus $\mathbb{T}^3$ for which, as proved by S. Tourville in her Ph.D. thesis [140], one can generalise the results of [22] after having systematically replaced the Fourier transform, the Littlewood-Paley analysis or the wavelets decomposition by the corresponding periodic versions [140,141].

On the other hand, if we try to solve the Navier-Stokes equations in an open set with more complicated boundary conditions (of the Dirichlet or Neumann type, for example), it is necessary to recall the Stokes equation with reminder, by making use of the associated semi-group operator. In other words, it is necessary to establish para-differential calculus on the open set Ω starting from the spectral resolution of the Laplacian. In 1962, T. Kato and H. Fujita introduced the fractional powers of the semi-group generator in L^2 and studied their domains. Several years later, F. B. Weissler in the half-space and Y. Giga and T. Miyakawa in bounded domains, considered the semi-group in its generality, in the Lebesgue spaces L^p. Finally in 1984, T. Kato published a famous paper obtaining the existence of mild values solutions without using the fractional powers method but involving only Hölder and Young inequalities.

After the introduction of the Besov spaces in the re-reading of Kato's theorem [22,23] especially in the uniqueness theorem of mild solutions in critical spaces [59] – results that, as we have already noted, had been originally formulated in $\mathbb{R}^3$ – it is very natural to go back to the case of an arbitrary domain $\Omega \subset \mathbb{R}^3$, going in an inverse historical direction to the one that led to T. Kato's paper of 1984.

Among the possible generalisations, three seem to be more useful for the applications: an open bounded set Ω_b, the half space $\mathbb{R}^3_+$ and finally an exterior domain Ω_e, that is the complementary of a compact set.

In his Ph.D. thesis in 1996 supervised by R. Beals at Yale University, K.A. Voss studied the half-space $\mathbb{R}^3_+$ problem aiming to prove the existence of self-similar solutions [144]. In collaboration with F. Planchon and M.E. Schonbek [33], we considered the same question by adapting a well-known work by S. Ukai [142] and giving an explicit formula for the solutions of the

Stokes equations in $\mathbb{R}^3_+$; in particular we reformulated and generalised part of Voss' work.

Finally, N. Depauw considered in his Ph.D. thesis the case of an exterior domain Ω_e, the complementary of a compact set with smooth boundaries [47]. An additional difficulty that arises here is the necessity of introducing the notion of a homogeneous Besov space on Ω_e. Having resolved this problem, N. Depauw showed as in [22,23], that a sequence of functions of norms uniformly bounded from below in $L^3(\Omega_e)$ can be arbitrarily small in the Besov space, thereby formulating Kato's theorem in an exterior domain for highly oscillating functions. Moreover, as we have already noted in Section 1, N. Depauw generalised the theorem on the uniqueness of mild solutions in $L^n(\Omega_e)$, $n > 3$ and $\Omega_e \subset \mathbb{R}^n$, the case $n = 3$ [48] having been obtained after the simplification of the proof given by Y. Meyer [103].

6   Other Equations

In the study of the Navier-Stokes equations in [22], we always replaced the bilinear term by its simplified scalar version, in such a way that the particular structure of these equations seemed not to play a role and our techniques seemed to be well-suited to other evolution partial differential equations, in particular the semi-linear parabolic ones of the form $\partial_t v - \Delta v = P(D)F(v)$, where $P(D)$ and F are respectively a pseudo-differential operator and a nonlinear given function.

F. Ribaud, was the first to study this type of generalisations in a systematic way. In his Ph.D. thesis of 1996 [124], he considered the case of a pseudo-differential operator $P(D)$ of order $d \in [0, 2[$ and constant coefficients and a function $F(x)$ whose nonlinearity looks like $|x|^\alpha$ or $x|x|^{\alpha-1}$, $\alpha > 1$. Among the equations that verify these conditions, we find the heat equation as well as Burgers' and of course the Navier-Stokes equations. Drawing his inspiration from [22], F. Ribaud considers the Cauchy problem associated to these equations for initial data in the Sobolev space of fractional order $H_p^s(\mathbb{R}^n)$. The techniques involved in his proofs allows him to treat not only the case for which the Sobolev embedding does not take place, say when $s \geq \max(0, \frac{n}{p} - \frac{n}{\alpha})$ [127], but also the more difficult case $s < 0$ for which the term $F(v)$ is *a priori* not well-defined in the distribution sense [126].

P. Biler also considers a parabolic type equation and, by making use of the *ad hoc* functional spaces introduced in [22,26], he obtains for this equation, the existence of self-similar solutions [15,16]. More exactly, Biler studies the Cauchy problem for the nonlocal parabolic equation $v_t = \Delta v + \nabla \cdot (v\nabla\varphi)$ in $\mathbb{R}^n \times \mathbb{R}^+$ associated to the initial condition $v(x, 0) = v_0(x)$ where the coefficient $\nabla\varphi$ is determined from v, via the potential $\varphi = E_n * v$, E_n being the fundamental solution of the Laplacian equation in $\mathbb{R}^n$. Biler gets in this way the local existence (for arbitrary data) as well as global existence (for small data) under the minimal hypothesis of regularity on the initial data $v_0(x)$ in terms of a Morrey-Campanato norm. Finally, he succeeds in constructing self-similar solutions for homogeneous initial data of degree -2.

G. Karch finds parts of the previous results in his study of the Cauchy problem for the parabolic equation $v_t = \Delta v + B(v, v)(t)$ where $B(v, v)$ is a bilinear operator verifying for a certain $b < 2$ the invariance property $B(f_\lambda, g_\lambda) = \lambda^b (B(f, g))_\lambda$ for any $\lambda > 0$ and any functions f and g (here $h_\lambda(x) = h(\lambda x)$) [69]. His results (existence, uniqueness and self-similar solutions) apply to the diffusion-convection equation ($b = 1$), the Navier-Stokes ones ($b = 1$), Debye's system ($b = 0$), the nonlocal parabolic equation considered by P. Biler ($b = 0$) as well as the one studied by F. Ribaud ($b = 1$).

In her Ph.D. thesis [140], defended in 1998 at Washington University of St. Louis, S. Tourville studied the Navier-Stokes equations with hyperdissipation, in other words with the linear diffusion term Δv replaced by the more general expression $-(-\Delta)^d v$ with $d \geq 1$. By making use of the techniques introduced in [22], S. Tourville proves some existence and uniqueness results of local and global solutions, justifying in this way some phenomena well-known from the numerical point of view.

The joint works of T. Cazenave and F.B. Weissler [35–37] follow the same direction. These works analyse the existence and uniqueness of global solutions for the nonlinear Schrödinger equation $iv_t + \Delta v = \gamma |v|^\alpha v$ with initial data $v(0) = \phi$ small with respect to the norm $\|\phi\| = \sup_{t>0} t^\beta \|S(t)\phi\|_{L^{\alpha+2}(\mathbb{R}^n)}$ where $S(t)$ is the Schrödinger group, $\beta = \frac{4-(n-2)\alpha}{2\alpha(\alpha+2)}$, γ is a real and $\alpha > 0$. A complete characterisation of the functional space for which the previous norm is finite seems difficult, since the space is not even stable when passing to the conjugate, as F. Oru proved shortly after [114]. This difficulty did not appear in the Navier-Stokes equations and this is why T. Cazenave and F.B. Weissler limit themselves to the case of special homogeneous initial data. In this way they obtain the existence of self-similar solutions for α in the interval $\alpha_0 < \alpha < 4/(n - 2)$, $n \geq 3$, where α_0 is the positive square root of the equation $(n - 2)\alpha^2 + (n - 4)\alpha - 4 = 0$ [35,36]. In fact, by a direct calculation, they prove that the functions $\phi(x) = \frac{P_k(x)}{|x|^{p+k}}$ (where $\operatorname{Re} p = \frac{2}{\alpha}$ and P_k is a harmonic polynomial of degree k) have a finite norm $\|\phi\|$, finally obtaining the existence of self-similar solutions. But a better result can be obtained, as was pointed out later by F. Oru [114]. In fact, by using some elementary techniques of harmonic analysis, Oru proves that one can take for ϕ any homogeneous function of degree $-p$, $\operatorname{Re} p = \frac{2}{\alpha}$, infinitely differentiable outside zero. Later, F. Ribaud and A. Youssfy [128] find that a finite regularity for ϕ is sufficient. Another result contained in F. Oru's Ph.D. thesis concerns the optimality of the condition $\alpha_0 < \alpha < 4/(n - 2)$. More precisely, if $\alpha < \alpha_0$ or if $\alpha > 4/(n - 2)$, the only tempered distribution ϕ whose norm $\|\phi\|$ is finite is the trivial one, whereas if $\alpha = \alpha_0$ or if $\alpha = 4/(n - 2)$ there is no homogeneous distribution of finite norm and accordingly, no self-similar solution in such a space. Finally, for the Schrödinger equation other results in the same direction have recently been announced by F. Planchon [120,121], who considers the Cauchy problem with initial data in the space $\dot{B}_2^{\frac{n}{2}-\frac{2}{\alpha},\infty}(\mathbb{R}^3)$, $\alpha > \frac{4}{n}$, and by G. Furioli [57].

Global solutions, in particular self-similar, were also obtained by F. Ribaud and A. Youssfi [129–131], using similar techniques to the ones introduced for the Schrödinger equation. In the same direction we also quote the successive works of F. Planchon [122] and H. Pecher [115].

Finally, the existence of self-similar solutions for a system of equations with dissipative terms, that contain as a particular case the Navier-Stokes equations for incompressible flows, the semi-linear heat equation, the Ginzburg-Landau and Cahn-Hilliard ones, was given by H.A. Biagioni, L. Cadeddu and T. Gramchev [13,14].

7   Perspectives and Open Problems

In spite of the progress and the multitude of results obtained in the last five years, some important problems are still without answer. We are going to present here some of the more significant ones.

7.1   The Uniqueness of the Solutions

Not only do we not dispose of an *existence* result of mild solutions for the Navier-Stokes equations with initial data in a sub-critical space, but their *uniqueness* is also unknown. This is the case for instance for the spaces $L^p(\mathbb{R}^3)$, $1 \leq p < 3$. With regard to this point, in Section 3, we opted in favour of the nonuniqueness of the solutions by means of a conjecture: the existence of forward regular self-similar solutions $\frac{1}{\sqrt{t}}V(\frac{x}{\sqrt{t}})$ with $V \in L^p(\mathbb{R}^3)$, $1 \leq p < 3$.

7.2   The Critical Spaces

A question, left open in [22] is still waiting for an answer. We are thinking of the Besov space $\dot{B}_\infty^{-1,\infty}(\mathbb{R}^3)$, corresponding to the largest one in the family of (increasing with q) Besov spaces $\dot{B}_q^{-1+\frac{3}{q},\infty}(\mathbb{R}^3)$, in which for $3 \leq q < \infty$ it is possible to prove the existence and uniqueness of mild solutions for Navier-Stokes. The importance of knowing whether this is also true for $q = \infty$ comes from the property that any critical space is embedded in $\dot{B}_\infty^{-1,\infty}(\mathbb{R}^3)$ as pointed out by Y. Meyer [103] (see also [53], [3], [114]). The closest result in this direction is certainly the one due to H. Koch and D. Tataru, concerning the space $BMO^{-1}(\mathbb{R}^3)$ [77]. Moreover, in her Ph.D. thesis [56] G. Furioli announces a generalisation of this result to the functional spaces E verifying, among others, the following property on the product between functions, say:

$$\|fg\|_E \leq C(\|f\|_E\|g\|_{L^\infty} + \|g\|_E\|f\|_{L^\infty}) \quad \forall f, g \in E \cap L^\infty,$$

as is the case for the Sobolev space $\dot{H}^s(\mathbb{R}^3)$ if $s \geq 0$ and the Hölder one $\dot{C}^\alpha(\mathbb{R}^3)$ if $\alpha > 0$.

In passing, let us note here that Stephen Montgomery-Smith, during the 6th Paseky school, proved a blow up result in the space $\dot{B}_\infty^{-1,\infty}(\mathbb{R}^3)$, for a modified ("cheap" in his terminology) Navier-Stokes system [108].

7.3   The Self-Similar Solutions

The *existence* of forward regular self-similar solutions of the form $\frac{1}{\sqrt{t}}V(\frac{x}{\sqrt{t}})$ with $V \in L^p(\mathbb{R}^3)$ and $3 < p < \infty$ was given in [22,26]. But no result is known if $1 \leq p \leq 3$ or $p = \infty$. Moreover, the work of the Czech school, in particular the paper of T.P. Tsai [139], shows the *nonexistence* of nontrivial backward singular self-similar solutions of the type $\frac{1}{\sqrt{2a(T-t)}}V(\frac{x}{\sqrt{2a(T-t)}})$ with $a > 0$, $V \in L^p(\mathbb{R}^3)$ and $3 \leq p < \infty$ and that, if $p = \infty$, then V is a constant. What happens if $1 \leq p < 3$? What happens if we consider V in the Le Jan-Sznitman space?

7.4   The Reynolds Numbers

In the super-critical case, *e.g.* $L^p(\mathbb{R}^3)$, $3 < p \leq \infty$, we know the existence and uniqueness of a local mild solution of the Navier-Stokes equations, but it is not clear whether such a solution can be extended to a global one. If one could get a uniform estimate of the type $\|v(t)\|_p \leq g(t)$, with $g(t) \in L^\infty_{\text{loc}}(\mathbb{R})$, the solution would of course be global. This happens to be the case, as we have already recalled, for the super-critical Sobolev space $\dot{H}^1(\mathbb{R}^3)$. Furthermore, the noninvariance of the $L^p(\mathbb{R}^3)$ norm, $p \neq 3$, ensures that such a global result would not depend on the size of the initial data, say of the Reynolds number $\frac{\|v_0\|_p}{\nu}$. But, as in the case of $\dot{H}^1(\mathbb{R}^3)$, the smallness of a Besov norm, or some oscillations, could be present.

Concerning the space $L^\infty(\mathbb{R}^3)$, let us remark that the existence and uniqueness of a local solution were obtained only recently in [22,25] by using the simplified structure of the bilinear term. In fact, as pointed out in a different proof by Y. Giga and his students in Sapporo [65], the Leray-Hopf projection $\mathbb{P}$ is not bounded in $L^\infty(\mathbb{R}^3)$, nor in $L^1(\mathbb{R}^3)$.

What happens in the critical case $p = 3$, when the Reynolds number is large, say $\frac{\|v_0\|_3}{\nu} \gg \delta$? Loss of analyticity for "turbulent" solutions (Leray), infinite number of bifurcation points (Landau-Hopf), or chaotic behaviour and strange attractors (Ruelle-Takens)? The situation remains inextricable and highly controversial. Maybe the three previous phenomenological theories could contribute to a description of turbulence in a complementary rather than competitive way. In fact, in all the examples of singularities known so far for other nonlinear equations, the loss of uniqueness and stability is also verified at the same time. Moreover, far from having reached unanimous consent, none of these conjectures was rigorously proved for the Navier-Stokes equations, even if they were all observed either for some model equations (Boussinesq, Burgers, Lorenz) or for some particular flows (Couette, Bénard).

Finally, the sub-critical case $1 \leq p < 3$ is the most difficult one, because, as we have already pointed out, neither the existence nor the uniqueness are known for the mild Navier-Stokes system in this space. If $2 \leq p < 3$, then the solutions are not stable [102], which implies that they cannot arise from a fixed point algorithm, whereas if $1 \leq p < 2$ a supplementary condition must to be imposed *a priori* on the solution $v \in \mathcal{C}([0, T); L^p(\mathbb{R}^3))$, in order to define, in the distribution sense, the product $v \otimes v$.

7.5   Other Equations

Some evolution equations very close to the Navier-Stokes ones for incompressible viscous fluids (INS) may be studied by the same techniques we described in the previous pages. I am thinking in particular of the Boltzmann equation of rarefied gases (B), the Euler equations of ideal incompressible fluids (IE) and compressible fluids (CE) and finally the compressible Navier-Stokes equations (CNS).

Now, as explained by C. Cercignani [38], it is possible, under some physical assumptions, to obtain the equations (INS), (CNS), (IE) and (CE) from (B), the mathematical correctness of such a reasoning being a challenging problem [5-8,4,98].

As far as the Boltzmann equation is concerned, the research of self-similar solutions seems to be an open problem, except for the results obtained by F. Golse [66]. Furthermore, we know of very few papers that make use of the dyadic decomposition of Littlewood-Paley for the study of the collision operator in the Boltzmann equation, as was done by R.J. DiPerna, P.-L. Lions and Y. Meyer [49].

This is not the case for the Euler equations of perfect fluids, thanks to the papers of J.-Y. Chemin [43] as well as a series of results obtained recently by M. Vishik [145–147] and H. Kozono and Y. Taniuchi [85,86].

Finally, for the compressible Navier-Stokes equations, except for the papers of R. Danchin [46], methods arising from the harmonical analysis such as the Littlewood-Paley decomposition, seem to not have been systematically exploited.

Acknowledgements

This paper contains the themes of a series of lectures I gave at the sixth Paseky school, in September 1999. I wish to express my gratitude to Julian Lange for his careful rereading of the manuscript.

References

1. H. Amann: On the strong solvability of the Navier-Stokes equations, J. Math. Fluid Mech. **2** (2000), 16–98.
2. H. Amann: Remarks on the strong solvability of the Navier-Stokes equations, preprint (2000), 1–6.
3. P. Auscher et P. Tchamitchian: Espaces critiques pour le système des équations de Navier-Stokes incompressibles, preprint de l'Université de Picardie Jules Verne (1999).
4. C. Bardos: From molecules to turbulence. An overview of multiscale analysis in fluid dynamics, Advances topics in theoretical fluid mechanics, J.Málek, J. Nečas, M. Rokyta (eds), Pitman Research Notes in Mathematics Series, Longman **392** (1998).
5. C. Bardos, F. Golse and D. Lavermore: Fluid dynamical limits of kinetic equations, I: Formal derivation, J. Stat. Physics **63** (1991), 323–344.
6. C. Bardos, F. Golse and D. Lavermore: Fluid dynamical limits of kinetic equations, II: Convergence Proofs, Comm. Pures et Appl. Math. **46** (1993), 667–753.
7. C. Bardos, F. Golse and D. Lavermore: Acoustic and Stokes limits for the Boltzmann equation, C.R.A.S.P. **327** (1998), 323–328.
8. C. Bardos and S. Ukai: The classical incompressible Navier-Stokes limit of the Boltzmann equation, Math. Mod. Meth. Appl. Sc. **1 (2)** (1991), 235–257.
9. O. A. Barraza: Self-similar solutions in weak L^p spaces of the Navier-Stokes equations, Rev. Mat. Iberoamericana **12** (1996), 411–439.
10. O. A. Barraza: Regularity and stability for the solutions of the Navier-Stokes equations in Lorentz spaces, Nonlinear Anal. Ser. A: Theory Methods **35 (6)** (1999), 747–764.
11. M. Ben-Artzi: Global solutions of two-dimensional Navier-Stokes and Euler equation, Arch. Rat. Mech. Anal. **128** (1994), 329–358.
12. O. V. Besov: Investigation of a family of functional spaces connected with embedding and extension theorems, Trudy Mat. Inst. Steklov **60** (1961) (in Russian).
13. H. A. Biagioni, L. Cadeddu and T. Gramchev: Semilinear parabolic equations with singular initial data in anisotropic weighted spaces, Diff. and Integral Equations **12** (1999), 613–636.
14. H. A. Biagioni and T. Gramchev: Evolution PDE with elliptic dissipative terms: critical index for singular initial data, self-similar solutions and analytic regularity, C. R. Acad. Sci. Paris Sér. I Math. **327 (1)** (1998), 41–46.
15. P. Biler: The Cauchy problem and self-similar solutions for a nonlinear parabolic equation, Studia Mathematica **114** (1995), 181–205.

16. P. Biler: Local and global solutions of a nonlinear nonlocal parabolic problem (Warsaw, 1994), Gakuto Internat. Ser. Math. Sci. Appl., 7, Gakkōtosho, Tokyo (1996), 49–66.

17. J.-M. Bony: Calcul symbolique et propagation des singularités pour les équations aux dérivées partielles non linéaires, Ann. Sci. Ecole Norm. Sup. **14 (4)** (1981), 209-246.

18. H. Brezis: Remarks on the preceding paper by M. Ben-Artzi, Arch. Rat. Mech. Anal. **128 (4)** (1994), 359–360.

19. H. Brezis and T. Cazenave: A nonlinear heat equation with singular initial data, J. Anal. Math. **68** (1996), 277–304.

20. C. P. Calderón: Existence of weak solutions for the Navier-Stokes equations with initial data in L^p, Trans. Amer. Math. Soc. **318 (1)** (1990), 179–200. Addendum, ibid. 201–207.

21. C. P. Calderón: Initial values of solutions of the Navier-Stokes equations, Proc. Amer. Math. Soc. **117 (3)** (1993), 761–766.

22. M. Cannone: Ondelettes, Paraproduits et Navier-Stokes, Diderot Editeur (1995).

23. M. Cannone: A generalisation of a theorem by Kato on Navier-Stokes equations, Rev. Mat. Iberoamericana **13 (3)** (1997), 515–541.

24. M. Cannone: Nombres de Reynolds, stabilité et Navier-Stokes, Evolution Equations: Existence and Singularities, Banach Center Publications, Institute of Mathematics, Polish Academy of Sciences, Warszawa **52** (2000).

25. M. Cannone and Y. Meyer: Littlewood-Paley decomposition and the Navier-Stokes equations, Meth. and Appl. of Anal. **2** (1995), 307-319.

26. M. Cannone, Y. Meyer et F. Planchon: Solutions auto-similaires des équations de Navier-Stokes in $\mathbb{R}^3$, Exposé n. VIII, Séminaire X-EDP, Ecole Polytechnique, (1994).

27. M. Cannone and F. Planchon: Self-similar solutions of the Navier-Stokes equations in $\mathbb{R}^3$, Comm. Part. Diff. Eq. **21** (1996), 179-193.

28. M. Cannone and F. Planchon: On the regularity of the bilinear term for solutions of the incompressible Navier-Stokes equations in $\mathbb{R}^3$, Rev. Mat. Iberoamericana (to appear) (2000).

29. M. Cannone and F. Planchon: Maximal function inequalities and Navier-Stokes equations, Investigations on the structure of solutions to partial differential equations (Japanese) (Kyōto, 1997). Sūrikaisekikenkyūsho Kōkyūroku **1036** (1998), 139–159.

30. M. Cannone and F. Planchon: On the non stationary Navier-Stokes equations with an external force, Adv. in Diff. Eq. **4 (5)** (1999), 697–730.

31. M. Cannone et F. Planchon: Fonctions de Lyapunov pour les équations de Navier-Stokes, Exposé n. VIII, Séminaire X-EDP, Ecole Polytechnique (2000), 1–7.

32. M. Cannone and F. Planchon: More Lyapunov functions for the Navier-Stokes equations, submitted (2000).

33. M. Cannone, F. Planchon and M. E. Schonbek: Navier-Stokes equations in the half space, Comm. Part. Diff. Equat. (to appear) (2000).

34. M. Cannone, F. Planchon and M. E. Schonbek: Navier-Stokes equations in an exterior domain, (in preparation) (2000).

35. T. Cazenave and F. B. Weissler: Asymptotically self-similar global solutions of the nonlinear Schrödinger and heat equations, Math. Zeit. **228 (1)** (1998), 83–120.

36. T. Cazenave and F. B. Weissler: More self-similar solutions of the nonlinear Schrödinger equation, NoDEA Nonlinear Differential Equations Appl. **5 (3)** (1998), 355–365.

37. T. Cazenave and F. B. Weissler: Scattering theory and self-similar solutions for the nonlinear Schrödinger equation, SIAM J. Math. Anal. **31** (2000), no. 3, 625–650 (electronic).

38. C. Cercignani: The Boltzmann equation and its applications, Springer-Verlag, Berlin (1988).

39. J.-Y. Chemin: Remarques sur l'existence globale pour le système de Navier-Stokes incompressible, SIAM J. Math. Anal. **23** (1992), 20–28.

40. J.-Y. Chemin: About Navier-Stokes system, Publ. Lab. Anal. Num. (Univ. Paris 6) **R 96023** (1996), 1–43.

41. J.-Y. Chemin: Sur l'unicité dans le système de Navier-Stokes tridimensionnel, Exposé n. XXIV, Séminaire X-EDP, Ecole Polytechnique, (1996-1997).

42. J.-Y. Chemin: Théorèmes d'unicité pour le système de Navier-Stokes tridimensionnel, Jour. d'Anal. Math. **77** (1997), 27–50.

43. J.-Y. Chemin: Fluides parfaits incompressibles, Astérisque, **230** 1995; *Perfect incompressible fluids.* Translated from the 1995 French original by Isabelle Gallagher and Dragos Iftimie. Oxford Lecture Series in Mathematics and its Applications, 14., The Clarendon Press, Oxford University Press, New York (1998).

44. Z.-M. Chen and Z. Xin: Homogeneity Criterion on the Navier-Stokes equations in the whole spaces, preprint (1999).

45. P. Constantin: A few results and open problems regarding incompressible fluids, Notices of the AMS **42 (6)**, 658–663.

46. R. Danchin: Existence globale dans des espaces critiques pour le système de Navier-Stokes compressible, C.R.A.S.P. **328 (8)** (1999), 649–652.

47. N. Depauw: Solutions peu régulières des équations d'Euler et Navier-Stokes incompressibles sur un domaine à bord, Ph.D. Thesis, Université de Paris Nord (1998).

48. N. Depauw: Personal communication (1998).

49. R. J. DiPerna, P.-L. Lions and Y. Meyer: L^p-regularity of velocity averages, Ann. I.H.P. Anal. Non Lin. **8** (1991), 271–287.

50. P. Federbush: Navier and Stokes meet the wavelet, Comm. Math. Phys. **155** (1993), 219–248.

51. P. Federbush: Navier and Stokes meet the wavelet. II, Mathematical quantum theory. I. Field theory and many-body theory (Vancouver, BC, 1993), CRM Proc. Lecture Notes, Amer. Math. Soc., Providence, RI **7** (1994), 163–169.

52. C. Foias and R. Temam: Self-similar Universal Homogeneous Statistical Solutions of the Navier-Stokes Equations, Comm. Math. Phys. **90** (1983), 187–206.

53. M. Frazier, B. Jawerth and G. Weiss: Littlewood-Paley theory and the study of function spaces, CBMS Regional Conference Series in Mathematics, **79**, AMS, Providence.

54. J. Fourier: Théorie analytique de la chaleur, 1822, published in *Œuvres de Fourier* **1** (1888) Gauthiers-Villar et fils, Paris.

55. H. Fujita and T. Kato: On the Navier-Stokes initial value problem I, Arch. Rat. Mech. Anal. **16** (1964), 269–315.

56. G. Furioli: Applications de l'analyse harmonique réelle à l'étude des équations de Navier-Stokes et de Schrödinger non linéaire, Ph.D. Thesis, Université d'Orsay (1999).

57. G. Furioli: Solutions auto-similaires pour les équations de Schrödinger non linéaires, en préparation (1999).

58. G. Furioli, P.-G. Lemarié-Rieusset et E. Terraneo: Sur l'unicité dans $L^3(\mathbb{R}^3)$ des solutions mild des équations de Navier-Stokes, C. R. Acad. Sci. Paris Sér. I Math. **325 (12)** (1997), 1253–1256.

59. G. Furioli, P.-G. Lemarié-Rieusset et E. Terraneo: Unicité dans $L^3(\mathbb{R}^3)$ et d'autres espaces fonctionnels limites pour Navier-Stokes, Rev. Mat. Iberoamericana (to appear) (2000).

60. Y. Giga: Solutions for semilinear parabolic equations in L^p and regularity of weak solutions of the Navier-Stokes system, J. Diff. Eq. **62** (1986), 186–212.

61. Y. Giga: Review of the paper by H. Brezis, "Remarks on the preceding paper by M. Ben-Artzi" Mathematical Reviews **96h:35149** (1996) [see, for the revised form, at the address http://klymene.mpim-bonn.mpg.de:80/msnpr-html/review_search.html].

62. Y. Giga and M.-H. Giga: Nonlinear Partial Differential Equations–Asymptotic Behaviour of Solutions and Self-Similar Solutions (in Japanese), Kyōritsu Shuppan, Tokyo, 1999.

63. Y. Giga and T. Miyakawa: Solutions in L^r of the Navier-Stokes initial value problem, Arch. Rat. Mech. Anal. **89** (1985), 267–281.

64. Y. Giga and T. Miyakawa: Navier-Stokes flows in $\mathbb{R}^3$ with measures as initial vorticity and the Morrey spaces, Comm. PDE **14** (1989), 577–618.

65. Y. Giga, K. Inui and S. Matsui: On the Cauchy problem for the Navier-Stokes equations with nondecaying initial data, Quaderni di Matematica, vol. **4**, Advances in Fluid Dynamics, edited by P. Maremonti (1999).

66. F. Golse: On the self-similar solutions of the Broadwell model for a discrete velocity gas, Comm. Part. Diff. Eq. **12** (1987), 315–326.

67. A. Haraux and F.B. Weissler: Non-uniqueness for a semi-linear initial value problem, Indiana Univ. Math. J. **31** (1985), 167–189.

68. D. D. Joseph: Stability of fluid motions (2 vols), Springer Tracts in Natural Philosophy, Berlin, Springer Verlag (1976).

69. G. Karch: Scaling in nonlinear parabolic equations, Jour. of Math. Anal. Appl. **234** (1999), 534–558.

70. T. Kato: Strong L^p solutions of the Navier-Stokes equations in $\mathbb{R}^m$ with applications to weak solutions, Math. Zeit. **187** (1984), 471–480.

71. T. Kato: Liapunov functions and monotonicity in the Navier-Stokes equation, Lecture Notes in Mathematics **1450** (1990), 53–63.

72. T. Kato: Well-posedness nitsuite, (in Japanese), Sūgaku **48 (3)** (1996), 298–300.

73. T. Kato and H. Fujita: On the non-stationary Navier-Stokes system, Rend. Sem. Mat. Univ. Padova **32** (1962), 243–260.

74. T. Kato and G. Ponce: The Navier-Stokes equations with weak initial data, Int. Math. Res. Notes **10** (1994), 435–444.

75. T. Kawanago: Stability of global strong solutions of the Navier-Stokes equations, Nonlinear evolution equations and their applications, (Japanese) Sūrikaisekikenkyūsho Kōkyūroku **913** (1995), 141–147.

76. T. Kawanago: Stability estimate of strong solutions for the Navier-Stokes system and its applications, Electron. J. Differential Equations (electronic) (see: http://ejde.math.swt.edu/Volumes/1998/15-Kawanago/abstr.html) **15** (1998), 1–23.

77. H. Koch and D. Tataru: Well-posedness for the Navier-Stokes equations, preprint (see: http://www.math.nwu.edu/~tataru/nas.html) (1999), 1–14.
78. H. Kozono and M. Yamazaki: Semilinear heat equations and the Navier-Stokes equation with distributions as initial data, C. R. Acad. Sci. Paris Sér. I Math. **317** (1993), 1127–1132.
79. H. Kozono and M. Yamazaki: Semilinear heat equations and the Navier-Stokes equation with distributions in new function spaces as initial data, Comm. P.D.E. **19** (1994), 959–1014.
80. H. Kozono and M. Yamazaki: The Navier-Stokes equation with distributions as initial data and application to self-similar solutions, New trends in microlocal analysis (Tokyo, 1995) (1997), 125–141.
81. H. Kozono and M. Yamazaki: The stability of small stationary solutions in Morrey spaces of the Navier-Stokes equation, Ind. Univ. Math. Journ. **44** (1995), 1307–1336.
82. H. Kozono and M. Yamazaki: The exterior problem for the non-stationary Navier-Stokes equation with data in the space $L^{n,\infty}$, C. R. Acad. Sci. Paris Sér. I Math. **320 (6)** (1995), 685–690.
83. H. Kozono and M. Yamazaki: Exterior problem for the stationary Navier-Stokes equations in the Lorentz space, Math. Ann. **310 (2)** (1998), 279–305.
84. H. Kozono and M. Yamazaki: On a larger class of stable solutions to the Navier-Stokes equations in exterior domains, Math. Z. **228** (1998), 751–785.
85. H. Kozono and Y. Taniuchi: Bilinear estimates in BMO and the Navier-Stokes equations, preprint (1999).
86. H. Kozono and Y. Taniuchi: Limiting case of the Sobolev inequality in BMO, with application to the Euler equation, preprint (1999).
87. L. D. Landau and E. M. Lifchitz: Fluid mechanics, Translated from the Russian by J. B. Sykes and W. H. Reid. Course of Theoretical Physics, Vol. 6, Pergamon Press, London-Paris-Frankfurt; Addison-Wesley Publishing Co., Inc., Reading, Mass. (1959).
88. Y. Le Jan et A. S. Sznitman: Cascades aléatoires et équations de Navier-Stokes, C. R. Acad. Sci. Paris **324 (7)** (1997), 823–826.
89. Y. Le Jan and A. S. Sznitman: Stochastic cascades and 3-dimensional Navier-Stokes equations, Probab. Theory Related Fields **109 (3)** (1997), 343–366.
90. P.-G. Lemarié-Rieusset: Some remarks on the Navier-Stokes equations in $\mathbb{R}^3$, J. Math. Phys **39 (8)** (1998), 4108–4118.
91. P.-G. Lemarié-Rieusset: Solutions faibles d'énergie infinie pour les équations de Navier-Stokes dans $\mathbb{R}^3$, C. R. Acad. Sci. Paris Sér. I **328 (8)** (1999), 1133–1138.
92. P.-G. Lemarié-Rieusset: Weak infinite-energy solutions for the Navier-Stokes equations in $\mathbb{R}^3$, preprint (1999).
93. P.-G. Lemarié-Rieusset: Cinq petits théorèmes d'unicité L^3 des solutions des équations de Navier-Stokes dans $\mathbb{R}^3$, Notes distribuées au Colloque Equations de Navier-Stokes et Analyse Microlocale, CIRM, Luminy, Marseille, (9–13 novembre 1999), 1–9.
94. J. Leray: Etudes de diverses équations intégrales non linéaires et de quelques problèmes que pose l'hydrodynamique, J. Math. Pures et Appl. **12** (1933), 1–82.
95. J. Leray: Sur le mouvement d'un liquide visqueux emplissant l'espace, Acta Math. **63** (1934), 193–248.

96. J. Leray: Aspects de la mécanique théorique des fluides, La Vie des Sciences, Comptes Rendus, Série Générale, II, **11 (4)** (1994), 287–290.

97. P.-L. Lions et N. Masmoudi: Unicité des solutions faibles de Navier-Stokes dans $L^N(\Omega)$, C. R. Acad. Sci. Paris Sér. I Math. **327** (1998), 491–496.

98. P.-L. Lions and N. Masmoudi: From Boltzmann equations to Fluid Mechanics equations, CIRM, Luminy, Marseille, (20–24, mars 2000).

99. J. Málek, J. Nečas, M. Pokorný and M. E. Schonbek: On possible singular solutions to the Navier-Stokes equations, Math. Nachr. **199** (1999), 97–114.

100. R. May: Thèse de Doctorat de l'Université d'Evry Val d'Essonne (en préparation, 2000).

101. Y. Meyer: Large time behavior and self-similar solutions of some semilinear diffusion equations, Harmonic analysis and partial differential equations (Chicago, IL, 1996), 241–261, Chicago Lectures in Math., Univ. Chicago Press, Chicago, IL, (1999).

102. Y. Meyer: New estimates for Navier-Stokes equations (manuscript), Colloque en l'honneur de J.-L. Lions à l'occasion de son 70e anniversaire, Paris 26–27 mai 1998, Auditorium du CNRS, 3 rue Michel-Ange, Paris XVIe (1998).

103. Y. Meyer: Wavelets, paraproducts and Navier-Stokes equations, Current developments in Mathematics 1996, International Press, 105–212 Cambridge, MA 02238-2872 (1999).

104. T. Miyakawa: Hardy spaces of solenoidal vector fields, with applications to the Navier-Stokes equations, Kyushu J. Math. **50 (1)** (1996), 1–64.

105. T. Miyakawa: Application of Hardy space techniques to the time-decay problem for incompressible Navier-Stokes flows in $\mathbb{R}^n$, Funkcial. Ekvac. **41 (3)** (1998), 383–434.

106. T. Miyakawa: On space-time decay properties of nonstationary incompressible Navier-Stokes Flows in $\mathbb{R}^n$, preprint (1999).

107. S. Monniaux: Uniqueness of mild solutions of the Navier-Stokes equation and maximal L^p-regularity, C. R. Acad. Sci. Paris Sér. I Math. **328 (8)** (1999), 663–668.

108. S. Montgomery-Smith: Finite time blow up for a Navier-Stokes like equation, preprint (see http://www.math.missouri.edu/~/stephen) (2000).

109. C. L. M. H. Navier: Mémoire sur les lois du mouvement des fluides, Mém. Acad. Sci. Inst. France **6** (1822), 389-440.

110. J. Nečas, M. Růžička et V. Šverák: Sur une remarque de J. Leray concernant la construction de solutions singulières des équations de Navier-Stokes, C. R. Acad. Sci. Paris Sér. I Math. **323 (3)** (1996), 245–249.

111. J. Nečas, M. Růžička and V. Šverák: On Leray self-similar solutions of the Navier-Stokes equations, Acta Math. **176 (2)** (1996), 283–294.

112. H. Okamoto: Exact solutions of the Navier-Stokes equations via Leray's scheme, Proceedings of Miniconference of Partial Differential Equations and Applications. Seoul Nat. Univ., Seoul, 1997. Lecture Notes Ser. **38** (1997), 1–30.

113. H. Okamoto: Exact solutions of the Navier-Stokes equations via Leray's scheme, Japan J. Indust. Appl. Math. **14 (2)** (1997), 169–197.

114. F. Oru: Rôle des oscillations dans quelques problèmes d'analyse non linéaire, Ph.D. Thesis, ENS Cachan (1998).

115. H. Pecher: Self-similar and asymptotically self-similar solutions of nonlinear wave equations, Math. Ann. **316** (2000), no. 2, 259–281.

116. F. Planchon: Solutions Globales et Comportement Asymptotique pour les Equations de Navier-Stokes, Ph.D. Thesis, Ecole Polytechnique, France (1996).

117. F. Planchon: Global strong solutions in Sobolev or Lebesgue spaces for the incompressible Navier-Stokes equations in $\mathbb{R}^3$, Ann. Inst. H. Poinc. **13** (1996), 319–336.

118. F. Planchon: Convergence de solutions des équations de Navier-Stokes vers des solutions auto-similaires, Exposé n. III, Séminaire X-EDP, Ecole Polytechnique (1996).

119. F. Planchon: Asymptotic behavior of global solutions to the Navier-Stokes equations in $\mathbb{R}^3$, Rev. Mat. Iberoamericana **14** (1) (1998), 71–93.

120. F. Planchon: Solutions auto-similaires et espaces de données initiales pour l'équation de Schrödinger, C. R. Acad. Sci. Paris **328** (1999), 1157–1162.

121. F. Planchon: On the Cauchy problem in Besov space for a non-linear Schrödinger equation, preprint (1999).

122. F. Planchon: Self-similar solutions and Besov spaces for semi-linear Schrödinger and wave equations. Journées "Equations aux Dérivées Partielles" (Saint-Jean-de-Monts, 1999), Exp. No. IX, 11 pp., Univ. Nantes, Nantes (1999).

123. F. Planchon: Sur un inégalité de type Poincaré, C. R. Acad. Sci. Paris Sér. I Math. **330** (2000), no. 1, 21–23.

124. F. Ribaud: Analyse de Littlewood-Paley pour la Résolution d'Equations Paraboliques Semi-linéaires, Ph.D. Thesis, Université d'Orsay, France (1996).

125. F. Ribaud: Problème de Cauchy pour les équations paraboliques semi-linéaires avec données dans $H_p^s(\mathbb{R}^n)$, C. R. Acad. Sci. Paris **322** (1) (1996), 25–30.

126. F. Ribaud: Semilinear parabolic equations with distributions as initial data, Discrete Contin. Dynam. Systems **3**(3) (1997), 305–316.

127. F. Ribaud: Cauchy problem for semilinear parabolic equations with initial data in $H_p^s(\mathbb{R}^n)$ spaces, Rev. Mat. Iberoamericana **14** (1) (1998), 1–46.

128. F. Ribaud and A. Youssfi: Regular and self-similar solutions of nonlinear Schrödinger equations, J. Math. Pures Appl. **77** (10) (1998), 1065–1079.

129. F. Ribaud et A. Youssfi: Solutions globales et solutions auto-similaires de l'équation des ondes non linéaire, C. R. Acad. Sci. Paris **329** (1) (1999), 33–36.

130. F. Ribaud and A. Youssfi: Self-similar solutions of the nonlinear wave equation, (to appear) (1999).

131. F. Ribaud and A. Youssfi: Global solutions and self-similar solutions of nonlinear wave equation, preprint (1999).

132. G. Rosen: Navier-Stokes initial value problem for boundary-free incompressible fluid flow, Jour. Phys. Fluid **13** (1970), 2891–2903.

133. S. Smale: Mathematical problems for the next century, The Mathematical Intelligencer **20** (2) (1998), 7–15.

134. M. Taylor: Analysis on Morrey spaces and applications to the Navier-Stokes and other evolution equations, Comm. Part. Diff. Eq. **17** (1992), 1407–1456.

135. R. Temam: Some developments on Navier-Stokes equations in the second half of the 20th century, Development of Mathematics 1950-2000, J.-P. Pier (ed.) (2000).

136. E. Terraneo: Applications de certains espaces fonctionnels de l'analyse harmonique réelle aux équations de Navier-Stokes et de la chaleur non-linéaire, Ph.D. Thesis, Université d'Evry Val d'Essonne (1999).

137. E. Terraneo: Sur la non-unicité des solutions faibles de l'équation de la chaleur non linéaire avec non-linéarité u^3, C. R. Acad. Sci. Paris Sér. I Math. **328 (9)** (1999), 759–762.

138. G. Tian and Z. Xin: One-point singular solutions to the Navier-Stokes equations, Topol. Methods Nonlinear Anal. **11 (1)** (1998), 135–145.

139. T. P. Tsai: On Leray's self-similar solutions of the Navier-Stokes equations satisfying local energy estimates, Arch. Rational Mech. Anal **143 (1)** (1998), 29–51. Erratum, ibid. **147** (1999), 363.

140. S. Tourville: An analysis of Numerical Methods for Solving the two-dimensional Navier-Stokes Equations, Ph. D. Thesis, Washington University of St. Louis, MO (1998).

141. S. Tourville: Existence and uniqueness of solutions for a modified Navier-Stokes equation in R^2, Comm. Partial Differential Equations **23 (1–2)** (1998), 97–121.

142. S. Ukai: A solution formula for the Stokes equation in $\mathbb{R}^n_+$, Comm. in Pure and Appl. Math. **15** (1987), 611–621.

143. W. von Wahl: The equations of Navier-Stokes and abstract parabolic equations, Aspect der Mathematik, Vieweg & Sohn, Wiesbaden (1985).

144. K. A. Voss: Self-similar Solutions of the Navier-Stokes Equations, Ph.D. Thesis, Yale University (1996).

145. M. Vishik: Hydrodynamics in Besov Spaces, Arch. Rational Mech. Anal. **145** (1998), 197–214.

146. M. Vishik: Incompressible flows of an ideal fluid with vorticity in bordeline spaces of Besov type, Ann. Sci. Ecole Norm. Sup. (4) 32 (1999), no. 6, 769–812.

147. M. Vishik: Incompressible flows of an ideal fluid with unbounded vorticity, preprint, IHES/M /98/75 (1998).

148. F. B. Weissler: The Navier-Stokes initial value problem in L^p, Arch. Rat. Mech. Anal. **74** (1981), 219–230.

149. M. Wiegner: The Navier-Stokes equations – A neverending challenge?, Jahresber. Deutsch. Math.-Verein, **101** (1999), 1–25.

150. M. Yamazaki: Shuju no kansūkūkan ni okeru Navier-Stokes houteishiki, (in Japanese) Sūgaku **51 (3)** (1999), 291–308 (English translation to appear).

151. V. I. Yudovich: The Linearization Method in Hydrodynamical Stability Theory, Trans. Math. Mon. Amer. Math. Soc. Providence, RI **74** (1984).

The Dynamical Systems Approach to the Navier-Stokes Equations of Compressible Fluids

Eduard Feireisl*

Institute of Mathematics AV ČR, Žitná 25, 115 67 Praha 1, Czech Republic
e-mail: feireisl@math.cas.cz

Abstract. We develop a dynamical system theory for the Navier-Stokes equations of isentropic compressible fluid flows in three space dimensions. Such a theory is based on a priori estimates, asymptotic compactness of solutions and an exact description of propagation of oscillations in the density component. The abstract theory yields several results concerning the long-time behaviour of solutions, in particular, the convergence towards a stationary state of a potential flow is proved.

> ... *what we call objective reality is, in the last analysis, that which is common to several thinking beings, and could be common to all; this common part, we will see, can be nothing but the harmony expressed by mathematical laws.*
>
> H. Poincaré, *La valeur de la science*, p. 9.

1 The Model and its Basic Properties

The mathematical subject we call a dynamical system is completely characterized by its *state* and rules called the *dynamics* for determining the state which corresponds to a given present state at a specified future time. Once the dynamic is given, it is the task of dynamical systems theory to study the patterns of how states change in the long run. In an attempt to predict the long-time behaviour of dynamical systems we encounter several difficulties related to chaos, bifurcation, and sensitivity to initial data.

Since the time of Euler, partial differential equations have been used to describe the movements of continuous media like fluids or vibrations of solids. Consider $\Omega \subset R^3$ a spatial domain filled with a fluid. We shall assume its motion is characterized by the *velocity* $u = u(t, x)$ of the particle moving through $x \in \Omega$ at *time* $t \in I \subset R$. Moreover, at each time t, we shall suppose the fluid has a well-defined *density* $\varrho = \varrho(t, x)$.

The time evolution of ϱ, u is determined by two fundamental principles of classical mechanics:

* The work was supported by Grant 201/98/1450 GA ČR.

conservation of mass:

$$\frac{\partial \varrho}{\partial t} + \text{div}(\varrho \boldsymbol{u}) = 0 \; ; \tag{1}$$

balance of momenta:

$$\frac{\partial \varrho \boldsymbol{u}}{\partial t} + \text{div}(\varrho \boldsymbol{u} \otimes \boldsymbol{u}) + \nabla p = \text{div } \mathbb{T}(\boldsymbol{u}) + \varrho \boldsymbol{f} \tag{2}$$

$$\text{for all } t \in I, \, x \in \Omega \; .$$

We shall suppose the fluid is Newtonian, i.e., the *stress tensor* $\mathbb{T}$ is given by the formula

$$\mathbb{T}(\boldsymbol{u}) = \mu\big[\nabla \boldsymbol{u} + (\nabla \boldsymbol{u})^T\big] + \lambda \text{ div } \boldsymbol{u} \text{ Id}$$

where λ, μ are the *viscosity coefficients* assumed constant and satisfying

$$\mu > 0, \; \lambda + \mu \geq 0 \; .$$

We concentrate on *barotropic* fluids where the *pressure* p and the density ϱ are functionally dependent and the relation between them is given by the *equation of state*:

$$p = p(\varrho), \; p \in C[0, \infty) \cap C^1(0, \infty), \; p'(\varrho) > 0 \text{ for all } \varrho > 0 \; .$$

The fluid is driven by an *external force* of density $\boldsymbol{f} = \boldsymbol{f}(t, x)$.

The boundary $\partial \Omega$ will be smooth and solid in the sense the fluid cannot cross it, i.e., $\boldsymbol{u}(t, x).\mathbf{n}(x) = 0$ for all $t \in I$, $x \in \partial \Omega$ where $\mathbf{n}$ denotes the outer normal vector. Moreover, the fluid being viscous, it seem reasonable to suppose the the tangential component of $\boldsymbol{u}$ vanishes on $\partial \Omega$ as well. In other words, we impose

the no-slip boundary conditions:

$$\boldsymbol{u}(t, x) = \boldsymbol{0} \text{ for all } t \in I, \, x \in \partial \Omega \; . \tag{3}$$

Thus from the dynamical systems point of view, the state of the fluid at a given time $t \in I$ is characterized by its density $\varrho(t)$ and the momenta $(\varrho \boldsymbol{u})(t)$ while the dynamics is determined by the equations (1) and (2) complemented by the boundary conditions (3). Let us point out, at the very beginning of our study, that it is a major open problem to prove or disprove that *classical solutions* of the problem (1)–(3) exist for all time.

Taking the scalar product of (2) and u we obtain an equation for the *kinetic energy* density $\mathcal{E}_{\mathrm{kin}}$:

$$\frac{\partial \mathcal{E}_{\mathrm{kin}}}{\partial t} + \mathrm{div}\left(\mathcal{E}_{\mathrm{kin}} u + p u\right) - p\, \mathrm{div}\ u = \mathrm{div}\left(\mathbb{T}(u).u\right) - \mathbb{T}(u) : u + \varrho f.u \quad (4)$$

where

$$\mathcal{E}_{\mathrm{kin}} = \frac{1}{2}\varrho|u|^2 \ .$$

Similarly, multiplying (1) by $b'(\varrho)$ we deduce

$$\frac{\partial b(\varrho)}{\partial t} + \mathrm{div}(b(\varrho)u) + (b'(\varrho)\varrho - b(\varrho))\mathrm{div}\ u = 0 \qquad (5)$$

for any differentiable function b. The relations (4) and (5) yield an equation satisfied by the *total energy* $\mathcal{E}$:

$$\frac{\partial \mathcal{E}}{\partial t} + \mathrm{div}\left(\mathcal{E}u + pu\right) = \mathrm{div}\left(\mathbb{T}(u).u\right) - \mathbb{T}(u) : u + \varrho f.u \qquad (6)$$

where

$$\mathcal{E} = \mathcal{E}_{\mathrm{kin}} + \mathcal{E}_{\mathrm{pot}} = \frac{1}{2}\varrho|u|^2 + P(\varrho)$$

with P satisfying

$$P'(z)z - P(z) = p(z) \text{ for all } z > 0 \ .$$

The relations (5) and (6) give rise to *a priori* estimates which form the background of the existence theory providing weak (distributional) solutions of the problem. As we will see, it is precisely the total energy $\mathcal{E}$ giving the right "norm" on the "state" space we shall work in.

To simplify the presentation, the following hypotheses will be assumed throughout the whole text:

[**D**] $\Omega \subset R^3$ is a *bounded* domain with the Lipschitz boundary $\partial\Omega$;

[**P**] $p(\varrho) = a\varrho^\gamma$ with $a > 0$, i.e., the flow is isentropic with the *adiabatic constant* $\gamma > 1$;

[**F**] the driving force density f is a bounded measurable function,

$$\mathrm{ess}\ \sup_{t \in I,\ x \in \Omega}\ |f(t,x)| \leq B_f \ .$$

Accordingly, we have $P(\varrho) = \frac{a}{\gamma-1}\varrho^\gamma$ and the relation (6), integrated with respect to the space variable, gives rise to the *energy inequality*:

$$\frac{dE}{dt} + \int_\Omega \mu |\nabla \boldsymbol{u}|^2 + (\lambda + \mu) |\text{div } \boldsymbol{u}|^2 \, dx \le \int_\Omega \varrho \boldsymbol{f}.\boldsymbol{u} \, dx \quad \text{for } t \in I \qquad (7)$$

where

$$E = E[\varrho, (\varrho \boldsymbol{u})] = \int_\Omega \frac{1}{2} \varrho |\boldsymbol{u}|^2 + \frac{a}{\gamma - 1} \varrho^\gamma \, dx \ .$$

2 The Class of Finite Energy Weak Solutions

We shall deal with *finite energy weak solutions* of the problem (1)–(3), specifically,

[FES1] the density ϱ is a nonnegative function belonging to the space

$$\varrho \in L^\infty_{\text{loc}}(I; L^\gamma(\Omega));$$

[FES2] the components of the velocity u^i, $i = 1, 2, 3$ satisfy

$$u^i \in L^2_{\text{loc}}\left(I; W^{1,2}_0(\Omega)\right), \ i = 1, 2, 3;$$

[FES3] the energy $E = E[\varrho, \boldsymbol{u}]$, is locally integrable on I and satisfies the energy inequality (7) in $\mathcal{D}'(I)$;

[FES4] the equations (1) and (2) hold in $\mathcal{D}'(I \times \Omega)$;

[FES5] the density ϱ is square integrable up to the boundary, i.e.,

$$\varrho \in L^2_{\text{loc}}(I; L^2(\Omega)) \ .$$

The existence of finite energy weak solutions on the time interval $I = R^+ = (0, \infty)$, $\Omega \subset R^N$ with prescribed initial density $\varrho(0)$ and momenta $(\varrho \boldsymbol{u})(0)$ was proved by Lions [27] under the hypothesis $\gamma \ge \gamma(N)$. The square integrability up to the boundary of weak solutions (see [FES5]) was shown in [16] and [28]. Combining the technique of [27] with the results of [16] and [28], one can show that the critical value of the adiabatic constant γ for which one can prove existence of finite energy weak solutions for $\Omega \subset R^3$ is $\gamma(3) = \frac{9}{5}$.

The conditions [FES1], [FES5] yield a rigorous justification of an obvious but in the class of weak solutions only formal statement, namely, the continuity equation

$$\frac{\partial \varrho}{\partial t} + \text{div}(\varrho \boldsymbol{u}) = 0 \text{ holds in } \mathcal{D}'(I \times R^3) \qquad (8)$$

provided ϱ, $\boldsymbol{u}$ were prolonged by zero outside Ω (see [16, Lemma 3.1]). In particular, the total mass m is a constant of motion, i.e.,

$$m = \int_\Omega \varrho(t, x) \, dx \text{ is independent of } t \in I \ .$$

Rescaling the constants a, μ, and λ as the case may be we shall always assume $m = 1$.

The modern theory of dynamical systems applied to evolutionary partial differential equations as (1)–(3) represents the solutions of these equations in terms of a semiflow defined on a suitable phase space and possessing a global attractor compact in this space (cf. the monographs Babin and Vishik [2], Hale [19], Temam [35], Ladyzhenskaya [23] etc.). While this approach has found a widespread applicability to various equations of mathematical physics, there are several features of the problem (1)–(3) that make its use rather delicate. To begin, it is not known whether or not the finite energy weak solutions are uniquely determined by their initial values. Obviously, this question is closely related to the global regularity problem for (1) - (3). Typically for strongly nonlinear equations one can find either local regular solutions whose higher order norms may possibly "blow up" in a finite time or globally defined weak solutions enjoying only low regularity far from being sufficient to ensure uniqueness. One faces a similar gap between regularity and uniqueness as for the incompressible Navier-Stokes equations where the question of global regularity of solutions emanating from smooth initial data or uniqueness of the weak solutions constructed by Leray is still an outstanding open problem. Of course, everything turns relatively easy when the data are both small and smooth (cf. Matsumura and Nishida [31]).

Note, however, that the situation becomes much simpler when the space dimension is one. The question of existence and uniqueness for (1)–(3) has been largely settled (cf. the monograph Antontsev et al. [1], Hoff [20], Serre [33] etc.) while the existence of global compact attractors in the isothermal case was established recently by Hoff and Ziane [22].

There have been several attempts to bypass the possible lack of uniqueness and regularity of global in time solutions. Foias and Temam [17] showed the existence of a universal attracting set for the incompressible Navier-Stokes equations in three space dimensions, compact in the weak topology of the related phase space. More recently, Sell [32] developed a method of constructing a global attractor for the same problem considering the whole trajectories as "data" and replacing the solution semigroup by simple time shifts. Similar ideas form the base of the method of short trajectories introduced by Málek and Nečas [29] and further elaborated by Málek and Pražák [30].

One quickly encounters difficulties in trying to build a dynamical systems theory for (1)–(3) based on weak solutions. First of all, to determine the state of the system at any specific time, one needs both ϱ, the momenta ϱu and, possibly, the total energy E to be well-defined for any $t \in I$. A priori, the weak solutions are determined up to a set of measure zero in I. Fortunately, making use of the above hypotheses, one observes first that $E \in L^\infty_{\mathrm{loc}}(I)$ and, by virtue of (1) and (2),

$$\varrho \in C(J; L^\gamma_{\mathrm{weak}}(\Omega)), \quad \varrho u^i \in C(J; L^{p_1}_{\mathrm{weak}}(\Omega)) \text{ with } p_1 = \frac{2\gamma}{\gamma + 1} \qquad (9)$$

for any compact subinterval $J \subset I$. Moreover, the fact that the continuity equation holds in $\mathcal{D}'(I \times R^3)$ (cf. (8)) makes it possible to employ the regularizing machinery in the spirit of DiPerna and Lions [9] to deduce that (8) holds in the sense of renormalized solutions, specifically,

$$\frac{\partial b(\varrho)}{\partial t} + \operatorname{div}(b(\varrho)\boldsymbol{u}) + \left(b'(\varrho)\varrho - b(\varrho)\right) \operatorname{div} \boldsymbol{u} = 0 \text{ in } \mathcal{D}'(I \times R^3) \qquad (10)$$

for any globally Lipschitz function $b \in C^1(R)$ (cf. Lions [26, Lemma 2.3]). In addition, one can show by the same technique that

$$\varrho \in C(J; L^\alpha(\Omega)) \text{ for any } 1 \le \alpha < \gamma \text{ and any compact interval } J \subset I . \quad (11)$$

The above relations enable to justify the seemingly obvious observation that

$$(\varrho\boldsymbol{u})(t, x) = 0 \text{ for a.e. } x \in V(t) = \{x \mid \varrho(t, x) = 0\} \text{ for any } t \in I. \qquad (12)$$

Indeed one can use the Cauchy-Schwartz inequality to obtain

$$\int_{V(t)} |(\varrho u^i)(t)| \, dx \le \liminf_{t_n \to t} \int_{V(t)} |(\varrho u^i)(t_n)| \, dx$$

$$\le \liminf_{t_n \to t} \left(\int_{V(t)} \varrho(t_n) \, dx \right)^{\frac{1}{2}} \sqrt{2E(t_n)} = 0$$

for a suitable chosen sequence $t_n \to t$.

Now, redefining the total energy on a set of measure zero if necessary, we can set

$$E(t) = E[\varrho, (\varrho\boldsymbol{u})](t) = \frac{1}{2} \int_{\varrho(t)>0} \frac{|(\varrho\boldsymbol{u})|^2}{\varrho}(t) \, dx$$

$$+ \frac{a}{\gamma - 1} \int_\Omega \varrho^\gamma(t) \, dx \text{ for any } t \in I . \qquad (13)$$

Lemma 1. *The kinetic energy*

$$E_{\text{kin}} = \frac{1}{2} \int_{\varrho>0} \frac{|\boldsymbol{q}|^2}{\varrho} dx, \text{ considered as a function of } \varrho, \, \boldsymbol{q} ,$$

$$E_{\text{kin}} : L^1(\Omega) \times [L^p_{\text{weak}}(\Omega)]^3 \mapsto R \cup \{\infty\} ,$$

is (sequentially) lower semicontinuous for any $1 < p < \infty$.

Proof. We have

$$E_{\text{kin}}[\varrho, \boldsymbol{q}] = \frac{1}{2} \int_{\varrho>0} \frac{|\boldsymbol{q}|^2}{\varrho} \, dx = \lim_{\varepsilon \to 0+} \frac{1}{2} \int_{\varrho>0} \frac{|\boldsymbol{q}|^2}{\varrho + \varepsilon} \, dx$$

$$= \frac{1}{2} \lim_{\varepsilon \to 0+} \sum_{i=1}^3 \int_\Omega \left(\frac{h_\varepsilon(\varrho) \, q^i}{\sqrt{\varrho + \varepsilon}} \right)^2 dx$$

where $0 \leq h_\varepsilon \leq 1$, $h_\varepsilon(z) = 0$ if $z \leq 0$ $h_\varepsilon \nearrow 1$ for $z > 0$. On the other hand, for any sequence $\varrho_n \to \varrho$ in $L^1(\Omega)$, $q_n \to q$ weakly in $L^p(\Omega)$, one gets

$$\sum_{i=1}^{3} \int_\Omega \left(\frac{h_\varepsilon(\varrho)q^i}{\sqrt{\varrho+\varepsilon}}\right)^2(t) \, dx \leq \liminf_{n\to\infty} \sum_{i=1}^{3} \int_\Omega \left(\frac{h_\varepsilon(\varrho_n)q_n^i}{\sqrt{\varrho_n+\varepsilon}}\right)^2 \, dx$$

$$\leq \liminf_{n\to\infty} \int_{\varrho_n>0} \frac{|q_n|^2}{\varrho_n} \, dx$$

$$= 2 \liminf_{n\to\infty} E_{\text{kin}}[\varrho_n, q_n] \text{ for any } \varepsilon > 0 \ .$$

$\square$

Combining the conclusion of Lemma 1 with (9) and (10) we obtain the following:

Corollary 2. *For any finite energy weak solution ϱ, u, the total energy $E = E(t)$, defined for any $t \in I$ by (13), is lower semicontinuous on I.*

3   Concentrations in the Pressure Term

For any bounded energy weak solution of (1) and (2), the pressure term $p = p(\varrho)$ belongs only to the space $L^\infty(I; L^1(\Omega))$. Moreover, in view of the results of Lions [27], concentrations cannot occur on any compact subset of Ω. We shall show that one can control the possible concentration effects up to the boundary. To this end, we introduce the operator $\mathcal{B} = [\mathcal{B}_1, \mathcal{B}_2, \mathcal{B}_3]$ enjoying the properties:

- the operator

$$\mathcal{B} : \left\{g \in L^p(\Omega) \mid \int_\Omega g = 0\right\} \mapsto [W_0^{1,p}(\Omega)]^3$$

  is a bounded linear operator, i.e.,

$$\|\mathcal{B}[g]\|_{W_0^{1,p}(\Omega)} \leq c(p)\|g\|_{L^p(\Omega)} \text{ for any } 1 < p < \infty;$$

- the function $\mathbf{v} = \mathcal{B}[g]$ solves the problem

$$\text{div } \mathbf{v} = g \text{ in } \Omega, \ \mathbf{v}|_{\partial\Omega} = \mathbf{0};$$

- if, moreover, g can be written in the form $g = \text{div } \mathbf{h}$ for a certain $\mathbf{h} \in [L^r(\Omega)]^3$, $\mathbf{h}.\mathbf{n}|_{\partial\Omega} = 0$, then

$$\|\mathcal{B}[g]\|_{L^r(\Omega)} \leq c(r)\|\mathbf{h}\|_{L^r(\Omega)}$$

  for arbitrary $1 < r < \infty$.

The operator $\mathcal{B}$ was introduced by Bogovskii [4]. A complete proof of the above mentioned properties may be found in Galdi [18, Theorem 3.3] or Borchers and Sohr [5, Proof of Theorem 2.4].

Now, one can use $\mathcal{B}$ to construct multipliers of the form

$$\varphi_i(t, x) = \psi(t)\mathcal{B}_i\left[S_\varepsilon[b(\varrho)] - \oint_\Omega S_\varepsilon[b(\varrho)]\, dx\right], \quad i = 1, 2, 3, \ \psi \in \mathcal{D}(I)$$

where

$$b \in C^1(R), \ b(z) = z^{1/5} \text{ for } z \geq 1 \ .$$

and $S_\varepsilon[v] = \vartheta_\varepsilon * v$ where ϑ_ε is a regularizing sequence. The quantities φ_i, $i = 1, 2, 3$ may be used as test functions for the equations in (2) and, after a bit lengthy but straightforward computation, one arrives at the following formula:

$$
\begin{aligned}
a\int_I\int_\Omega \psi\varrho^\gamma S_\varepsilon[b(\varrho)]\, dx\, dt &= \int_I \psi\left(\int_\Omega a\varrho^\gamma\, dx\right)\left(\oint_\Omega S_\varepsilon[b(\varrho)]\, dx\right) dt \\
&\quad + (\lambda+\mu)\int_I\int_\Omega \psi\, S_\varepsilon[b(\varrho)] \operatorname{div} \boldsymbol{u}\, dx\, dt \\
&\quad - \int_I\int_\Omega \psi_t\, \varrho u^i\, \mathcal{B}_i\Big\{S_\varepsilon[b(\varrho)] - \oint_\Omega S_\varepsilon[b(\varrho)]dx\Big\}\, dx\, dt \\
&\quad + \mu\int_I\int_\Omega \psi\partial_{x_j}u^i\, \partial_{x_j}\mathcal{B}_i\Big\{S_\varepsilon[b(\varrho)] - \oint_\Omega S_\varepsilon[b(\varrho)]\, dx\Big\}\, dx\, dt \\
&\quad - \int_I\int_\Omega \psi\varrho u^i u^j\, \partial_{x_j}\mathcal{B}_i\Big\{S_\varepsilon[b(\varrho)] - \oint_\Omega S_\varepsilon[b(\varrho)]\, dx\Big\}\, dx\, dt \\
&\quad + \int_I\int_\Omega \psi\varrho u^i\, \mathcal{B}_i\Big\{S_\varepsilon\big[(b(\varrho) - b'(\varrho)\varrho) \operatorname{div} \boldsymbol{u}\big] \\
&\quad - \oint_\Omega S_\varepsilon\big[(b(\varrho) - b'(\varrho)\varrho)\operatorname{div} \boldsymbol{u}\big]\, dx\Big\}dx\, dt \\
&\quad + \int_I\int_\Omega \psi\varrho u^i\, \mathcal{B}_i\Big\{r_\varepsilon - \oint_\Omega r_\varepsilon\, dx\Big\}\, dx \\
&\quad - \int_I\int_\Omega \psi\varrho u^i\, \mathcal{B}_i\Big\{\operatorname{div}\big(S_\varepsilon[b(\varrho)]\boldsymbol{u}\big)\Big\}\, dx\, dt \\
&\quad - \int_I\int_\Omega \psi\varrho f_i\, \mathcal{B}_i\Big\{S_\varepsilon[b(\varrho)] - \oint_\Omega S_\varepsilon[b(\varrho)]\, dx\Big\}\, dx\, dt
\end{aligned}
\tag{14}
$$

where

$$r_\varepsilon \to 0 \text{ in } L^2_{\text{loc}}(I; L^{18/11}(\Omega)) \text{ as } \varepsilon \to 0+$$

(see [13, Section 3]).

Making use of the properties of $\mathcal{B}$ one has $\mathcal{B}_i\{r_\varepsilon\} \to 0$ in $L^2_{\text{loc}}(I; L^{18/5}(\Omega))$ while ϱu_i is bounded in $L^2_{\text{loc}}(I; L^{18/13}(\Omega))$ provided $\gamma \geq \frac{9}{5}$. Consequently, one can pass to the limit for $\varepsilon \to 0+$ in (14) and to use the energy estimates together with the Hölder inequality to obtain the following conclusion (see [16] for details):

Proposition 3. *Under the hypotheses* [D], [P], [F], *let* ϱ, u *be a finite energy weak solution of the problem (1)–(3) defined on an open time interval* $I \subset R$. *Let* $\gamma \geq \frac{9}{5}$.

Then

$$\int_J \int_\Omega |\varrho|^{\gamma+1/5} \, dx \, dt \leq c = c(E_J, B_f, |J|) \text{ for any compact } J \subset I$$

where we have denoted

$$E_J = \sup_{t \in J} E(t) \ .$$

In particular, under the hypotheses of Proposition 3, we have bounds on the L^2–norm of the density up to the boundary. These estimates are in fact necessary when proving *existence* of finite energy weak solutions if $9/5 \leq \gamma < 2$.

Finite energy weak solutions enjoying, in addition to [FES1]–[FES5], the regularity properties stated in Proposition 3 can be constructed by the method of Lions [28] making use of the regularity properties of the pressure in the Stokes problem.

4   The Effective Viscous Flux, Paracommutators

There is an important quantity $p(\varrho) - (\lambda + 2\mu) \operatorname{div} u$ called the effective viscous flux. It was pointed out by Hoff [21] and Serre [34] that it is in a certain sense more regular than its components. The main discovery of Lions [27] asserts that the space oscillations (if any) of this quantity are "orthogonal" to those of the density for any bounded family of finite energy weak solutions. This is a remarkable property which plays the main role in the existence proof given in [27].

To explain this phenomenon, consider a sequence ϱ_n, u_n solving (1) and (2) with $f = f_n$ in $\mathcal{D}'(I \times \Omega)$. Assume

$$f_n \to f \text{ weakly star in } [L^\infty(I \times \Omega)]^3,$$

$$\varrho_n \to \overline{\varrho} \text{ weakly star in } L^\infty(I; L^\gamma(\Omega)), \gamma > \tfrac{3}{2},$$
$$\text{and weakly in } L^2(I \times \Omega),$$

$$p(\varrho_n) \to \overline{p(\varrho)} \text{ weakly } L^1(I \times \Omega), \tag{15}$$

$$T_k(\varrho) \to \overline{T_k(\varrho)} \text{ weakly star in } L^\infty(I \times \Omega),$$

$$u_n \to \overline{u} \text{ weakly in } [L^2(I; W_0^{1,2}(\Omega))]^3 \ .$$

Here, and in what follows, the bar denotes a weak limit and T_k stands for the cut-off function

$$T_k(z) = \operatorname{sgn}(z) \min\{|z|, k\}, \ k = 1, 2, \ldots$$

The result mentioned above reads as follows:

Proposition 4. *Let ϱ_n, $\boldsymbol{u}_n$ be a sequence solving (1) and (2) with $\boldsymbol{f} = \boldsymbol{f}_n$ in $\mathcal{D}'(I \times \Omega)$ and satisfying (15). Moreover, assume the energy $E[\varrho_n, \boldsymbol{u}_n]$ is bounded independently of n.*
   Then

$$\overline{p(\varrho)T_k(\varrho)} - (\lambda + 2\mu)\,\overline{\operatorname{div}\boldsymbol{u}\,T_k(\varrho)} = \overline{p(\varrho)}\,\overline{T_k(\varrho)} - (\lambda + 2\mu)\,\operatorname{div}\overline{\boldsymbol{u}}\,\overline{T_k(\varrho)},$$

$k = 1, 2, \dots.$

Proposition 4 is the heart of the existence theory of Lions [27, Chapter 1]. The proof in [27, Appendix B] is based on hidden regularity of the commutators

$$v\mathcal{R}_{i,j}[w] - \mathcal{R}_{i,j}[vw] \text{ with } \mathcal{R}_{i,j} = \partial_{x_i}\Delta^{-1}\partial_{x_j}$$

where $v \in W^{1,2}$, $w \in L^p$, $1/p + 1/2 < 1$ (see Coifman and Meyer [7]). Note that this approach requires a slight modification when $\gamma \leq 3$.

Here, we give a different and more elementary proof based on weak continuity of the bilinear operator

$$\mathcal{Q}_{i,j}[v, w] = v\mathcal{R}_{i,j}[w] - w\mathcal{R}_{i,j}[v].$$

(cf. [15, Lemma 4.1]). Such an approach is more "compensated compactness like" in the spirit of Li et al. [24].

Proof (of Proposition 4). Since ϱ_n satisfies (1) in $\mathcal{D}'(I \times \Omega)$, (15) implies

$$\varrho_n \to \overline{\varrho} \text{ in } C(J; L^\gamma_{\text{weak}}(\Omega)) \text{ for any compact } J \subset I$$

and, consequently,

$$\varrho_n u_n^i \to \overline{\varrho}\,\overline{u}^i \text{ weakly in } L^2(I; L^{\frac{6\gamma}{7+6}}), \ i = 1, 2, 3 .$$

Thus $\overline{\varrho}$, $\overline{\boldsymbol{u}}$ solve (1) in $\mathcal{D}'(I \times \Omega)$. Moreover, since ϱ_n, $\overline{\varrho}$, $\nabla\boldsymbol{u}_n$, $\nabla\overline{\boldsymbol{u}}$ belong to $L^2(I \times \Omega)$, the equation (1) holds also in the sense of renormalized solutions, in particular,

$$\frac{\partial T_k(\varrho_n)}{\partial t} + \operatorname{div}\Big(T_k(\varrho_n)\boldsymbol{u}_n\Big) - k\,\operatorname{sgn}^+(\varrho_n - k)\,\operatorname{div}\boldsymbol{u}_n = 0 \qquad (16)$$

in $\mathcal{D}'(I \times \Omega)$, and

$$\frac{\partial \overline{T_k(\varrho)}}{\partial t} + \operatorname{div}\Big(\overline{T_k(\varrho)}\,\overline{\boldsymbol{u}}\Big) - k\,\overline{\operatorname{sgn}^+(\varrho - k)\,\operatorname{div}\boldsymbol{u}} = 0 \text{ in } \mathcal{D}'(I \times \Omega) . \qquad (17)$$

Moreover, since the energy is bounded, we have

$$\varrho_n\boldsymbol{u}_n \to \overline{\varrho}\,\overline{\boldsymbol{u}} \text{ weakly star in } L^\infty(I; L^{\frac{2\gamma}{\gamma+1}}(\Omega))$$

and, making use of (2),

$$\partial_t(\varrho_n \boldsymbol{u}_n) \text{ bounded in } L^2(I; W^{-2,2}(\Omega)).$$

Consequently,

$$\varrho_n u_n^i \to \overline{\varrho}\, \overline{u}^i \text{ in } C(J; L_{\text{weak}}^{\frac{2\gamma}{\gamma+2}}(\Omega)), \ J \subset I \text{ compact} \qquad (18)$$

and, since $\gamma > 3/2$,

$$\varrho_n u_n^i u_n^j \to \overline{\varrho}\, \overline{u}^i\, \overline{u}^j \text{ in, say, } \mathcal{D}'(I \times \Omega) \ .$$

Thus in the limit we get

$$\partial_t(\overline{\varrho}\, \overline{u}) + \text{div}(\overline{\varrho}\, \overline{u} \otimes \overline{u}) + \nabla \overline{p} = \text{div } \mathbb{T}(\overline{u}) + \overline{\varrho f} \text{ in } \mathcal{D}'(\Omega) \ . \qquad (19)$$

Now, let $\mathcal{H}_j = \Delta^{-1}\partial_{x_j}$, $j = 1, 2, 3$ be a differential operator whose Fourier symbol is $\tilde{\mathcal{H}}_j(\xi) = \frac{-i\xi_j}{|\xi|^2}$. We use the quantities

$$\psi(t)\phi_1(x)\mathcal{H}\{\phi_2(x)T_k(\varrho_n)\}, \ \psi \in \mathcal{D}(0,T), \ \phi_i \in \mathcal{D}(\Omega), \ i = 1, 2$$

as test function for (2). Making use of (16), we arrive at the following formula:

$$\int\int \psi\phi_1\phi_2\Big(p(\varrho_n) - (\lambda + 2\mu)\text{div}\boldsymbol{u}_n\Big)T_k(\varrho_n) \, dx \, dt =$$

$$\mathbf{M}_n \begin{cases} \int\int \psi\, \mathbb{T}(\boldsymbol{u}_n) : \Big[\nabla\phi_1 \otimes \mathcal{H}\{\phi_2 T_k(\varrho_n)\}\Big] \, dx \, dt \\[2mm] \quad - \int\int \psi\, p(\varrho_n)\nabla\phi_1.\mathcal{H}\{\phi_2 T_k(\varrho_n)\} \, dx \, dt \\[2mm] \quad - \int\int \psi\, \varrho_n[\boldsymbol{u}_n \otimes \boldsymbol{u}_n] : [\nabla\phi_1 \otimes \mathcal{H}\{\phi_2 T_k(\varrho_n)\}] \, dx \, dt \\[2mm] \quad - \int\int \psi\phi_1\varrho_n\boldsymbol{u}_n.\mathcal{H}\{T_k(\varrho_n)\boldsymbol{u}_n.\nabla\phi_2\} \, dx \, dt \\[2mm] \quad -2\mu\int\int \psi\, [\nabla\phi_1 \otimes \boldsymbol{u}_n] : [\nabla\mathcal{H}\{\phi_2 T_k(\varrho_n)\}] \, dx \, dt \\[2mm] \quad -k\int\int \psi\phi_1\varrho_n\boldsymbol{u}_n.\mathcal{H}\{\phi_2 \, \text{sgn}^+(\varrho_n - k)\text{div}\boldsymbol{u}_n\} \, dx \, dt \\[2mm] \quad - \int\int \psi_t\phi_1\varrho_n\boldsymbol{u}_n.\mathcal{H}\{\phi_2 T_k(\varrho_n)\} \, dx \, dt \\[2mm] \quad - \int\int \psi\phi_1\varrho_n\boldsymbol{f}_n.\mathcal{H}\{\phi_2 T_k(\varrho_n)\} \, dx \, dt \end{cases}$$

$$+ \sum_{i,j}\int\int \psi\, u_n^j\, \mathcal{Q}_{i,j}\Big\{\phi_1\varrho_n u_n^i \,\Big|\, \phi_2 T_k(\varrho_n)\Big\} \, dx \, dt$$

where the bilinear form $\mathcal{Q}_{i,j}$ is defined as

$$\mathcal{Q}_{i,j}\{v \mid w\} = v\, \partial_{x_i}\mathcal{H}_j\{w\} - w\, \partial_{x_i}\mathcal{H}_j\{v\} = v\, \mathcal{R}_{i,j}\{w\} - w\mathcal{R}_{i,j}\{v\} \ .$$

Next, taking the quantities $\psi\phi_1\mathcal{H}\{\phi_2\overline{T_k(\varrho)}\}$ as test functions for (19) and making use of (17), we obtain

$$\int\int \psi\phi_1\phi_2\Big(\overline{p}-(\lambda+2\mu)\mathrm{div}\overline{u}\Big)\overline{T_k(\varrho)}\ \mathrm{d}x\ \mathrm{d}t =$$

$$M\begin{cases} \int\int\psi\ \mathbb{T}(\overline{u}):\Big[\nabla\phi_1\otimes\mathcal{H}\{\phi_2\overline{T_k(\varrho)}\}\Big]\ \mathrm{d}x\ \mathrm{d}t- \\[2mm] \quad -\int\int\psi\ \overline{p}\nabla\phi_1.\mathcal{H}\{\phi_2\overline{T_k(\varrho)}\}\ \mathrm{d}x\ \mathrm{d}t \\[2mm] \quad -\int\int\psi\ \overline{\varrho[u\otimes u]}:[\nabla\phi_1\otimes\mathcal{H}\{\phi_2\overline{T_k(\varrho)}\}]\ \mathrm{d}x\ \mathrm{d}t- \\[2mm] \quad -\int\int\psi\phi_1\overline{\varrho}\ \overline{u}.\mathcal{H}\{\overline{T_k(\varrho)}\ \overline{u}.\nabla\phi_2\}\ \mathrm{d}x\ \mathrm{d}t \\[2mm] -2\mu\int\int\psi\ [\nabla\phi_1\otimes\overline{u}]:[\nabla\mathcal{H}\{\phi_2\overline{T_k(\varrho)}\}]\ \mathrm{d}x\ \mathrm{d}t \\[2mm] -k\int\int\psi\phi_1\overline{\varrho}\ \overline{u}.\mathcal{H}\{\phi_2\ \overline{\mathrm{sgn}^+(\varrho-k)\mathrm{div}u}\}\ \mathrm{d}x\ \mathrm{d}t \\[2mm] \quad -\int\int\psi_t\phi_1\overline{\varrho}\ \overline{u}.\mathcal{H}\{\phi_2\overline{T_k(\varrho)}\}\ \mathrm{d}x\ \mathrm{d}t \\[2mm] \quad -\int\int\psi\phi_1\overline{\varrho f}.\mathcal{H}\{\phi_2\overline{T_k(\varrho)}\}\ \mathrm{d}x\ \mathrm{d}t \end{cases}$$

$$+\sum_{i,j}\int\int\psi\ \overline{u}^j\ \mathcal{Q}_{i,j}\Big\{\phi_1\overline{\varrho}\ \overline{u}^i\ \Big|\ \phi_2\overline{T_k(\varrho)}\Big\}\ \mathrm{d}x\ \mathrm{d}t\ .$$

First of all, one can use the smoothing properties of the operator $\mathcal{H}$ and the standard Sobolev embedding relations to deduce that $M_n \to M$ (see [15] for details) . Consequently, it remains to pass to the limit in the term containing the bilinear form $\mathcal{Q}_{i,j}$. By virtue of (15) and (18), and

$$T_k(\varrho_n) \to \overline{T_k(\varrho)} \text{ in } C(J;L^q_{\mathrm{weak}}(\Omega)) \text{ for any } q \geq 1\ ,$$

which may be deduced from (16), the desired conclusion will follow from the next compensated compactness like result:

Lemma 5. *Let v_n, w_n be two sequences,* supp v_n, supp $w_n \subset \Omega$, *such that*

$$v_n \to v \text{ weakly in } L^p(R^3), \ w_n \to w \text{ weakly in } L^q(R^3),$$

$$\frac{1}{p}+\frac{1}{q}=\frac{1}{r}<1\ .$$

Then

$$\mathcal{Q}_{i,j}\{v_n\mid w_n\} \to \mathcal{Q}_{i,j}\{v\mid w\} \text{ weakly in } L^r(\Omega)\ . \tag{20}$$

Remark 6. If $p = q = 2$, the above assertion may be viewed as a particular case of Coifman et al. [6, Theorem 5.1].

Proof (i) Assume first that

$$v_n \to v, \ w_n \to w \text{ weakly in } L^2(R^3) \ .$$

Consider the Fourier transforms $\tilde{v}_n, \tilde{w}_n$. Accordingly, we have

$$\tilde{v}_n, \tilde{w}_n \text{ bounded in } L^\infty(R^3) \text{ independently of } n$$

and

$$\tilde{v}_n(\xi) \to \tilde{v}(\xi), \ \tilde{w}_n(\xi) \to \tilde{w}(\xi) \text{ for all } \xi \ . \tag{21}$$

Denoting the quantities in (20) by Q_n, Q respectively and taking the Fourier transform of Q_n we get

$$\tilde{Q}_n(\xi) = \int_{R^3} \left[\frac{(\xi - \eta)_i (\xi - \eta)_j}{|\xi - \eta|^2} - \frac{\eta_i \eta_j}{|\eta|^2} \right] \tilde{v}_n(\xi - \eta) \tilde{w}_n(\eta) \ d\eta \ .$$

Consequently, to show the convergence of Q_n to Q in, say, $\mathcal{D}'(\Omega)$, we have to prove

$$\tilde{Q}_n(\xi) \to \tilde{Q}(\xi) \text{ for all } \xi \in \mathbb{R}^3 \ .$$

Thus, by virtue of (21), it is enough to show

$$\lim_{|\eta| \to \infty} \left[\frac{(\xi - \eta)_i (\xi - \eta_j)}{|\xi - \eta|^2} - \frac{\eta_i \eta_j}{|\eta|^2} \right] = 0 \text{ for any fixed } \xi \in R^3 \ .$$

Arguing by contradiction we find a sequence η^m such that

$$|\eta^m| \to \infty \text{ and } \left| \frac{(\xi - \eta^m)_i (\xi - \eta^m)^j}{|\xi - \eta^m|^2} - \frac{\eta_j^m \eta_i^m}{|\eta^m|^2} \right| \geq k > 0 \ . \tag{22}$$

Now observe first that necessarily

$$\eta_i^m, \eta_j^m \to \infty \ ,$$

and, consequently,

$$\left| \frac{(\xi - \eta^m)_i (\xi - \eta^m)_j}{|\xi - \eta^m|^2} - \frac{\eta_i^m \eta_j^m}{|\eta|^2} \right| = \left| \frac{\eta_i^m \eta_j^m}{|\eta^m|^2} \right| \left| \frac{(\frac{\xi}{\eta^m} - 1)_i (\frac{\xi}{\eta^m} - 1)_j}{|\frac{\xi}{\eta^m} - 1|^2} - 1 \right| \to 0$$

in contrast with (22). Thus we have proved that (20) holds in $\mathcal{D}'(\Omega)$ provided $p = q = 2$.

(ii) In the general case, assume e.g. that $p > 2$ and replace w_n by $T_k(w_n)$. Thus

$$\|T_k(w_n) - w_n\|_{L^s(R^3)}^s \leq k^{s-q} \sup_n \|w_n\|_{L^q(R^3)} \ ,$$

and, consequently,

$$\|\overline{T_k(w)} - w\|^s_{L^s(R^3)} \le k^{s-q} \sup \|w_n\|_{L^q(R^3)}$$

where

$$s = p' < q, \quad \frac{1}{p} + \frac{1}{s} = 1 \ .$$

Finally, we write

$$\mathcal{Q}_{i,j}\{v_n \mid w_n\} - \mathcal{Q}_{i,j}\{v \mid w\} = \mathcal{Q}_{i,j}\{v_n \mid T_k(w_n)\} - \mathcal{Q}_{i,j}\{v \mid \overline{T_k(w)}\}$$
$$+ \Big[v_n \mathcal{R}_{i,j}[w_n - T_k(w_n)] - (w_n - T_k(w_n))\mathcal{R}_{i,j}[v_n] \Big]$$
$$+ \Big[v\mathcal{R}_{i,j}[\overline{T_k(w)} - w] - (\overline{T_k(w)} - w)\mathcal{R}_{i,j}[v] \Big]$$

where the first term on the right-hand side converges to zero in $\mathcal{D}'(\Omega)$ because of what we have proved in part **(i)** and the rest is uniformly small for large k in view of the estimates above and the continuity of $\mathcal{R}_{i,j}$ in the L^p−spaces.

$\square$

We have proved Proposition 4.

5   Bounded Absorbing Sets

We shall see that the total energy E is the right quantity to play the role of a "norm" of finite energy weak solutions. If the driving force f is uniformly bounded as in [**F**], the "dynamical system" generated by the finite energy weak solutions of the problem (1)–(3) is ultimately bounded or dissipative in the sense of Levinson with respect to the energy "norm". Specifically, we report the following result:

Theorem 7. *Assume $I = (a, b)$. Let the hypotheses* [**D**], [**P**], *and* [**F**] *be satisfied with $\gamma > \frac{5}{3}$.*
Then there exists a constant E_∞, depending solely on B_f, with the following property: Given E_0, there exists a time $T = T(E_0)$ such that

$$E[\varrho, \boldsymbol{u}](t) \le E_\infty \text{ for all } t \in I, \quad T + a < t < b,$$

for any ϱ, $\boldsymbol{u}$ – a finite energy weak solution of the problem (1)–(3) satisfying

$$\limsup_{t \to a+} E[\varrho, \boldsymbol{u}](t) \le E_0 \ .$$

The just stated property is an evidence of the dissipative nature of the system (1), (2). In finite-dimensional setting, J.E.Billoti and J.P.LaSalle proposed it as a definition of dissipativity. Unfortunately, however, some difficulties inherent to infinite-dimensional dynamical systems make it, in that case, less appropriate.

From the mathematical point of view, the existence of an ultimate bound on the energy, or, equivalently, the existence of a bounded absorbing set, reduces to the problem of finding suitable *a priori* estimates. We only sketch the proof referring to [13, Theorem 1.1] for details.

To begin, observe that, by virtue of the energy inequality (7), the total energy E, redefined on a set of measure zero if necessary, has locally bounded variation on I and

$$E(t+) = \operatorname{ess} \lim_{s \to t+} E(t) \le \operatorname{ess} \lim_{s \to t-} E(t) = E(t-) \text{ for any } t \in I .$$

Moreover,

$$E(t_2-) \le \left(1 + E(t_1+)\right) e^{\sqrt{2}B} f^{(t_2 - t_1)} - 1 \text{ for all } 0 < t_1 < t_2 . \tag{23}$$

It can be shown that for any given $T > 0$ the energy inequality (7) gives the following alternatives: Either

$$E((T+1)-) \le E(T+) - 1 \tag{24}$$

or

$$\left. \begin{array}{l} \int_T^{T+1} \|u(t)\|^2_{W_0^{1,2}(\Omega)} \, dt \le c_1 \left(1 + \int_T^{T+1} \|\varrho(t)\|_{L^{\frac{3}{2}}(\Omega)} \, dt\right), \\[2ex] E(t+) \le c_2 \left(1 + \int_T^{T+1} \|\varrho(s)\|^\gamma_{L^\gamma(\Omega)} \, ds\right) \text{ for any } t \in [0, T], \\[2ex] \qquad\qquad \text{and} \\[2ex] \int_T^{T+1} \int_\Omega \varrho^{\gamma + \theta} \le c_3 \left[\left(1 + \sup_{t \in [T, T+1]} \|\varrho(t)\|_{L^\gamma(\Omega)}\right) \int_T^{T+1} \|u\|^2_{W_0^{1,2}(\Omega)} \right. \\[2ex] \left. \qquad\qquad + \sup_{t \in [T, T+1]} \|\sqrt{\varrho(t)} u(t)\|_{L^2(\Omega)} \right] \end{array} \right\} \tag{25}$$

where $\theta = \min\{\frac{1}{4}, \frac{2\gamma - 3}{3\gamma}\}$ and the constants c_1, c_2, c_3 depend solely on γ and B_f. The proof of (25), in particular the last estimate, is of the same nature as the estimates in Section 3 and uses the same multipliers based on the operator $\mathcal{B} \approx \operatorname{div}^{-1}$.

The estimates (25) combined with interpolation arguments imply boundedness of E on the interval $[T, T + 1]$ by a constant depending solely on B_f (and on γ) and, consequently, the conclusion of Theorem 7 can be deduced from (23)–(25).

6   Propagation of Oscillations

The issue we shall address now is the ultimate compactness of bounded trajectories or, equivalently, the problem of propagation of spatial oscillations of
the density. To begin, note that any reasonable "solution operator" we might
associate to the finite energy weak solutions cannot be compact, at least
with respect to the density component, due to the hyperbolic character of
the continuity equation (1). In view of the observations of Lions [25], one can
anticipate that possible oscillations of the density will propagate in time. On
the other hand, if one is able to control the oscillations at some specific time,
they should not appear later. Unlike "compensated compactness" statements
similar to Proposition 4, the problem is not of local character and one must
take care also of the disturbances emanating from the boundary though, by
virtue of (3), there are no in-going characteristics but there is still a competition between the speed of decay of the velocity u in the proximity of $\partial\Omega$
and possible concentrations of the density ϱ in the same area. Fortunately,
the oscillations cannot be "created near the boundary" provided ϱ satisfies
(7). This is apparently related to the fact that the "driving" term $\varrho\,\mathrm{div}u$ in
(1) is then integrable (see Section 3 and the discussion in [16]).

The propagation of oscillations was studied by Serre [34]. It was shown
that the amplitude of the Young measure characterizing the oscillations of
the density is a nonincreasing function of time. The proof is complete in
the dimension $N = 1$ and formal for $N \geq 2$ assuming a priori the validity
of Proposition 4 proved later by Lions [27]. In fact, this property plays an
important role in the existence theory developed in [27].

We shall see that the amplitude of the oscillations is *decreasing* in time
at uniform rate. Such a result can be used to prove the existence of a global
attracting set for the density component.

Consider a sequence of finite energy weak solutions ϱ_n, u_n of (1)–(3) on
$I \subset R$ driven by f_n satisfying [F] uniformly in $n = 1, 2, \dots$ Moreover, suppose
that

$$E[\varrho_n, (\varrho_n u_n)](t) \leq E_I \quad \text{for all } t \in I \text{ and all } n = 1, 2, \dots$$

Thus we have (at least for subsequences)

$$\varrho_n \to \overline{\varrho} \quad \text{in } C(J; L^\gamma_{\text{weak}}(\Omega)) \ ,$$

$$u_n \to \overline{u} \quad \text{weakly in } L^2(I; W_0^{1,2}(\Omega)) \ ,$$

$$T_k(\varrho_n) \to \overline{T_k(\varrho)} \quad \text{weakly star in } L^\infty(J \times \Omega) \ ,$$

and, by virtue of Proposition 3,

$$\varrho_n \to \overline{\varrho} \quad \text{weakly in } L^2(J \times \Omega) \ ,$$

$$p(\varrho_n) \to \overline{p(\varrho)} \quad \text{weakly in, say, } L^1(J \times \Omega)$$

on any compact intarval $J \subset I$. Moreover, $\overline{\varrho}$, $\overline{u}$ solve the continuity equation

$$\frac{\partial \bar{\varrho}}{\partial t} + \mathrm{div}(\bar{\varrho}\,\bar{u}) = 0 \qquad (26)$$

both in $\mathcal{D}'(I \times R^3)$ and in the sense of renormalized solutions.

In particular, $\bar{\varrho} \in C(J; L^{\alpha}(\Omega))$ for any $1 \le \alpha < \gamma$, $J \subset I$ compact, and

$$\frac{\partial M_k(\bar{\varrho})}{\partial t} + \mathrm{div}(M_k(\bar{\varrho})\,\bar{u}) + T_k(\bar{\varrho})\,\mathrm{div}\,\bar{u} = 0 \text{ in } \mathcal{D}'(I \times R^3) \qquad (27)$$

where

$$M_k(z) = \begin{cases} z\log(z) \text{ for } 0 \le z < k \\[2mm] k\log(k) + (1 + \log(k))(z - k) \text{ for } z \ge k \ . \end{cases}$$

To describe the oscillations, we employ the standard concept of defect measure. As we will see that the right candidate is the quantity

$$\nu(t, x) = \overline{\varrho\log(\varrho)} - \bar{\varrho}\log(\bar{\varrho}) \ .$$

We shall show that the function

$$D(t) = \int_{\Omega} \nu(t, x) \, dx, \ t \in I$$

is decreasing in time at a uniform rate independent of the particular choice of the sequence ϱ_n, u_n. To carry out this idea, one has to approximate the function $z\log(z)$ by $M_k(z)$ because of the low inegrability of the pressure term.

Since ϱ_n, u_n satisfy (1) in the sense of renormalized solutions, one can deduce that

$$\frac{\partial M_k(\varrho_n)}{\partial t} + \mathrm{div}(M_k(\varrho_n)\,u_n) + T_k(\varrho_n)\,\mathrm{div}\,u_n = 0 \text{ in } \mathcal{D}'(I \times R^3); \qquad (28)$$

whence

$$M_k(\varrho_n) \to \overline{M_k(\varrho)} \text{ in } C(J; L^{\gamma}_{\mathrm{weak}}(\Omega)) \qquad (29)$$

where

$$\sup_{t \in J} \|\overline{M_k(\varrho)}(t)\|_{L^{\alpha}(\Omega)} \le c(\alpha) \text{ for } 1 \le \alpha < \gamma \text{ independently of } k = 1, 2, \ldots$$

Approximating $z\log(z)$ by $M_k(z)$ one can deduce from (29) that

$$\varrho_n\log(\varrho_n) \to \overline{\varrho\log(\varrho)} \text{ in } C(J; L^{\alpha}_{\mathrm{weak}}(\Omega)) \text{ for any } 1 \le \alpha < \gamma \ .$$

Taking the difference of (28) and (27) and integrating by parts one gets

$$\int_{\Omega} [M_k(\varrho_n) - M_k(\bar{\varrho})](t_2) \, dx - \int_{\Omega} [M_k(\varrho_n) - M_k(\bar{\varrho})](t_1) \, dx$$

$$+ \int_{t_1}^{t_2} \int_{\Omega} T_k(\varrho_n) \, \mathrm{div}\,u_n \, dx \, dt = \int_{t_1}^{t_2} \int_{\Omega} T_k(\bar{\varrho}) \, \mathrm{div}\,\bar{u} \, dx \, dt,$$

whence, passing to the limit for $n \to \infty$,

$$\int_\Omega [\overline{M_k(\varrho)} - M_k(\overline{\varrho})](t_2) \ dx - \int_\Omega [\overline{M_k(\varrho)} - M_k(\overline{\varrho})](t_1) \ dx$$

$$+ \lim_{n\to\infty} \int_{t_1}^{t_2} \int_\Omega T_k(\varrho_n) \ \mathrm{div} \ u_n \ dx \ dt$$

$$= \int_{t_1}^{t_2} \int_\Omega T_k(\overline{\varrho}) \ \mathrm{div} \ \overline{u} \ dx \ dt$$

for any $t_1, t_2 \in J$. Consequently, by virtue of Proposition 4,

$$\int_\Omega [\overline{M_k(\varrho)} - M_k(\overline{\varrho})](t_2) \ dx - \int_\Omega [\overline{M_k(\varrho)} - M_k(\overline{\varrho})](t_1) \ dx$$

$$= \int_{t_1}^{t_2} \int_\Omega \left(T_k(\overline{\varrho}) - \overline{T_k(\varrho)} \right) \ \mathrm{div} \ \overline{u} \ dx \ dt \qquad (30)$$

$$- \lim_{n\to\infty} \int_{t_1}^{t_2} \int_\Omega \frac{1}{\lambda + 2\mu} \left(p(\varrho_n) T_k(\varrho_n) - \overline{p(\varrho)} \ \overline{T_k(\varrho)} \right) \ dx \ dt \ .$$

Now, we have

$$\lim_{n\to\infty} \int_{t_1}^{t_2} \int_\Omega \left(p(\varrho_n) T_k(\varrho_n) - \overline{p(\varrho)} \ \overline{T_k(\varrho)} \right) \ dx \ dt$$

$$= \lim_{n\to\infty} \int_{t_1}^{t_2} \int_\Omega \left(p(\varrho_n) - p(\overline{\varrho}) \right) \left(T_k(\varrho_n) - T_k(\overline{\varrho}) \right) \ dx \ dt$$

$$+ \int_{t_1}^{t_2} \int_\Omega \left(\overline{p(\varrho)} - p(\overline{\varrho}) \right) \left(T_k(\overline{\varrho}) - \overline{T_k(\varrho)} \right) \ dx \ dt$$

$$\geq \lim_{n\to\infty} \int_{t_1}^{t_2} \int_\Omega \left(p(\varrho_n) - p(\overline{\varrho}) \right) \left(T_k(\varrho_n) - T_k(\overline{\varrho}) \right) \ dx \ dt$$

as p is convex and T_k concave.

Seeing that

$$\int_\Omega \left(p(\varrho_n) - p(\overline{\varrho}) \right) \left(T_k(\varrho_n) - T_k(\overline{\varrho}) \right) \ dx \geq a \int_\Omega |T_k(\varrho_n) - T_k(\overline{\varrho})|^{\gamma+1} \ dx$$

$$\geq a|\Omega|^{1 - \frac{\gamma+1}{\alpha}} \|T_k(\varrho_n) - T_k(\overline{\varrho})\|^{\gamma+1}_{L^\alpha(\Omega)}$$

for any $1 \leq \alpha < \gamma$, one can get from (30) that

$$\int_\Omega [\overline{M_k(\varrho)} - M_k(\overline{\varrho})](t_2) \ dx - \int_\Omega [\overline{M_k(\varrho)} - M_k(\overline{\varrho})](t_1) \ dx$$

$$+ a|\Omega|^{1 - \frac{\gamma+1}{\alpha}} \lim_{n\to\infty} \int_{t_1}^{t_2} \|T_k(\varrho_n) - T_k(\overline{\varrho})\|^{\gamma+1}_{L^\alpha(\Omega)} \ dt$$

$$\leq \int_{t_1}^{t_2} \int_\Omega \left(T_k(\overline{\varrho}) - \overline{T_k(\varrho)} \right) \ \mathrm{div} \ \overline{u} \ dx \ dt \qquad (31)$$

at least for a suitable subsequence of ϱ_n. Passing to the limit for $k \to \infty$ in (31) we infer

$$D(t_2) - D(t_1) + a|\Omega|^{1 - \frac{\gamma+1}{\alpha}} \lim_{n \to \infty} \int_{t_1}^{t_2} \|\varrho_n - \overline{\varrho}\|_{L^\alpha(\Omega)}^{\gamma+1} \, dt \le 0 \qquad (32)$$

for any $t_1, t_2 \in J$ and any $1 \le \alpha < \gamma$.

At this stage, we need a technical assertion following from the structural properties of the function log.

Lemma 8. *Let* $\alpha \in (1, \gamma)$ *be fixed. Then there exists* $c = c(\alpha) > 0$ *such that for any* $y, z \ge 0$,

$$z \log(z) - y \log(y) \le (1 + \log^+(y))(z - y) + c(\alpha)\left(|z - y|^{\frac{1}{2}} + |z - y|^\alpha\right) .$$

Proof. If $z \ge y$ or $1 \le z \le y$, one has

$$z \log(z) - y \log(y) - (1 + \log^+(y))(z - y)$$
$$\le (\log^+(z) - \log^+(y))(z - y) \le \log(1 + |z - y|)|z - y|$$

On the other hand, in the case $z \le \min\{1, y\}$, we obtain

$$z \log(z) - y \log(y) - (1 + \log^+(y))(z - y) \le \log(1 + y - z)(z - y)$$
$$- \int_z^{\min\{1, y\}} \log(s) \, ds \le \log(1 + y - z)(z - y) - \int_0^{\min\{1, y-z\}} \log(s) \, ds$$
$$\le \log(1 + |z - y|)|z - y| + |z - y||\log(|z - y|)| + |z - y|$$

$\square$

According to Lemma 8 we can write

$$\int_\Omega \varrho_n \log(\varrho_n) - \overline{\varrho} \log(\overline{\varrho}) \, dx - \int_\Omega (1 + \log^+(\overline{\varrho}))(\varrho_n - \overline{\varrho}) \, dx$$
$$\le c(\alpha)\left(|\Omega|^{\frac{2\alpha-1}{2\alpha}} \|\varrho_n - \overline{\varrho}\|_{L^\alpha(\Omega)}^{2\alpha} + \|\varrho_n - \overline{\varrho}\|_{L^\alpha(\Omega)}^\alpha\right) \qquad (33)$$

which, together with (32), yields

$$D(t_2) - D(t_1) + \int_{t_1}^{t_2} \Phi(D(t)) \, dt \le 0 \text{ for any } t_1 \le t_2, \; t_1, t_2 \in I \qquad (34)$$

where Φ is a strictly increasing continuous function on R, $\Phi(0) = 0$.

Since D is continuous on I, one obtains from (34) that

$$D(t_2) \le \chi(t_2 - t_1), \text{ where } \chi'(t) + \Phi(\chi(t)) = 0, \; \chi(0) = D(t_1), \qquad (35)$$

in other words, the amplitude of possible oscillations decreases in time at uniform rate given by the function χ independent of the upper bound on the energy E_I and the norm of the forcing term B_f.

In other words, we have proved the the deffect measure D is decreasing in time with uniform rate given by the function Φ depending solely on the structural properties of the function log. This is a remarkable property and we are going to formulate it in a rigorous way. To this end, some notation is needed. For any bounded set B in $L^\alpha(\Omega)$, $\alpha > 1$, we define a *measure of noncompactness* β,

$$\beta(B) = \sup \left\{ s \mid s = \int_\Omega \overline{v \log(v)} - \overline{v} \log(\overline{v}) \, dx \text{ for some sequence} \right.$$

$$\left. v_n \in B, \ v_n \to \overline{v}, \ v_n \log(v_n) \to \overline{v \log(v)} \text{ weakly in } L^1(\Omega) \right\} .$$

To bypass the possible nonuniqueness of weak solutions we introduce the quantity $U(t_0, t)$ playing the role of the evolution operator related to the problem (1)–(3). Let $\mathcal{F}$ be a bounded subset of $L^\infty_{loc}(R; L^\infty(\Omega))$. We denote

$$U[E_0, \mathcal{F}](t_0, t) = \left\{ [\varrho, \boldsymbol{q}] \mid \varrho = \varrho(t), \ \boldsymbol{q} = \varrho \boldsymbol{u}(t) \text{ where } \varrho, \boldsymbol{u} \text{ is a finite energy} \right.$$

$$\text{weak solution of the problem (1)–(3) on an open}$$

$$\text{interval } I, (t_0, t] \subset I, \text{ with } \boldsymbol{f} \in \mathcal{F}, \text{ and such that}$$

$$\left. \limsup_{t \to t_0+} E[\varrho(t), (\varrho \boldsymbol{u})(t)] \leq E_0 \right\} .$$

Theorem 9. *In addition to the hypotheses* [D], [P], *assume that* $\gamma \geq \frac{9}{5}$ *and that* $\mathcal{F}$ *is a bounded subset of* $L^\infty_{loc}(R; L^\infty(\Omega))$.
Then

$$\beta \left(\{ \varrho \mid [\varrho, \boldsymbol{q}] \in U[E_0, \mathcal{F}](a, t_2) \} \right) \leq \chi(t_2 - t_1) \text{ for any } a < t_1 \leq t_2 ,$$

where the function χ *is the unique solution of the problem*

$$\chi'(t) + \Phi(\chi(t)) = 0, \ \chi(0) = \beta \left(\{ \varrho \mid [\varrho, \boldsymbol{q}] \in U[E_0, \mathcal{F}](a, t_1) \} \right), \quad (36)$$

with Φ *a strictly increasing continuous function on* R, $\Phi(0) = 0$, *independent of* E_0, $\mathcal{F}$.

Proof. Fix $a < b < t_1$ and let

$$s = \int_\Omega \overline{v \log(v)} - \overline{v} \log(\overline{v}) \, dx \text{ for a certain } [\overline{v}, \boldsymbol{q}] \in U[E_0, \mathcal{F}](a, t_2) .$$

Accordingly, there exists a sequence of finite energy weak solutions ϱ_n, $\boldsymbol{u}_n$, defined on I_n, $[b, t_2] \subset \cap I_n$, where $\boldsymbol{f}_n$ satisfy [F] with B_f and the energy $E[\varrho_n, (\varrho_n \boldsymbol{u}_n)]$ is bounded independently of n; such that

$$\varrho_n \to \overline{\varrho} \text{ in } C([b, t_2]; L^\gamma_{\text{weak}}(\Omega)), \quad \overline{\varrho}(t_2) = \overline{v},$$

$$\varrho_n \log(\varrho_n) \to \overline{\varrho \log(\varrho)} \text{ in } C([b, t_2]; L^1_{\text{weak}}(\Omega)), \quad \overline{\varrho \log(\varrho)}(t_2) = \overline{v \log(v)},$$

where, by virtue of Proposition 3,

$$\overline{\varrho} \in L^2((b, t_2) \times \Omega) . \tag{37}$$

Moreover, the relation (35) yields

$$\int_\Omega \left(\overline{\varrho \log(\varrho)} - \overline{\varrho} \log(\overline{\varrho}) \right)(t) \, dx \leq \chi(t - t_1) \text{ for all } t \in [b, t_2)$$

with χ as in (36).

Consequently, to complete the proof, it is enough to show that the function $t \mapsto \int_\Omega \overline{\varrho} \log(\overline{\varrho})(t) \, dx$ is (right) continuous at $t = t_2$. To this end, one observes first that $\overline{\varrho}$ satisfies the continuity equation (1) in $D'((b, t_2) \times R^3)$ with a certain $\overline{\boldsymbol{u}} \in L^2((b, t_2); W_0^{1,2}(\Omega))$. Now, following DiPerna and Lions [9] one can use the standard regularization kernels to obtain

$$\overline{\varrho}^\varepsilon_t + \text{div}(\overline{\varrho}^\varepsilon \overline{\boldsymbol{u}}) = r^\varepsilon \text{ on } (b, t_2) \times R$$

for the regularized function $\overline{\varrho}^\varepsilon$ where, by virtue of (37), $r^\varepsilon \to 0$ in $L^1((b, t_2) \times R^3)$. Consequently, since we already know that $\overline{\varrho} \in C([b, t_2]; L^\gamma_{\text{weak}}(\Omega))$ we can pass to the limit for $\varepsilon \to 0$ to infer that $\overline{\varrho} \in C([b, t_2]; L^\alpha(\Omega))$ for any $1 \leq \alpha < \gamma$. $\qquad \square$

7   Complete Bounded Trajectories

We start with the concept of the so-called short trajectory in the spirit of Málek and Nečas [29]:

$$U^s[E_0, \mathcal{F}](t_0, t)$$

$$= \Big\{ [\varrho(\tau), \boldsymbol{q}(\tau)], \ \tau \in [0, 1] \ \Big| \ \varrho(\tau) = \varrho(t + \tau), \boldsymbol{q}(\tau) = (\varrho \boldsymbol{u})(t + \tau)$$

where $\varrho, \boldsymbol{u}$ is a finite energy weak solution of the problem (1)–(3) on an open interval I, $(t_0, t + 1] \subset I$, with $\boldsymbol{f} \in \mathcal{F}$, and such that

$$\limsup_{t \to t_0} E(t) \leq E_0 \Big\} .$$

The following result is a slight modification of [12, Theorem 1.1].

Proposition 10. *Assume* [D], [P] *with* $\gamma \geq \frac{9}{5}$. *Let* $\mathcal{F}$ *be bounded in* $L^\infty(R \times \Omega)$. *Assume* $[\varrho_n, \boldsymbol{q}_n] \in U^s[E_0, \mathcal{F}](a, t_n)$ *for a certain sequence* $t_n \to \infty$.
Then there is a subsequence (not relabeled) such that

$$\varrho_n \to \bar{\varrho} \text{ in } L^\gamma((0,1) \times \Omega) \text{ and in } C([0,1]; L^\alpha(\Omega)) \text{ for } 1 \leq \alpha < \gamma, \qquad (38)$$

$$\boldsymbol{q}_n \to (\bar{\varrho}\,\bar{\boldsymbol{u}}) \text{ in } L^p((0,1) \times \Omega) \text{ and in } C([0,1]; L_{\text{weak}}^{\frac{2\gamma}{\gamma+1}}(\Omega)) \qquad (39)$$

for any $1 \leq p < \frac{2\gamma}{\gamma+1}$, *and*

$$E[\varrho_n, \boldsymbol{q}_n] \to E[\bar{\varrho}, (\bar{\varrho}\,\bar{\boldsymbol{u}})] \text{ in } L^1(0,1), \qquad (40)$$

where $\bar{\varrho}$, $\bar{\boldsymbol{u}}$ *is a finite energy weak solution of the problem (1)–(3) defined on the whole real line* $I = R$ *such that* $E \in L^\infty(R)$ *and* $\boldsymbol{f} \in \mathcal{F}^+$ *where*

$$\mathcal{F}^+ = \left\{ \boldsymbol{f} \ \middle| \ \boldsymbol{f} = \lim_{\tau_n \to \infty} \boldsymbol{h}_n(. + \tau_n) \text{ weak star in } L^\infty(R \times \Omega) \right.$$
$$\left. \text{for a certain } \boldsymbol{h}_n \in \mathcal{F} \text{ and } \tau_n \to \infty \right\} .$$

Proof. It follows from the estimates stated in Theorem 7 and the fact ϱ_n satisfy (8) and (10), that

$$\varrho_n \to \bar{\varrho} \text{ in } C([0,1]; L_{\text{weak}}^\gamma(\Omega)), \ \varrho_n^\alpha \to h \text{ in } C([0,1]; L_{\text{weak}}^1(\Omega)), \ 1 \leq \alpha < \gamma$$

where, by virtue of Theorem 9, $h = \bar{\varrho}^\alpha$. Consequently, making use of the bounds in Proposition 3, we infer that (38) holds and, moreover, the function $\bar{\varrho} \in L^2((0,1) \times \Omega)$.

Now, by the same token,

$$\boldsymbol{q}_n = (\varrho_n \boldsymbol{u}_n) \to (\bar{\varrho}\,\bar{\boldsymbol{u}}) \text{ in } C([0,1]; L_{\text{weak}}^{\frac{2\gamma}{\gamma+1}}(\Omega)) \qquad (41)$$

with

$$\boldsymbol{u}_n \to \bar{\boldsymbol{u}} \text{ weakly in } L^2((0,1); W_0^{1,2}(\Omega)) \qquad (42)$$

where $\bar{\varrho}$, $\bar{\boldsymbol{u}}$ is a finite energy weak solution of the problem (1)–(3) on the interval $I = (0,1)$ with $\boldsymbol{f} \in \mathcal{F}^+$ and such that $\sup_{t \in (0,1)} E(t) \leq E_\infty$. The constant E_∞ is that from Theorem 7 depending solely on the norm of $\mathcal{F}$.

To show the strong convergence of the momenta, one has to write

$$\varrho_n \boldsymbol{u}_n = (\varrho_n)^{\frac{1}{2}} \, (\varrho_n)^{\frac{1}{2}} \boldsymbol{u}_n$$

where

$$(\varrho_n)^{\frac{1}{2}} \to (\bar{\varrho})^{\frac{1}{2}} \text{ in } L^2((0,1) \times \Omega),$$
$$(\varrho_n)^{\frac{1}{2}} \boldsymbol{u}_n \to (\bar{\varrho})^{\frac{1}{2}} \bar{\boldsymbol{u}} \text{ weakly in } L^2((0,1) \times \Omega) .$$

Moreover, by virtue of (41) and (42), one has

$$\varrho_n u_n^i u_n^j \to \overline{\varrho}\, \overline{u}^i\, \overline{u}^j \text{ weakly in, say, } L^1((0,1) \times \Omega), \ i,j = 1,2,3,$$

in particular,

$$\|(\varrho_n)^{\frac{1}{2}} u_n\|_{L^2((0,1)\times\Omega)}^2 = \int_0^1 \int_\Omega \varrho_n |u_n|^2 \ dx \ dt$$

$$\to \int_0^1 \int_\Omega \overline{\varrho}\ |\overline{u}|^2 \ dx \ dt = \|(\overline{\varrho})^{\frac{1}{2}}\ \overline{u}\|_{L^2((0,1)\times\Omega)}^2 \tag{43}$$

which yields strong convergence of the quantity $(\varrho_n)^{\frac{1}{2}} u_n$ in $L^2((0,1) \times \Omega)$ and thus completes the proof of (39). Moreover, the energy being a function whose variation is bounded on compact time intervals of unit length, (43) implies (40).

To conclude the proof, one has only to observe that replacing the sequence t_n by $t_n + 1$, $t_n - 1$, the same result can be obtained taking any compact subinterval of R instead of $[0,1]$. In other words, $\overline{\varrho}$, $\overline{u}$ solve the problem (1)–(3) on $I = R$. □

Proposition 10 shows the importance of the *complete bounded trajectories*, i.e., the finite energy weak solutions defined on $I = R$ whose energy E is uniformly bounded on R. Let us define

$$\mathcal{A}^s[\mathcal{F}] = \Big\{ [\varrho(\tau), q(\tau)], \ \tau \in [0,1] \ \Big| \ \varrho, \ q = (\varrho u) \text{ is a finite energy weak solution}$$
of the problem (1)–(3) on the interval $I = R$ with $f \in \mathcal{F}^+$ and $E \in L^\infty(R)\Big\}$.

The next statement is a straighforward consequence of Proposition 10 (see also [10]).

Theorem 11. *Let* [D], [P] *be satisfied with* $\gamma \geq \frac{9}{5}$. *Let* $\mathcal{F}$ *be a bounded subset of* $L^\infty(R \times \Omega)$.
Then the set $\mathcal{A}^s[\mathcal{F}]$ *is compact in* $L^\gamma((0,1) \times \Omega) \times [L^p((0,1) \times \Omega)]^3$ *and*

$$\sup_{[\varrho,q] \in U[E_0,\mathcal{F}](t_0,t)} \Big[\inf_{[\overline{\varrho},\overline{q}] \in \mathcal{A}^s[\mathcal{F}]} \Big(\|\varrho - \overline{\varrho}\|_{L^\gamma((0,1)\times\Omega)}$$

$$+ \|q - \overline{q}\|_{L^p((0,1)\times\Omega)} \Big) \Big] \to 0 \text{ as } t \to \infty$$

for any $1 \leq p < \frac{2\gamma}{\gamma+1}$.

Theorem 11 says that the set $\mathcal{A}^s[\mathcal{F}]$ is a global attractor on the space of "short" trajectories.

8   Limit Sets

Theorem 11 represents the right tool to develop a dynamical systems theory for solutions of the problem (1)–(3). Assume ϱ, $\boldsymbol{u}$ is a finite energy weak solution defined on a half-line, say, $I = (0, \infty) = R^+$. We introduce the $\omega-$*limit set* for the density:

$$\omega_d[\varrho, \boldsymbol{u}] = \cap_{s>0} \, \mathrm{cl}_{L^1(\Omega)} \, \{\cup_{t \geq s} \varrho(t)\} \ .$$

Proposition 12. *Let the hypotheses* [D], [P], [F] *hold on* $I = R^+$ *with* $\gamma \geq \frac{9}{5}$. *Let* ϱ, $\boldsymbol{u}$ *be a finite energy weak solution of the problem (1)–(3). Then the* $\omega-$*limit set* $\omega_d[\varrho, \boldsymbol{u}]$ *enjoys the following properties:*

- ω_d *is contained in a bounded ball in* $L^\gamma(\Omega)$ *whose radius depends solely on* B_f;
- ω_d *is compact in* $L^1(\Omega)$ *(and thus in any* $L^\alpha(\Omega)$ *with* $1 \leq \alpha < \gamma$*);*
- ω_d *attracts the trajectory* $\varrho(t)$ *in* $L^1(\Omega)$, *i.e.,* $\mathrm{dist}_{L^1(\Omega)}[\varrho(t), \omega_d] \to 0$ *as* $t \to \infty$;
- ω_d *is connected in the topology of* $L^1(\Omega)$.

Proof. **(i)** It follows from Theorem 7 that ω_d is bounded in $L^\gamma(\Omega)$ by a constant depending on B_f. Moreover, by definition, ω_d is closed in $L^1(\Omega)$.

Now, consider a sequence $\{\varrho_n\} \subset \omega_d$. Consequently, there exists a sequence of times $t_n \to \infty$ such that $\|\varrho_n - \varrho(t_n)\|_{L^1(\Omega)} \to 0$ as $n \to \infty$. On the other hand, by virtue of Proposition 10, there is a subsequence of $\{\varrho(t_n)\}$ which is precompact in $L^1(\Omega)$ and so is the corresponding subsequence of $\{\varrho_n\}$, i.e., ω_d is precompact in $L^1(\Omega)$.

(ii) To show that ω_d attracts the trajectory $\varrho(t)$ assume the contrary, i.e., there exists a sequence $t_n \to \infty$ such that

$$\|\varrho(t_n) - \overline{\varrho}\|_{L^1(\Omega)} \geq \delta > 0 \text{ for all } t_n \text{ and all } \overline{\varrho} \in \omega_d \ . \tag{44}$$

On the other hand, by virtue of Proposition 10, we can assume that

$$\varrho(t_n) \to \varrho_\infty \text{ in } L^1(\Omega) \text{ as } t_n \to \infty \ .$$

It is easy to see that $\varrho_\infty \in \omega_d$ in contrast with (44).

(iii) Finally, assume that ω_d has two distinct components ω_d^1, ω_d^2. As it is a compact set, there exist two disjoint open neighbourhoods U_1, U_2 such that

$$\emptyset \neq \omega_d^1 \subset U_1, \ \emptyset \neq \omega_d^2 \subset U_2, \ U_1 \cap U_2 = \emptyset \ .$$

On the other hand, the density $\varrho(t)$ is a continuous function of t with respect to the strong topology of $L^1(\Omega)$ and, at the same time, it is attracted by ω_d. Necessarily, one has either $\varrho(t) \subset U_1$ or $\varrho(t) \subset U_2$ for all t large enough. Thus either ω_d^1 or ω_d^2 must be empty. $\square$

Similar statement could be proved for the $\omega-$limit sets of the momenta $\omega_m[\varrho, \boldsymbol{u}]$ replacing the strong topology of $L^1(\Omega)$ by the weak one.

9   Potential Flows

We shall examine the flows driven by a potential force, i.e., we assume

$$\boldsymbol{f} = \boldsymbol{f}(x) = \nabla F(x) \ .$$

In this case, the term on the right-hand side of the energy inequality (7) can be rewritten as

$$\int_\Omega (\varrho \boldsymbol{u}).\nabla F \ \mathrm{d}x = \frac{dH}{dt} \ \text{where} \ H(t) = \int_\Omega \varrho F \ \mathrm{d}x,$$

and, consequently, (7) reads as follows:

$$\frac{d}{dt}\Big(E(t) - H(t) \Big) + \int_\Omega \mu |\nabla \boldsymbol{u}|^2 + (\lambda + 2\mu)|\mathrm{div}\boldsymbol{u}|^2 \ \mathrm{d}x \le 0 \ . \qquad (45)$$

We denote

$$EH_\infty = \text{ess}\lim_{t \to \infty}\left[E(t) - H(t) \right] \ .$$

By virtue of (45), the integral

$$\int_1^\infty \int_\Omega \|\boldsymbol{u}\|^2_{W^{1,2}(\Omega)} \ \mathrm{d}x \ \mathrm{d}t \ \text{is convergent,} \qquad (46)$$

in particular,

$$\lim_{T \to \infty} \int_T^{T+1} E_{\text{kin}}(t) \ \mathrm{d}t = 0$$

and

$$\lim_{T \to \infty} \int_T^{T+1} \int_\Omega \frac{a}{\gamma - 1}\varrho^\gamma - \varrho F \ \mathrm{d}x \ \mathrm{d}t = EH_\infty \ . \qquad (47)$$

On the other hand, by virtue of Proposition 10, any sequence $t_n \to \infty$ contains a subsequence such that

$$\int_{t_n}^{t_n+1} \|\varrho(t) - \varrho_s\|_{L^\gamma(\Omega)} \ \mathrm{d}t \to 0$$

where, in view of (46) and (47), ϱ_s is a solution of the *stationary problem*

$$a\nabla \varrho_s^\gamma = \varrho_s \ \nabla F \ \text{in} \ \Omega, \ \int_\Omega \varrho_s \ \mathrm{d}x = 1, \ \int_\Omega \frac{a}{\gamma - 1}\varrho_s^\gamma - \varrho_s F \ \mathrm{d}x = EH_\infty \ . \quad (48)$$

Making use of Proposition 10 again we conclude that the $\omega-$limit set ω_d introduced in Section 8 is contained in the set of solutions of the stationary problem (48):

$$\omega_d[\varrho, \boldsymbol{u}] \subset \Big\{ \varrho_s \mid \varrho_s \text{ solves (48)} \Big\} \ . \qquad (49)$$

Since E_{pot} is lower semi-continuous with respect to t, one gets

$$
\begin{aligned}
EH_\infty &= \int_\Omega \frac{a}{\gamma-1}\varrho_s^\gamma - \varrho_s F \ dx \\
&\leq \liminf_{t\to\infty} \int_\Omega \frac{a}{\gamma-1}\varrho^\gamma(t) - \varrho(t)F \ dx \\
&\leq \limsup_{t\to\infty} \int_\Omega \frac{a}{\gamma-1}\varrho^\gamma(t) - \varrho(t)F \ dx \\
&\leq \operatorname{ess}\lim_{t\to\infty} E(t) - H(t) = EH_\infty
\end{aligned}
$$

for arbitrary $\varrho_s \in \omega_d$. Consequently,

$$
\lim_{t\to\infty} \int_\Omega \frac{a}{\gamma-1}\varrho^\gamma(t) - \varrho(t)F \ dx = EH_\infty, \tag{50}
$$

and, in view of the lower-semicontinuity of the kinetic energy,

$$
\lim_{t\to\infty} E_{\text{kin}}(t) = 0 \ .
$$

Moreover, by virtue of (50), ω_d attracts the trajectories in the strong topology of the space $L^\gamma(\Omega)$:

Theorem 13. *Let assumptions* **[D]**, **[P]** *hold with $\gamma \geq \frac{9}{5}$ and let* $\mathbf{f} = \nabla F(x)$ *where the potential F is globally Lipschitz continuous on Ω. Let ϱ, $\mathbf{u}$ be a finite energy weak solution of the problem (1)–(3) defined on the time interval $I = R^+$.*
Then

$$
E_{\text{kin}}(t) = \int_{\varrho(t)>0} \frac{|\varrho\mathbf{u}|^2}{\varrho} \ dx \to 0 \ as \ t \to \infty
$$

and

$$
\varrho(t) \to \omega_d[\varrho, \mathbf{u}] \ strongly \ in \ L^\gamma(\Omega)
$$

where $\omega_d[\varrho, \mathbf{u}]$ is a compact (in the L^1-topology) and connected subset of the set of solutions to the stationary problem (48) with

$$
EH_\infty = \lim_{t\to\infty} E(t) \ .
$$

In view of Theorem 13, it is of interest to study the structure of the set of the static solutions, i.e., the solutions of the problem (48), in particular, whether or not they form a discrete set. If this is the case, any finite energy weak solution of (1)–(3) is *convergent* to a static state. A partial answer was obtained in the case of potentials with at most two "peaks" ([14, Theorem 1.2]):

Theorem 14. *Let $\Omega \subset R^3$ be an arbitrary domain. Assume F is locally Lipschitz continuous on Ω. Moreover, suppose Ω can be decomposed as*

$$\overline{\Omega} = \overline{\Omega}_1 \cup \overline{\Omega}_2, \ \Omega_1 \cap \Omega_2 = \emptyset$$

where Ω_i are two subdomains (one of them possibly empty) satisfying

$$[F > k] \cap \Omega_i \text{ is connected in } \Omega_i \text{ for } i = 1, 2 \text{ for any } k \in R \qquad (51)$$

where

$$[F > k] = \{x \in \Omega \mid F(x) > k\} \ .$$

Then, given EH_∞, the problem (48) admits at most two distinct solutions.

As an immediate consequence of Theorems 13, 14 we get the following conclusion:

Theorem 15. *Let* **[D]**, **[P]** *hold with $\gamma \geq \frac{9}{5}$. Let $\boldsymbol{f} = \boldsymbol{f}(x) = \nabla F(x)$ where F is globally Lipschitz on Ω. Moreover, assume that Ω can be decomposed in such a way that (51) holds.*
Then for any finite energy weak solution ϱ, $\boldsymbol{u}$ of the problem (1)–(3) defined on the time interval $I = R^+$, there exists a solution ϱ_s of the stationary problem (48) such that

$$\varrho(t) \to \varrho_s \text{ strongly in } L^\gamma(\Omega) \text{ as } t \to \infty,$$

$$E_{\text{kin}}(t) \to 0 \text{ as } t \to \infty \ .$$

10   Attractors

Let

$$\mathcal{A}[\mathcal{F}] = \left\{ [\varrho, \boldsymbol{q}] \ \middle| \ \varrho = \varrho(0), \ \boldsymbol{q} = (\varrho \boldsymbol{u})(0) \text{ where } \varrho, \boldsymbol{u} \text{ is a finite energy weak}\right.$$

$$\left. \text{solution of the problem (1)–(3) on } I = R \text{ with } \boldsymbol{f} \in \mathcal{F}^+ \text{ and } E \in L^\infty(R) \right\} .$$

The next statement shows that $\mathcal{A}[\mathcal{F}]$ is a global attractor in the sense of Foias and Temam [17] (cf. [11]) .

Theorem 16. *Suppose* [D], [P] *hold with* $\gamma \geq \frac{9}{5}$. *Let* $\mathcal{F}$ *be a bounded subset of* $L^{\infty}(R \times \Omega)$.
Then $\mathcal{A}[\mathcal{F}]$ *is compact in* $L^{\alpha}(\Omega) \times L_{\text{weak}}^{\frac{2\gamma}{\gamma+1}}(\Omega)$ *and*

$$\sup_{[\varrho, q] \in U[E_0, \mathcal{F}](t_0, t)} \left[\inf_{[\bar{\varrho}, \bar{q}] \in \mathcal{A}[\mathcal{F}]} \left(\|\varrho - \bar{\varrho}\|_{L^{\alpha}(\Omega)} + \left| \int_{\Omega} (q - \bar{q}).\phi \, dx \right| \right) \right] \to 0$$

as $t \to \infty$ *for any* $1 \leq \alpha < \gamma$ *and any* $\phi \in [L^{\frac{2\gamma}{\gamma-1}}(\Omega)]^3$.

Proof. The proof is a direct consequence of Proposition 10 and Theorem 11. $\qquad \square$

The last issue we want to address in this section is strong convergence of the density in $L^{\gamma}(\Omega)$ and of the momenta in $L^1(\Omega)$ in Theorem 16. To this end, we shall pursue the idea of Ball [3], namely, that weak convergence together with convergence of the energy yields strong convergence.

Theorem 17. *In addition to the hypotheses of Theorem 16, assume that the energy* $E = E[\varrho, q]$ *is (sequentially) continuous on* $\mathcal{A}[\mathcal{F}]$, *specifically, for any sequence*

$$\{[\bar{\varrho_n}, \bar{q}_n]\} \subset \mathcal{A}[\mathcal{F}] \text{ such that } \bar{\varrho}_n \to \bar{\varrho} \text{ in } L^1(\Omega), \ q_n \to q \text{ weakly in } L^1(\Omega)$$

one requires

$$E[\bar{\varrho}_n, \bar{q}_n] \to E[\bar{\varrho}, \bar{q}] \ . \tag{52}$$

Then

$$\sup_{[\varrho, q] \in U[E_0, \mathcal{F}](t_0, t)} \left[\inf_{[\bar{\varrho}, \bar{q}] \in \mathcal{A}[\mathcal{F}]} \left(\|\varrho - \bar{\varrho}\|_{L^{\gamma}(\Omega)} + \|q - \bar{q}\|_{L^1(\Omega)} \right) \right] \to 0$$

as $t \to \infty$.

The proof of Theorem 17 will follow from to next assertion where we denote by $\{\phi_n\}$ the countable dense subset of the separable space $[L^{\frac{2\gamma}{\gamma-1}}(\Omega)]^3$.

Lemma 18. *Under the hypotheses of Theorem 17, given* $\varepsilon > 0$ *there exists* $\delta > 0$ *such that*

$$\|\varrho - \bar{\varrho}\|_{L^1(\Omega)} + \max_{j \leq \delta^{-1}} \left| \int_{\Omega} (q - \bar{q}).\phi_j \, dx \right| < \delta, \ E[\varrho, q] < E[\bar{\varrho}, \bar{q}] + \delta \tag{53}$$

implies

$$\|\varrho - \bar{\varrho}\|_{L^\gamma(\Omega)} + \|q - \bar{q}\|_{L^1(\Omega)} < \varepsilon \qquad (54)$$

for all $[\bar{\varrho}, \bar{q}] \in \mathcal{A}[\mathcal{F}]$ and all ϱ, q such that $q = 0$ on the set where ϱ vanishes and $E[\varrho, q] \leq E_\infty$.

Proof. Arguing by contradiction and using the continuity of the energy on the attractor, one can construct a sequence such that

$$\varrho_n \to \bar{\varrho} \text{ in } L^\alpha(\Omega), \ 1 \leq \alpha < \gamma, \qquad q_n \to \bar{q} \text{ weakly in } L^{\frac{2\gamma}{\gamma+1}}(\Omega),$$

and, by virtue of Proposition 3,

$$E[\varrho_n, q_n] \to E[\bar{\varrho}, \bar{q}]$$

but

$$\|\varrho_n - \bar{\varrho}\|_{L^\gamma(\Omega)} + \|q_n - \bar{q}\|_{L^1(\Omega)} \geq \varepsilon_0 > 0$$

for a certain $[\bar{\varrho}, \bar{q}] \in \mathcal{A}[\mathcal{F}]$.

Since both the kinetic and potential energy are lower semicontinuous (cf. Lemma 1), one has, in fact

$$\varrho_n \to \bar{\varrho} \text{ in } L^\gamma(\Omega), \ q_n \to \bar{q} \text{ weakly in } L^{\frac{2\gamma}{\gamma+1}}(\Omega), \text{ and}$$

$$\int_{\varrho_n > 0} \frac{|q_n|^2}{\varrho_n} \, dx \to \int_{\varrho > 0} \frac{|q|^2}{\bar{\varrho}} \, dx$$

together with

$$\|q_n - \bar{q}\|_{L^1(\Omega)} \geq \varepsilon_0 > 0 . \qquad (55)$$

Now, one can write

$$q_n = (\varrho_n)^{\frac{1}{2}} 1_{\varrho_n > 0}(\varrho_n)^{-\frac{1}{2}} q_n$$

where

$$(\varrho_n)^{\frac{1}{2}} \to \bar{\varrho}^{\frac{1}{2}} \text{ in } L^2(\Omega), \ 1_{\varrho_n > 0}(\varrho_n)^{-\frac{1}{2}} q_n \to h \text{ weakly in } L^2(\Omega)$$

with

$$\bar{q} = \bar{\varrho}^{\frac{1}{2}} h \text{ on the set } \{\bar{\varrho} > 0\} = \{x \in \Omega \mid \bar{\varrho}(x) > 0\} .$$

On one hand,

$$\|1_{\varrho_n > 0}(\varrho_n)^{-\frac{1}{2}} q_n\|^2_{L^2\{\bar{\varrho} > 0\}} \leq \int_{\varrho_n > 0} \frac{|q_n|^2}{\varrho_n} \, dx \to \int_{\bar{\varrho} > 0} \frac{|\bar{q}|^2}{\bar{\varrho}} \, dx = \|h\|^2_{L^2\{\bar{\varrho} > 0\}},$$

i.e.,

$$q_n \to \bar{q} \text{ in } L^1(\{\bar{\varrho} > 0\}) . \qquad (56)$$

On the other hand,

$$\int_{\bar{\varrho} = 0} |q_n| \, dx \leq \left(\int_{\bar{\varrho} = 0} \varrho_n \, dx \right)^{\frac{1}{2}} \sqrt{E_\infty} \to 0,$$

i.e., $q_n \to 0$ in $L^1\{\bar{\varrho} = 0\}$ which, together with (56) contradicts to (55). □

Proof (of Theorem 17). We will argue by contradiction. Assume there are sequences $t_n \to \infty$ and ϱ_n, $\boldsymbol{u}_n$ - finite energy weak solutions of the problem (1)–(3) with $\boldsymbol{f}_n$ satisfying (4) on I_n, $(t_0, t_n] \subset I_n$ independently of n, such that, in accordance with Theorem 7,

$$E[\varrho_n, (\varrho_n \boldsymbol{u}_n)](t) \leq E_\infty \text{ for all } t \geq T(E_0, B_f),$$

and

$$\|\varrho(t_n) - \overline{\varrho}\|_{L^\gamma(\Omega)} + \|(\varrho_n \boldsymbol{u}_n)(t_n) - \overline{\boldsymbol{q}}\|_{L^1(\Omega)} \geq \kappa > 0 \text{ for all } [\overline{\varrho}, \overline{\boldsymbol{q}}] \in \mathcal{A}[\mathcal{F}] \ . \quad (57)$$

Making use of the energy inequality (7) and the lower semi-continuity of the energy stated in Lemma 1, we have

$$
\begin{aligned}
E[\varrho_n, (\varrho_n \boldsymbol{u}_n)](t_n) &\leq \sup_{t \in [t_n - \tau/2, t_n]} E[\varrho_n, \boldsymbol{u}_n](t) \\
&\leq \operatorname{ess\,sup}_{t \in [t_n - \tau/2, t_n]} E[\varrho_n, \boldsymbol{u}_n](t) \qquad (58) \\
&\leq \operatorname{ess\,inf}_{t \in [t_n - \tau, t_n - \tau/2]} E[\varrho_n, \boldsymbol{u}_n](t) + \tau B_f \sqrt{E_\infty} \ .
\end{aligned}
$$

Moreover, by virtue of Proposition 10, we can suppose there exists a global trajectory $\overline{\varrho}$, $\overline{\boldsymbol{u}}$ such that

$$\sup_{t \in [t_n - \tau, t_n]} \|\varrho_n(t) - \overline{\varrho}(t - t_n)\|_{L^1(\Omega)} \to 0 \text{ as } t_n \to \infty, \qquad (59)$$

$$\sup_{t \in [t_n - \tau, t_n]} \left| \int_\Omega (\boldsymbol{q}_n(t) - \overline{\boldsymbol{q}}(t - t_n)) . \phi_i \ \mathrm{d}x \right| \to 0 \text{ as } t_n \to \infty \qquad (60)$$

for all $i = 1, 2, \ldots$, and

$$\int_{t_n - \tau}^{t_n} \left| E[\varrho_n, (\varrho_n \boldsymbol{u}_n)](t) - E[\overline{\varrho}, (\overline{\varrho}\,\overline{\boldsymbol{u}})](t - t_n) \right| \ \mathrm{d}t \to 0 \text{ as } t_n \to \infty \ . \quad (61)$$

Now, choosing $\varepsilon = \frac{\kappa}{2}$ in the hypotheses of Lemma 18, one can take $\tau > 0$ in (58) so small and t_n in (59)–(61) so large that (53) is satisfied at least for one $[\overline{\varrho}(T_n), (\overline{\varrho}\,\overline{\boldsymbol{u}})(T_n)] \in \mathcal{A}[\mathcal{F}]$ for a certain $T_n \in (-\tau, 0)$. Consequently, we have (54) in contrast to (57). $\qquad \square$

Theorem 17 has been proved.

The hypothesis (52) on compactness of the attractor in Theorem 17 is of the same nature as the hypothesis of regularity of globally defined weak solutions for the incompressible Navier-Stokes equations made by Constantin, Foias and Temam [8].

References

1. S. N. Antontsev, A. V. Kazhikhov, and V. N. Monachov: *Krajevyje zadaci mechaniki neodnorodnych zidkostej.* Novosibirsk, 1983.
2. A. V. Babin and M. I. Vishik: *Attractors of evolution equations.* North-Holland, Amsterdam, 1992.
3. J. M. Ball: Continuity properties and global attractors of generalized semiflows and the Navier-Stokes equations. *J. Nonlinear Sci.*, **7**:475–502, 1997.
4. M. E. Bogovskii: Solution of some vector analysis problems connected with operators div and grad (in Russian). *Trudy Sem. S.L. Sobolev*, **80**(1):5–40, 1980.
5. W. Borchers and H. Sohr: On the equation $rotv = g$ and $divu = f$ with zero boundary conditions. *Hokkaido Math. J.*, **19**:67–87, 1990.
6. R. Coifman, P.-L. Lions, Y. Meyer, and S. Semmes: Compensated compactness and Hardy spaces. *J. Math. Pures Appl.*, **72**:247–286, 1993.
7. R. Coifman and Y. Meyer: On commutators of singular integrals and bilinear singular integrals. *Trans. Amer. Math. Soc.*, **212**:315–331, 1975.
8. P. Constantin, C. Foias, and R. Temam: *Attractors representing turbulent flows.* Mem. Amer. Math. Soc. 53, Providence, 1985.
9. R. J. DiPerna and P.-L. Lions: Ordinary differential equations, transport theory and Sobolev spaces. *Invent. Math.*, **98**:511–547, 1989.
10. E. Feireisl: Global attractors for the Navier-Stokes equations of three-dimensional compressible flow. *C. R. Acad. Sci. Paris, Sér. I*, 1999. Submitted.
11. E. Feireisl: Propagation of oscillations, complete trajectories and attractors for compressible flows. *Archive Rational Mech. Anal.*, 2000. Submitted.
12. E. Feireisl and H. Petzeltová: Asymptotic compactness of global trajectories generated by the Navier-Stokes equations of compressible fluid. *J. Differential Equations*, 1999. Submitted.
13. E. Feireisl and H. Petzeltová: Bounded absorbing sets for the Navier-Stokes equations of compressible fluid. *Commun. Partial Differential Equations*, 1999. To appear.
14. E. Feireisl and H. Petzeltová: Zero-velocity-limit solutions to the Navier-Stokes equations of compressible fluid revisited. *Navier-Stokes equations and applications, Proceedings, Ferrara*, 1999. Submitted.
15. E. Feireisl and H. Petzeltová: On compactness of solutions to the Navier-Stokes equations of compressible flow. *J. Differential Equations*, **163**(1):57–74, 2000.
16. E. Feireisl and H. Petzeltová: On integrability up to the boundary of the weak solutions of the Navier-Stokes equations of compressible flow. *Commun. Partial Differential Equations*, **25**(3-4):755–767, 2000.
17. C. Foias and R. Temam: The connection between the Navier-Stokes equations, dynamical systems and turbulence. *In Directions in Partial Differential Equations, Academic Press, New York*, pages 55–73, 1987.
18. G. P. Galdi: *An introduction to the mathematical theory of the Navier-Stokes equations, I.* Springer-Verlag, New York, 1994.
19. J. K. Hale: *Asymptotic behavior of dissipative systems.* Amer. Math. Soc., 1988.
20. D. Hoff: Global existence for 1D compressible, isentropic Navier-Stokes equations with large initial data. *Trans. Amer. Math. Soc.*, **303**:169–181, 1987.

21. D. Hoff: Global solutions of the Navier-Stokes equations for multidimensional compressible flow with discontinuous initial data. *J. Differential Equations*, **120**:215–254, 1995.

22. D. Hoff and M. Ziane: Compact attractors for the Navier-Stokes equations of one-dimensional compressible flow. *C.R. Acad. Sci. Paris, Sér I.*, **328**:239–244, 1999.

23. O. A. Ladyzhenskaya: *The mathematical theory of viscous incompressible flow.* Gordon and Breach, New York, 1969.

24. C. Li, A. McIntosh, K. Zhang, and Z. Wu: Compensated compactness, para-commutators and Hardy spaces. *J. Funct. Anal.*, **150**:289–306, 1997.

25. P.-L. Lions: Compacité des solutions des équations de Navier- Stokes compressible isentropiques. *C.R. Acad. Sci. Paris, Sér I.*, **317**:115–120, 1993.

26. P.-L. Lions: *Mathematical topics in fluid dynamics, Vol.1, Incompressible models.* Oxford Science Publication, Oxford, 1996.

27. P.-L. Lions: *Mathematical topics in fluid dynamics, Vol.2, Compressible models.* Oxford Science Publication, Oxford, 1998.

28. P.-L. Lions: Bornes sur la densité pour les équations de Navier- Stokes compressible isentropiques avec conditions aux limites de Dirichlet. *C.R. Acad. Sci. Paris, Sér I.*, **328**:659–662, 1999.

29. J. Málek and J. Nečas: A finite-dimensional attractor for the three dimensional flow of incompressible fluid. *J. Differential Equations*, **127**:498–518, 1996.

30. J. Málek and D. Pražák: Large time behavior via the method of l-trajectories. *J. Differential Equations*, 2000. Submitted.

31. A. Matsumura and T. Nishida: The initial value problem for the equations of motion of compressible and heat conductive fluids. *Comm. Math. Phys.*, **89**:445–464, 1983.

32. G. R. Sell: Global attractors for the three-dimensional Navier-Stokes equations. *J. Dynamics Differential Equations*, **8**(1):1–33, 1996.

33. D. Serre: Solutions faibles globales des équations de Navier-Stokes pour un fluide compressible. *C.R. Acad. Sci. Paris*, **303**:639–642, 1986.

34. D. Serre: Variation de grande amplitude pour la densité d'un fluid viscueux compressible. *Physica D*, **48**:113–128, 1991.

35. R. Temam: *Infinite-dimensional dynamical systems in mechanics and physics.* Springer-Verlag, New York, 1988.

Adaptive Wavelet Solvers for the Unsteady Incompressible Navier-Stokes Equations

Michael Griebel, Frank Koster*

Institut für Angewandte Mathematik, Universität Bonn
Wegelerstr. 6, D-53115 Bonn, Germany
e-mail: `griebel@iam.uni-bonn.de`, `koster@iam.uni-bonn.de`

Abstract. In this paper we describe adaptive wavelet-based solvers for the Navier-Stokes equations. Our approach employs a Petrov-Galerkin scheme with tensor products of Interpolet wavelets as ansatz functions. We present the fundamental algorithms for the adaptive evaluation of differential operators and non-linear terms. Furthermore, a simple but efficient preconditioning technique for the resulting linear systems is introduced. For the Navier-Stokes equations a Chorin-type projection method with a stabilized pressure discretization is used. Numerical examples demonstrate the efficiency of our approach.

1 Introduction

A main problem for the numerical solution of the unsteady Navier-Stokes equations is the large range of scales present in the solution. The most prominent estimate for that range is Kolmogorov's $Re^{3/4}$ law for isotropic, three-dimensional turbulence [28]. This leads to a huge number of degrees of freedom which are required to resolve *all* active scales. This number scales like $Re^{9/4}$ just for the spatial resolution. However, it is widely accepted that even in the turbulent regime the flow is governed by very localized, coherent structures. These structures are quasi-singularities in the flow which are characterized by strong gradients of the velocity, the vorticity and/or the pressure. One typical example are the essentially two-dimensional vortex rolls known for example from cigarette smoke. This suggests that the *intrinsic* complexity of turbulent flows is much smaller than that estimated by Kolmogorov's law. It should be possible to exploit this fact by means of *adaptive* solution methods.

The main advantage of wavelets are their localization in space and scale. Therefore, they are perfect candidates for adaptive methods. More precisely it can be shown that functions which have at most a few (quasi-)singularities can be approximated with a quite small number of degrees of freedom in terms of a wavelet basis. For a recent review on this topic we refer to the excellent paper [23]. The foundation for this property are the following two aspects: First, polynomials up to a certain degree are in the span of a wavelet basis. This explains their approximation power. Second, for a certain range of Sobolev spaces H^s the Sobolev-norms of wavelet expansions and sequence-norms of the expansion coefficients are equivalent.

* Research supported by the Deutsche Forschungsgemeinschaft, GR 1144/7-2.

These norm equivalences also play a role for an important practical problem: The efficient preconditioning of the large linear systems that have to be solved during the numerical integration of the Navier-Stokes equations. It is this unification of insight into approximation theory on one side and the efficient preconditioning of linear systems on the other side which makes up the beauty of wavelets and which explains to some extent the interest they have attracted in the last years.

However, wavelets are not for free. The associated algorithms for the evaluation of linear or non-linear terms are in general quite complicated. A main topic of this paper is the description of such basic algorithms for adaptive wavelet methods for the treatment of the Navier-Stokes equations. Of course, these algorithms are dependent on the wavelets at hand. There exist orthogonal Daubechies wavelets [19–21], biorthogonal spline wavelets [16] or Interpolets [22,26]. In this paper we will use Interpolets.

To treat the multivariate case, there are two different types of constructions by means of tensor products of univariate wavelets. The first one employs tensor products of wavelets and so-called scaling functions of the same level of refinement. This leads to a linear order of successively finer approximations spaces as in the univariate case, i.e. to a multi-resolution analysis (MRA) [21]. However, in this paper we will use another approach. Here, simple tensor products of wavelets of arbitrary levels of refinement are used similar to Fourier spectral methods. This technique does not lead to a linear order of approximation spaces, but instead to a d-dimensional array of approximation spaces (c.f. [53] sect. 5.4), where d denotes the dimension of the computational domain. Therefore, we will call this approach MRA-d. The main advantage of this approach is its excellent approximation property. For example, for the approximation of functions with bounded mixed derivatives a much smaller number of degrees of freedom is required compared to the classical MRA-approach. This property has been extensively studied in e.g. [24,34,62]. Another situation where the MRA-d-approach is superior to the MRA-approach is the approximation of strong anisotropic functions like the velocity or vorticity field in a turbulent boundary layer with its strong anisotropic coherent structures (so-called hair-pin vortices) [49].

A further advantage of our MRA-d-approach is its algorithmical simplicity. For the discretization of PDEs we use a Petrov-Galerkin scheme with specially chosen test functions. Then, the adaptive evaluation of discrete differential operators completely boils down to the univariate case. We furthermore use univariate Interpolets as basic building blocks in the MRA-d-approach. Then also the adaptive evaluation of non-linear terms essentially boils down to the univariate adaptive wavelet transform and its inverse.

Hence, in our opinion the MRA-d-approach with Interpolet wavelets and a Petrov-Galerkin discretization is the simplest way to carry the adaptive univariate wavelet setup over to the adaptive multivariate case. This also allows for very efficient data-structures and algorithms. Furthermore, code developed for the univariate case can directly be reused.

The efficient solution of the resulting linear systems is another important issue. To solve this problem we use a simple preconditioning technique based on a diagonal scaling together with the lifting scheme [13,60]. This leads to $O(1)$ condition numbers of the linear systems of equations and, therefore, to efficient iterative solution methods.

For the discretization of time-dependent problems we will use a common time-stepping method. This method is combined with an algorithm for the selection of the adaptive basis in space. Here, the main idea is that locations where adaptive refinement is required evolve smoothly in time. Therefore, adaptive refinement for a new time slab is required in the vicinity of regions which were refined in the previous time slab.

All these ingredients (MRA-d, Petrov-Galerkin, Interpolets, time stepping methods and adaptive basis selection) form the building blocks of an adaptive wavelet-based Navier-Stokes solver. According to our 'efficient, but as simple as possible' strategy we here employ a Chorin-type projection method. This method decouples the momentum and continuity equation and leads to a Poisson equation for the pressure in each time step. We present two numerical experiments which show that our wavelet-based Navier-Stokes solver works very well and yields accurate and reliable results.

This paper is organized as follows. In the following section we introduce the setting of univariate biorthogonal wavelets and approximation spaces for the example of Interpolets. Furthermore, we derive a general algebraic form of the wavelet transform algorithms and we briefly sketch how wavelets on the interval are built. In the third section we carry our setting over to the adaptive case. Then, we also present the basic adaptive algorithms for the wavelet transform, for its inverse, for the adaptive evaluation of non-linear terms and for the matrix-vector multiplication with the (generalized) stiffness matrix. In addition we also briefly introduce the wavelet-finite-difference technique as an alternative to the Petrov-Galerkin approach. The multivariate MRA-d-approach is described in section 4. There we show how the previous adaptive algorithms for the univariate case can be used in the multivariate case as well. In section 5 we briefly present what is known about consistency and stability of the Petrov-Galerkin method associated to MRA-d-Interpolets. The problem of efficient preconditioning of linear systems which arise from elliptic equations is treated in section 6. Here, we introduce two simple preconditioners which exploit the norm-equivalences mentioned earlier. Furthermore, an analysis of these preconditioners is given for the non-adaptive case. After these preparations we describe in section 7 how adaptive wavelet methods can be applied to time-dependent problems such as convection-diffusion equations. Here, we describe how in a time-stepping method the adaptive basis is selected from time step to time step. Section 8 deals with the correct treatment of the continuity equation and with the stabilization of the pressure discretization. The main idea is to use a Chorin type projection method together with a consistent discretization of the pressure-Poisson equation. For stabilization reasons we introduce an upwind-downwind discretization for the pressure gradient and the divergence operator, respectively. Then, we give numerical examples where our adaptive Interpolet Petrov-Galerkin method developed so far is applied to two complicated two-dimensional flow problems. The first is the interaction and merger of three vortices and the second is the evolution of a shear layer.

In order to simplify the reading and understanding we give a detailed table of symbols and notation at the end of this paper.

2   Univariate Interpolets

In this section we introduce the scaling functions, the wavelets and their dual counterparts for the Interpolet family [22,27]. Furthermore, basic notation is set up. To this end, we first consider spaces of functions defined on $\mathbb{R}$. Interpolets of order N ($N \in \mathbb{N}$, N even) are then defined via a single scaling function ϕ with the following properties:

• (Scaling equation:) The function ϕ is the solution of

$$\phi(x) = \sum_{s \in \mathbb{Z}} h_s \phi(2x - s) \ . \tag{1}$$

The mask coefficients $\mathbf{h} := \{h_s\}_{s \in \mathbb{Z}}$ are given by $h_0 = 1$, $h_s = h_{-s}$ ($s \in \mathbb{Z}$) and

$$\begin{pmatrix} 1^0 & 3^0 & \dots & (N-1)^0 \\ 1^2 & 3^2 & \dots & (N-1)^2 \\ \vdots & \vdots & \vdots & \vdots \\ 1^{N-2} & 3^{N-2} & \dots & (N-1)^{N-2} \end{pmatrix} \cdot \begin{pmatrix} h_1 \\ h_3 \\ \vdots \\ h_{N-1} \end{pmatrix} = \begin{pmatrix} 1/2 \\ 0 \\ \vdots \\ 0 \end{pmatrix} \ . \tag{2}$$

All other h_s are zero.

• (Compact support:)

$$\text{supp } \phi = [-N+1, N-1] \ , \quad \text{supp } \mathbf{h} = \{-N+1, ..., N-1\} \ . \tag{3}$$

• (Polynomial exactness:) In a pointwise sense, polynomials of degree less than N can be written as linear combination of translates of ϕ, e.g.

$$\forall \, 0 \leq i < N \ : \ \forall x \in \mathbb{R} \ : \ x^i = \sum_{s \in \mathbb{Z}} s^i \phi(x - s) \ . \tag{4}$$

• (Interpolation property:) Let $\delta(x)$ be the Dirac functional. It should hold

$$\forall s \in \mathbb{Z} \ : \ \phi(s) = \delta(s) \ . \tag{5}$$

This is a special property of the Interpolet family. The scaling functions of other wavelet types usually do not fulfill this relation.

• (Dual scaling functions:) Let $\tilde{\phi}(x) := \delta(x)$. Then (5) can also be written as a biorthogonality relation

$$\langle \tilde{\phi}(x - s) \ , \ \phi(x - t) \rangle = \delta(s - t) \ . \tag{6}$$

This means that translates $\{\tilde{\phi}(. - s)\}_{s \in \mathbb{Z}}$ of the Dirac functional are the dual scaling functions to $\{\phi(. - s)\}_{s \in \mathbb{Z}}$. Clearly the function $\tilde{\phi}$ also satisfies a scaling equation since

$$\tilde{\phi}(x) = \sum_{s \in \mathbb{Z}} \tilde{h}_s \tilde{\phi}(2x - s) \quad \text{where} \quad \tilde{h}_s = \delta(s) \ . \tag{7}$$

- (*Complementary spaces:*) The odd translates $\phi(2x-(2t+1))$ can now be considered as so-called wavelets (see the discussion below and section 2.1). We have

$$\text{span}\{\phi(x - s)\}_{s \in \mathbb{Z}} \oplus \text{span}\{\phi(2x - (2t + 1))\}_{t \in \mathbb{Z}} = \text{span}\{\phi(2x - s)\}_{s \in \mathbb{Z}}. \tag{8}$$

This is the main property which distincts Interpolet wavelets from other wavelet families. Since the wavelets are interpolating again, the algorithms for the evaluation of non-linear terms become particularly simple. In the remainder of this paper we will denote wavelets with properties (1) – (8) as Interpolets or Interpolet wavelets [22,26].

2.1 Multi-Resolution Analysis

Equipped with functions $\phi(x)$ and $\tilde{\phi}(x)$ which possess the above properties (1)–(8), we now define a multi-resolution analysis of $L^2(\mathbb{R})$. To this end, we define approximation spaces V^l spanned by translates and dilates of ϕ, i.e.

$$V^l = \text{span}\{\phi^{(l,s)}\}_{s \in \mathbb{Z}} \ , \quad l \in \mathbb{N} \quad \text{where} \quad \phi^{(l,s)}(x) := \phi(2^l x - s) \ . \tag{9}$$

Because of (1) the V^l form a ladder of successively finer approximation spaces

$$... \subset V^l \subset V^{l+1} \subset ... \subset L^2(\mathbb{R}) \ . \tag{10}$$

The properties of polynomial exactness (4) and compact support (3) together with a Bramble-Hilbert argument yield a classical Jackson-type approximation estimate

$$\inf_{u^l \in V^l} \|u - u^l\|_0 \le c 2^{-lN} \|u^{(N)}\|_0 \ ,$$

where $u^{(N)}$ is the Nth derivative of u. Of course the *local* approximation quality on e.g. $\text{supp} \, \phi^{(l,s)}$ depends only on $\int_{\mathcal{N}} |u^{(N)}|^2 \, dx$ for a neighbourhood $\mathcal{N}$ of $\text{supp} \, \phi^{(l,s)}$. Furthermore it is clear that in regions where $u^{(N)}$ is quite small, there is no significant difference between the best approximation $u^l \in V^l$ and the best approximation $u^{l-1} \in V^{l-1}$. This is the main idea behind any multi-scale representation which attempts to decompose the approximation spaces V^l into the sum of complementary spaces W_l. There we have

$$V^L = V^{l_0} \oplus W_{l_0+1} \oplus W_{l_0+2} \oplus .. \oplus W_L \quad \text{where} \quad W_l = V^{l-1} \oplus V^l \ .$$

Here, V^{l_0} is a certain coarsest approximation space which is later on related to wavelets on the interval and which can not be further decomposed (see subsection 2.3). To simplify the notation we set

$$W_{l_0} := V^{l_0} , \quad \text{i.e. } V^L = \oplus_{l=l_0}^L W_l .$$

In the above-mentioned situation of local smoothness in the neighbourhood of $\operatorname{supp} \phi^{(l,s)}$ we can assume the following: If l is sufficiently large then the local contributions of W_l are quite small and may be discarded without introducing significant errors. This means that an approximate representation of a function $u \in V^L$ will be much sparser in terms of basis functions of W_l than the single-scale representation in terms of $\phi^{(L,s)}$.

At the present stage neither the spaces W_l nor an appropriate basis of W_l were determined yet. Property (8) suggests a simple definition of W_l:

$$W_l := \operatorname{span}\{\psi_{(l,t)}\}_{t \in \mathbf{Z}} , \quad \text{where} \quad \psi_{(l,t)} = \begin{cases} \phi^{(l,2t+1)} & \text{if } l > l_0 \\ \phi^{(l_0,t)} & \text{if } l = l_0 . \end{cases} \tag{11}$$

First of all, $\{\psi_{(l,t)}\}_{t \in \mathbf{Z}}$ is indeed a basis of W_l and $V^l = W_l \oplus V^{l-1}$ is a direct sum. In the following the basis functions $\psi_{(l,s)}$ will be called wavelets and in case of the Interpolet family alternatively Interpolets or Interpolet wavelets. Figure 1 depicts the different bases for Interpolets in the case $N = 6$.

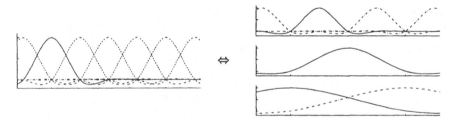

Fig. 1. Left: single-scale basis. Right: multi-scale basis.

2.2   Wavelet Transform

With a specific choice of scaling functions, wavelets and their duals the complete multi-resolution analysis is determined. It then remains to give algorithms which calculate the multi-scale representation from a single-scale representation. To simplify the explanation we will use the following notation in the remainder of this paper,

– *row* vectors of functions

$$\Phi^l := \{\phi^{(l,s)}\}_{s \in \mathbf{Z}} , \quad \Psi_l := \{\psi_{(l,t)}\}_{t \in \mathbf{Z}} , \tag{12}$$

– *column* vectors of coefficient sequences

$$\mathbf{u}^l := \{u^{(l,s)}\}_{s \in \mathbf{Z}} , \quad \mathbf{u}_l := \{u_{(l,t)}\}_{t \in \mathbf{Z}} . \tag{13}$$

With this notation we get the following representations of $u \in V^L$

$$u = \sum_{s \in \mathbb{Z}} u^{(L,s)} \phi^{(L,s)} = \Phi^L \cdot \mathbf{u}^L$$

$$= \sum_{l=l_0}^{L} \sum_{t \in \mathbb{Z}} u_{(l,t)} \psi_{(l,t)} = \sum_{l=l_0}^{L} \Psi_l \cdot \mathbf{u}_l \; .$$

Note that here and in the following superscript indices are used for coefficients and functions in a scaling function representation and subscript indices are applied for coefficients and functions in a wavelet representation.

Now, the wavelet transform has to compute the coefficients $\{\mathbf{u}_l\}_l$ with respect to the multi-scale basis $\{\Psi_l\}_l$ from the coefficients $\mathbf{u}^L$ with respect to the single-scale basis Φ^L. In case of Interpolets there is a very simple interpretation of the scheme: First of all, note that $\mathbf{u}^L$ are the nodal values of u in the grid points $\{2^{-L}s\}_{s \in \mathbb{Z}}$ because of (5). Now consider one step of the wavelet transform, i.e.

$$\Phi^L \cdot \mathbf{u}^L \overset{!}{=} \Phi^{L-1} \cdot \mathbf{u}^{L-1} + \Psi_L \cdot \mathbf{u}_L \tag{14}$$

which is associated to the splitting $V^L = V^{L-1} \oplus W_L$. To calculate for example $u_{(L,t)}$ we pick the odd node $2^{-L}(2t+1)$ and his neighbouring evenly numbered nodes $\mathcal{N} := 2^{-L}\{2t - N + 2,\ 2t - N + 4,\ ...,\ 2t + N - 2\}$. We calculate the polynomial $p_\mathcal{N}$ of degree $N - 1$ which interpolates at $\{\mathcal{N}, u(\mathcal{N})\}$. Then, the coefficient $u_{(L,t)}$ is given by

$$u_{(L,t)} := (u - p_\mathcal{N})(2^{-L}(2t + 1)) \; , \tag{15}$$

see Figure 2 for an example. This is done for all $t \in \mathbb{Z}$. Because of (5) the auxiliary coefficients $\{u^{(L-1,s)}\}_{s \in \mathbb{Z}}$ are given by $u^{(L-1,s)} = u^{(L,2s)}$.

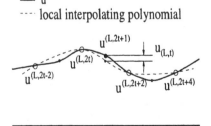

Fig. 2. Calculation of wavelet coefficients

To set up the above algorithm more formally it is appropriate to introduce the complete machinery of dual scaling functions and wavelets. Analogously to the decomposition $V^L = \oplus_l W_l$, we can decompose the dual space as

$$\tilde{V}_L = \text{span}\{\tilde{\phi}^{(L,s)}\}_{s \in \mathbb{Z}} = \oplus_{l=l_0}^L \tilde{W}_l$$

where $\tilde{W}_l = \text{span}\{\tilde{\psi}_{(l,t)}\}$ and the dual wavelets $\tilde{\psi}_{(l,t)}$ are linear combinations of the elements of $\tilde{\Phi}^l$. More precisely we have

$$\tilde{\psi}_{(l,t)}(x) = \sum_{r \in \mathbb{Z}} \tilde{g}_{r-2t} \tilde{\phi}^{(l,r)}(x) , \tag{16}$$

where the mask coefficients $\tilde{g}_r$ are defined by

$$\tilde{g}_r := (-1)^{1-r} h_{1-r} . \tag{17}$$

This choice ensures that we have the following *biorthogonality relations*

$$\langle \tilde{\Phi}^l, \Phi^l \rangle = I , \quad \langle \tilde{\Psi}^l, \Phi^{l-1} \rangle = 0 ,$$
$$\langle \tilde{\Psi}^l, \Psi^l \rangle = I , \quad \langle \tilde{\Phi}^{l-1}, \Psi^l \rangle = 0 .$$

An immediate consequence of these relations are the general biorthogonality relations of the wavelets and their duals

$$\langle \tilde{\psi}_{(k,s)} , \psi_{(l,t)} \rangle = \delta(k-l)\delta(s-t) . \tag{18}$$

We also can recast the scaling relations (1), (7), (11) and (16) as matrix-vector products

$$\Phi^{l-1} = \Phi^l \cdot H^l , \quad \tilde{\Phi}^{l-1} = \tilde{\Phi}^l \cdot \tilde{H}^l , \quad \Psi^l = \Phi^l \cdot G^l , \quad \tilde{\Psi}^l = \tilde{\Phi}^l \cdot \tilde{G}^l , \tag{19}$$

where H^l, $\tilde{H}^l$, G^l and $\tilde{G}^l$ are sparse matrices, $(H^l)_{r,s} = h_{r-2s}$, $(\tilde{H}^l)_{r,s} = \tilde{h}_{r-2s}$, $(G^l)_{r,s} = \delta(r-(2s+1))$ and $(\tilde{G}^l)_{r,s} = \tilde{g}_{r-2s}$. In terms of these matrices the biorthogonality relations read

$$(H^l, G^l) = (\tilde{H}^l, \tilde{G}^l)^{-T} . \tag{20}$$

Furthermore, we have the following algebraic representation of one step of the wavelet transform

$$\Phi^L \cdot u^L = \Phi^{L-1} \cdot \langle \tilde{\Phi}^{L-1}, \Phi^L \rangle \cdot u^L + \Psi^L \cdot \langle \tilde{\Psi}^L, \Phi^L \rangle \cdot u^L$$
$$= \Phi^{L-1} \cdot (\tilde{H}^L)^T \cdot u^L + \Psi^L \cdot (\tilde{G}^L)^T \cdot u^L ,$$

and, consequently, we get from (14)

$$u^{L-1} = (\tilde{H}^L)^T \cdot u^L , \quad u_L = (\tilde{G}^L)^T \cdot u^L .$$

The complete wavelet transform and its inverse read

algorithm (WT)	algorithm (IWT)
given u^L	given u^{l_0} and $\{u_l\}_{l=l_0+1}^l$
for $l = L$ downto $l_0 + 1$	for $l = l_0 + 1$ to L
$\quad u_l = (\tilde{G}^l)^T \cdot u^l$	$\quad u^l = G^l \cdot u_l + H^l \cdot u^{l-1}$
$\quad u^{l-1} = (\tilde{H}^l)^T \cdot u^l$	end
end	

2.3   Wavelets on the Interval

The MRA for functions defined on $\mathbb{R}$ can not be used for the solution of PDEs on finite domains. To this end the above setting must be modified to function spaces defined on bounded domains. We consider the domain $[0, 1]$ and briefly describe such a construction for Interpolets. A detailed description of a similar approach for orthogonal wavelets on the interval is given in [45].

The main idea is the following: We have to find scaling functions defined on $[0, 1]$ which fulfill a scaling equation similar to (19) and which span the polynomials of degree less than N for $0 \leq x \leq 1$ to maintain the approximation properties.

To this end consider the scaling functions $\{\phi^{(l,s)}\}_{N \leq s \leq 2^l - N}$ for sufficiently large l, i.e. $l \geq l_0$, where

$$l_0 = \lceil log_2(2N) \rceil .$$

Obviously these scaling functions have their support completely in $[0, 1]$ and they also satisfy a scaling relation with $\{\phi^{(l+1,s)}\}_{N \leq s \leq 2^{l+1} - N}$. However, they do not span all polynomials of degree less than N in $[0, 1]$. The remedy is to augment $\{\phi^{(l,s)}\}_{N \leq s \leq 2^l - N}$ by additional scaling functions $\phi^{(l,0)}, .., \phi^{(l,N-1)}$ for the left boundary $x = 0$ and $\phi^{(l,2^l - N+1)}, .., \phi^{(l,2^l)}$ for the right boundary $x = 1$ which are no longer simple dilates and translates of ϕ, but instead are defined by

$$\left(\phi^{(l,0)}, .., \phi^{(l,N-1)}\right)(x) := \left(\phi(2^l x + N - 2), .., \phi(2^l x - N + 1)\right) \cdot A \qquad (21)$$

and

$$\left(\phi^{(l,2^l - N+1)}, .., \phi^{(l,2^l)}\right)(x) := \left(\phi(2^l x - 2^l - N + 1), .., \phi(2^l x - 2^l + N - 2)\right) \cdot B .$$

Here, the matrices A and B consist of N specially chosen eigenvectors of the matrix $\{h_{r-2s}\}_{-N+2 \leq r,s \leq N-1}$ such that, altogether, the boundary modified scaling functions $(0 \leq s < N, 2^l - N < s \leq 2^l)$ together with the ordinary translates and dilates $(N \leq s \leq 2^l - N)$ satisfy:

- (*Scaling equation:*)

$$\text{span}\{\phi^{(l,s)} \mid 0 \leq s \leq 2^l\} \subset \text{span}\{\phi^{(l+1,s)} \mid 0 \leq s \leq 2^{l+1}\} . \qquad (22)$$

- (*Polynomial exactness:*) For $0 \leq x \leq 1$ and $0 \leq i < N$ there holds

$$x^i = \sum_{s=0}^{2^l} s^i \phi^{(l,s)}(x) . \qquad (23)$$

- (*Interpolation property:*)

$$\forall 0 \leq s, t \leq 2^l \ : \ \phi^{(l,s)}(2^{-l} t) = \delta(s - t) . \qquad (24)$$

In case of a prescribed homogeneous Dirichlet boundary condition in for example the point $x = 0$ there exist just $N - 1$ modified scaling functions

$\phi^{(l,1)}, ..., \phi^{(l,N-1)}$. In this case A consists of only $N - 1$ eigenvectors of the matrix $\{h_{r-2s}\}_{-N+2 \leq r,s \leq N-1}$ and (23) is satisfied only for $1 \leq i < N$ and (24) only for $s, t \neq 0$.

Likewise, for a prescribed homogeneous Neumann boundary condition in for example $x = 0$ there exist just the $N - 1$ boundary modified scaling functions $\phi^{(l,0)}, \phi^{(l,2)}, ..., \phi^{(l,N-1)}$. This slightly different labeling $(0, 2, 3, .., N-1)$ instead of $(1, 2, .., N - 1)$ is motivated by the fact that here (23) holds for $i \in \{0, 2, ..., N - 1\}$ and (24) for $s, t \neq 1$. Analogous constructions hold for the point $x = 1$.

In order to simplify the further explanations we define row vectors of scaling functions again by

$$\Phi^l = \{\phi^{(l,s)}\}_{s \in S(l)} ,$$

with the index sets $S(l) = \{0, .., 2^l\}$ (or $\{1, 2, .., 2^l - 1\}$ or $\{0, 2, .., 2^l - 2, 2^l\}$) depending on the boundary conditions). The corresponding dual scaling functions are

$$\tilde{\Phi}^l := \{\tilde{\phi}^{(l,s)}\}_{s \in S(l)} \quad \text{where} \quad \tilde{\phi}^{(l,s)}(x) := \delta(2^l x - s) . \tag{25}$$

Similar to subsection 2.1, for $l > l_0$ the wavelets $\psi_{(l,t)}$ are essentially the odd numbered scaling functions $\phi^{(l,2t+1)}$

$$\Psi_l := \{\psi_{(l,t)}\}_{t \in T(l)} , \quad T(l) = \{0, ..., 2^{l-1} - 1\} \tag{26}$$

and

$$\psi_{(l,t)} := \begin{cases} \phi^{(l,2)} & \text{if } t = 0 \text{ and homog. Neumann b.c. in } x = 0 , \\ \phi^{(l,2^l-2)} & \text{if } t = 2^{l-1}-1 \text{ and homog. Neumann b.c. in } x = 1 , \\ \phi^{(l,2t+1)} & \text{else .} \end{cases}$$

$$\tag{27}$$

For $l = l_0$ we simply define

$$\psi_{(l_0,t)} := \phi^{(l_0,t)} , \quad \text{for } t \in T(l_0) := S(l_0) . \tag{28}$$

For the scaling relations we use the same notation as in (19), but now, the matrices H^l, $\tilde{H}^l$, G^l, $\tilde{G}^l$ are finite, e.g. H^l is a $\#S(l) \times \#S(l-1)$ matrix, etc. These matrices have essentially the same banded structure as their counterparts from subsection 2.2, but at the upper left and lower right corners there are blocks which contain the scaling relation coefficients for the boundary modified scaling functions and wavelets, respectively.

Periodic boundary conditions are even simpler. Here, we set

$$\phi^{(l,s)}(z) = \phi^{(l,s)}(x) , \quad \text{where } z := \begin{cases} x & : 0 \leq x \leq 1 \\ 1 + x & : x < 0 \\ x - 1 & : x > 1 \end{cases}$$

for $s \in S(l) = \{0, .., 2^l - 1\}$. The wavelets are $\psi^{(l, 2t+1)}$ for $t \in T(l) :=$ $\{0, .., 2^{l-1} - 1\}$. Thus, the scaling matrices have a simple periodic structure. For example for H^l we have

$$(H^l)_{(r,s)} = h_{mod(r - 2s, 2^l)} . \tag{29}$$

3 Adaptive Algorithms

In this section we will introduce algorithms for the adaptive wavelet transform, its inverse transform and the adaptive evaluation of constant coefficient linear differential operators. In section 4 we will see that the simple tensor product application of these univariate algorithms is enough to handle also the multivariate case.

Now assume that a function $u \in V^L$ is given in its wavelet representation

$$u = \sum_{l=l_0}^{L} \sum_{t \in T(l)} u_{(l,t)} \psi_{(l,t)} = \sum_{l=l_0}^{L} \Psi_l \cdot \mathbf{u}_l .$$

We already noted that the coefficients $u_{(l,t)}$ decay fast where u is locally smooth and decay less fast where u is locally less smooth. Thus, it may be advantageous to discard all coefficients with an absolute value smaller than some prescribed ϵ and to work with the resulting adapted and compressed function approximation instead. This motivates to use certain subsets of the wavelet basis $\bigcup_{l=l_0}^{L} \Psi_l$ for function representation. Analogously, the wavelet transform and its inverse can be generalized to the adaptive case. Also, further operations with functions like the addition, the multiplication of two functions or more general, the non-linear function evaluation can be considered directly for the adaptive setting. Furthermore, the discretization and application of a differential operator to a function can be treated in an adaptive fashion.

To this end, recall that $T(l)$ denotes the index set of the wavelets of level l and that $S(l)$ denotes the index set for the scaling functions. Now, an adaptive basis out of $\bigcup_{l=l_0}^{L} \Psi_l$ consists of

$$\bigcup_{l=l_0}^{L} \bigcup_{t \in \mathcal{T}(l)} \psi_{(l,t)} \quad \text{where} \quad \mathcal{T}(l) \subseteq T(l) . \tag{30}$$

Thus $\{\mathcal{T}(l)\}_{l=l_0}^{L}$ describes the set of active wavelet indices whose associated wavelets are 'active' in the adaptive wavelet basis. For the adaptive wavelet transforms we will also deal with subsets

$$\mathcal{S}(l) \subseteq S(l) , \text{ where in general } \mathcal{S}(l) \subset S(l) , \tag{31}$$

in order to handle adaptive versions of the auxiliary $\mathbf{u}^l$ vectors. Similar to subsection 2.2 we will use the following notation for row vectors of functions, column vectors of coefficient sequences and scalar products

$$
\begin{aligned}
\mathbf{u}^{\mathcal{S}(l)} &:= \{u^{(l,s)}\}_{s \in \mathcal{S}(l)} , \quad \Phi^{\mathcal{S}(l)} := \{\phi^{(l,s)}\}_{s \in \mathcal{S}(l)} , \\
\mathbf{u}_{\mathcal{T}(l)} &:= \{u_{(l,t)}\}_{t \in \mathcal{T}(l)} , \quad \Psi_{\mathcal{T}(l)} := \{\psi_{(l,t)}\}_{t \in \mathcal{T}(l)} , \\
\mathbf{u}_{\mathcal{T}} &:= \{\mathbf{u}_{\mathcal{T}(l)}\}_{l=l_0}^{L} , \quad \Psi_{\mathcal{T}} := \{\Psi_{\mathcal{T}(l)}\}_{l=l_0}^{L}
\end{aligned}
$$

$$
\text{and} \quad \Psi_{\mathcal{T}} \cdot \mathbf{u}_{\mathcal{T}} = \sum_{(l,t) \in \mathcal{T}} u_{(l,t)} \psi_{(l,t)} . \tag{32}
$$

In this section we will also use submatrices of H^l, $\tilde{H}^l$, G^l and $\tilde{G}^l$. E.g. for $\mathcal{S}(l) \subseteq S(l)$ and $\mathcal{T}(l) \subseteq T(l)$ we define

$$
G^l_{\mathcal{S}(l),\mathcal{T}(l)} := \left(G^l_{s,t}\right)_{s \in \mathcal{S}(l), t \in \mathcal{T}(l)} .
$$

These submatrices will be used for the algebraic description of the adaptive wavelet transforms where we will work only with the sparse vectors $\mathbf{u}_{\mathcal{T}(l)}$ or $\mathbf{u}^{\mathcal{S}(l)}$ instead of $\mathbf{u}_l$ or $\mathbf{u}^l$, respectively.

3.1 Adaptive Wavelet Transform

In the following we will generalize the wavelet transform of the previous section and its inverse to the adaptive setting. Let us first describe the basic idea:

Each of our wavelets $\psi_{(l,t)}$ is also a scaling function $\phi^{(l,s(t))}$ with a certain $s(t)$ which is $s(t) = 2t + 1$ (except for the case of homogeneous Neumann boundary conditions, see (27),(28)). Therefore, $\psi_{(l,t)}$ is one in $s(t)2^{-l}$ and vanishes in all other nodes of the dyadic grid $2^{-l}S(l)$. Thus, we can uniquely identify each of the wavelets with its associated point $s(t)2^{-l}$. Consequently, the adaptive grids

$$
\Omega_l := \bigcup_{k=l_0}^{l} \bigcup_{t \in \mathcal{T}(l)} s(t)2^{-k} , \quad l_0 \leq l \leq L
$$

are associated to our adaptive basis (30). To calculate the coefficient $u_{(l,t)}$ like in (15) we needed the nodal values of u in $s(t)2^{-l}$ and the surrounding even nodes $(s(t) + \{-N+1, -N+3, ..., N-1\})2^{-l}$. Thus, these nodes have to belong to Ω_{l-1} and therefore the corresponding wavelets $\psi_{(l',t')}$ $(l' \leq l-1)$ have to belong to the adaptive basis. If this condition is satisfied we can run the adaptive versions of the algorithms (WT) and (IWT) by just restricting them to the index sets $\mathcal{T}(l) \subseteq T(l)$ on the wavelet side, and yet to be properly defined $\mathcal{S}(l) \subseteq S(l)$ on the scaling function side, respectively.

In order to formalize this approach it is convenient to introduce the concept of generating systems. Furthermore we need general operators R and C which act on matrices and index sets. Let X, Y be index sets and A be a $X \times Y$ matrix $A = \{a_{x,y}\}_{x \in X, y \in Y}$. Recall that the set of subsets of X is denoted by $\mathcal{P}X$. Then we define

$$R(A,.) : \mathcal{P}X \to \mathcal{P}Y , R(A, M) := \{y \in Y \mid \exists x \in M : a_{x,y} \neq 0\}, \text{'R} \cong \text{rows'},$$
$$C(A,.) : \mathcal{P}Y \to \mathcal{P}X , C(A, M) := R(A^T, M) , \quad \text{'C} \cong \text{columns'.} \tag{33}$$

The operators R and C will be used to determine dependencies of 'active' wavelets $\psi_{(l,t)}$ on scaling functions $\phi^{(l,s)}$ and conversely.

Consider the following typical step of the wavelet transform **(WT)**

$$u_{(l,t)} = \left(\tilde{G}^l_{.,t}\right)^T \cdot \mathbf{u}^l . \tag{34}$$

In the adaptive case we are interested in a few coefficients $u_{(l,t)}$ only (more precisely: $t \in \mathcal{T}(l)$). In this case we can not effort to store/calculate the complete vector $\mathbf{u}^l$, since this would destroy the advantages of adaptivity. Now, the problem is which of the coefficients $u^{(l,s)}$ are really required to calculate the above term (34). The answer is given by the matrix $\tilde{G}^l$: We need $u^{(l,s)}$ for all $s \in C(\tilde{G}^l, \{t\})$. so-called generating system.

The main idea of so-called generating systems is that an *incomplete* representation of $u = \sum_l \Psi_{\mathcal{T}(l)} \cdot \mathbf{u}_{\mathcal{T}(l)}$ in terms of a scaling functions basis $\Phi^{\mathcal{S}(l)} \cdot \langle \tilde{\Phi}^{\mathcal{S}(l)}, u \rangle$ may be *locally* identical to u if l is *locally* the finest level. This can be exploited to evaluate non-linear functionals or even differential operators.

A generating system for $\bigcup_{l=l_0}^L \Psi_l$ is a set

$$G := \bigcup_{l=l_0}^L \Phi^{\mathcal{S}(l)} , \quad \mathcal{S}(l) \subseteq S(l), \quad \text{such that} \quad \text{span} \bigcup_{l=l_0}^L \Psi_l \subseteq \text{span } G .$$

The index sets $\mathcal{S}(l)$ shall be as small as possible (numerical efficiency) and they shall allow for the following adaptive wavelet transform:

---
algorithm (AWT)
given $\{\mathbf{u}^{\mathcal{S}(l)}\}_{l=l_0}^L$
for $l = L$ downto $l_0 + 1$
   $\mathbf{u}_{\mathcal{T}(l)} = \left(\tilde{G}^l_{\mathcal{S}(l),\mathcal{T}(l)}\right)^T \cdot \mathbf{u}^{\mathcal{S}(l)}$
   $\Delta(l-1) = \{t \in \mathcal{S}(l-1) \mid C(\tilde{H}^l, \{t\}) \subseteq \mathcal{S}(l)\}$
   $\mathbf{u}^{\Delta(l-1)} = \left(\tilde{H}^l_{\mathcal{S}(l),\Delta(l-1)}\right)^T \cdot \mathbf{u}^{\mathcal{S}(l)}$
end

The idea of the $\Delta(l-1)$ correction is the following: Assume that the values of $\mathbf{u}^{\mathcal{S}(l)}$ are only approximations of the true values $\langle \tilde{\Phi}^{\mathcal{S}(l)}, u \rangle$. Then, it is likely that the values of $\mathbf{u}^{\mathcal{S}(l)}$ are more accurate for large l than for small l. Thus, the $\Delta(l-1)$ correction uses the more accurate values of $\mathbf{u}^{\mathcal{S}(l)}$ to calculate an improved approximation of $\mathbf{u}^{\mathcal{S}(l-1)}$. From **(AWT)** we immediately obtain the condition

$$C(\tilde{G}^l, \mathcal{T}(l)) \subseteq \mathcal{S}(l) . \tag{35}$$

In the same fashion we carry the inverse wavelet transform over to the adaptive case:

---

algorithm (AIWT)
given $\{\mathbf{u}_{\mathcal{T}(l)}\}_{l=l_0}^{L}$
for $l = l_0 + 1$ to L
$\quad \mathbf{u}^{\mathcal{S}(l)} = G_{\mathcal{S}(l),\mathcal{T}(l)}^{l} \cdot \mathbf{u}_{\mathcal{T}(l)} \ + \ H_{\mathcal{S}(l),\mathcal{S}(l-1)}^{l} \cdot \mathbf{u}^{\mathcal{S}(l-1)}$
end

This algorithm shall give the same results $u^{(l,s)}$ for $s \in \mathcal{S}(l)$ as the non-adaptive algorithm (**IWT**). Thus, the second condition on $\mathcal{S}(l)$ is

$$R(H^l, \mathcal{S}(l)) \ \subseteq \ \mathcal{S}(l-1) . \tag{36}$$

Combining conditions (35) and (36) we end up with an algorithm for the recursive determination of $\mathcal{S}(l)$:

---

algorithm ($\mathcal{S}$)
given $\{\mathcal{T}(l)\}_{l=l_0}^{L}$
$\mathcal{S}(L) = C(\tilde{G}^L, \mathcal{T}(L))$
for $l = L - 1$ downto l_0
$\quad \mathcal{S}(l) = C(\tilde{G}^l, \mathcal{T}(l)) \ \bigcup \ R(H^{l+1}, \mathcal{S}(l+1))$
end

A short calculation shows that the amount of operations involved in (**AWT**), (**AIWT**) and ($\mathcal{S}$) is proportional to the number of active wavelets.

3.2   Adaptive Non-Linear Function Evaluation

The three above algorithms are not restricted to just Interpolet wavelets but can be used in connection with general biorthogonal wavelets to solve a difficult problem in adaptive wavelet methods: the evaluation of non-linear terms. In the following we will briefly sketch the main idea of the associated algorithm.

Assume that we have a quite accurate adaptive wavelet approximation $\boldsymbol{\Psi}_{\mathcal{T}} \cdot \mathbf{u}_{\mathcal{T}}$ of a function u. Now we are looking for a wavelet approximation $\boldsymbol{\Psi}_{\mathcal{T}} \cdot \mathbf{v}_{\mathcal{T}}$ of $f(u(x))$, where $f : \mathbb{R} \to \mathbb{R}$ is a given smooth function. Thus, the coefficients

$$v_{(l,t)} \ \approx \ \langle \tilde{\psi}_{(l,t)} \, , \, f \circ u \rangle$$

are needed. The main idea for the calculation of these terms is to use the generating system representation of $\boldsymbol{\Psi}_{\mathcal{T}} \cdot \mathbf{u}_{\mathcal{T}}$. To this end, first the corresponding coefficients $\mathbf{u}^{\mathcal{S}(l)}$ are calculated by algorithm (**AIWT**). These coefficients are approximations of $\langle \tilde{\phi}^{(l,s)} \, , \, u \rangle$

$$u^{(l,s)} \approx \langle \tilde{\phi}^{(l,s)} \, , \, u \rangle , \quad s \in S(l) . \tag{37}$$

All commonly used scaling functions have certain (quasi-) interpolation properties. More precisely, there is a $K \in \mathbb{N}$ such that for all $s \in S(l)$ there is a point $x(l,s)$ (in general $x(l,s) \approx s2^{-l}$) which satisfies

$$\langle \tilde{\phi}^{(l,s)} , u \rangle = u(x(l,s)) + O(2^{-lK}\frac{\partial^K u}{\partial x^K}) . \tag{38}$$

For the orthogonal Daubechies wavelets $K = \min(3,N)$, compare [61]. For Coiflets K also depends on N and is usually $\geq N$, see [51]. For Interpolets there is no restriction on K, since $\langle \tilde{\phi}^{(l,s)} , u \rangle = u(2^{-l}s)$. The above relation (38) applied to $f \circ u$ yields

$$f(u^{(l,s)}) \overset{(37)}{\approx} f(\langle \tilde{\phi}^{(l,s)} , u \rangle)$$

$$\overset{(38)}{=} f\big(u(x(l,s)) + O(2^{-lK}\frac{\partial^K u}{\partial x^K})\big)$$

$$\overset{f \in C^1}{=} f\big(u(x(l,s))\big) + O(2^{-lK}\frac{\partial^K u}{\partial x^K})$$

$$\overset{(38) \text{ for } f \circ u}{=} \langle \tilde{\phi}^{(l,s)} , f \circ u \rangle - O\big(2^{-lK}\frac{\partial^K f \circ u}{\partial x^K}\big) + O\big(2^{-lK}\frac{\partial^K u}{\partial x^K}\big) .$$

Therefore, if f is sufficiently smooth there holds

$$f(u^{(l,s)}) \approx \langle \tilde{\phi}^{(l,s)} , f \circ u \rangle + O\big(2^{-lK}\|u\|_{C^K(\text{supp}\tilde{\phi}^{(l,s)})}\big) \tag{39}$$

where

$$\|u\|_{C^K(\text{supp}\tilde{\phi}^{(l,s)})} := \sum_{i=0}^{K} \sup\{|\frac{\partial^i u(x)}{\partial x^i}| \mid x \in \text{supp}\tilde{\phi}^{(l,s)}\} .$$

If we would have sufficiently accurate approximations $v^{(l,s)}$ of the values $\langle \tilde{\phi}^{(l,s)} , f \circ u \rangle$, algorithm (AWT) applied to $\{v^{(l,s)}\}$ would calculate sufficiently accurate approximations $\mathbf{v}_{\mathcal{T}(l)}$.

Two cases are possible in (39). First, the $O(2^{-lK}\|u\|_{C^K})$-term is sufficiently small, i.e. it is of the same order as the error $u^{(l,s)} - \langle \tilde{\phi}^{(l,s)} , u \rangle$, see (37). Then, $f(u^{(l,s)})$ is the desired approximation $v^{(l,s)}$. In the other case the $O(2^{-lK}\|u\|_{C^K})$-term is not sufficiently small. But then, adaptive refinement in the vicinity of $\text{supp}\tilde{\phi}^{(l,s)}$ was already needed in the wavelet approximation process to obtain $\Psi_{\mathcal{T}} \cdot \mathbf{u}_{\mathcal{T}}$ with sufficient accuracy. Therefore, there are active wavelets in the vicinity of $\text{supp}\tilde{\phi}^{(l,s)}$ on higher levels than l in $\Psi_{\mathcal{T}}$. Thus, we may assume that $s \in \Delta(l)$, compare algorithm (AWT). Consequently, the approximation $v^{(l,s)}$ of $\langle \tilde{\phi}^{(l,s)} , f \circ u \rangle$ is not necessarily to be calculated by $f(u^{(l,s)})$, but can be calculated by the wavelet transform from sufficiently accurate values of higher levels. Now, the complete algorithm reads

---

algorithm (NL)

given $\mathbf{u}_{\mathcal{T}}$

Calculate $\{\mathbf{u}^{S(l)}\}_{l=l_0}^{L}$ by means of (AIWT)

Calculate $\{\mathbf{v}^{S(l)}\}_{l=l_0}^{L}$ by means of $v^{(l,s)} = f(u^{(l,s)})$

Calculate $\mathbf{v}_{\mathcal{T}}$ by means of (AWT) from $\{\mathbf{v}^{S(l)}\}_{l=l_0}^{L}$

An analysis of this algorithm for general biorthogonal wavelets is given in the paper [18].

Note that for Interpolets there holds

$$u^{(l,s)} = \langle \tilde{\phi}^{(l,s)} , u \rangle$$

and thus

$$f\left(u^{(l,s)}\right) = \langle \tilde{\phi}^{(l,s)} , f \circ u \rangle .$$

Therefore, the results $v_{(l,t)}$ of algorithm (AWT) in algorithm (NL) satisfy

$$v_{(l,t)} = \langle \tilde{\psi}_{(l,t)} , f \circ u \rangle$$

without any error. Thus, for Interpolets the exact values are obtained while for general biorthogonal wavelets one can only expect approximate values.

We will now introduce a further simplification for the Interpolet case which deals with the following problem: In general the number of degrees of freedom of the adaptive generating system is larger than the number of degrees of freedom of the adaptive Interpolet basis. But, if both numbers would be the same then we could carry out the multivariate analogue of (AWT) and (AIWT) 'in place' by the successive application of (AWT) or (AIWT) with respect to each coordinate direction. Now, consider the special properties of the generating system representation for the Interpolet case. Here, many of the coefficients $u^{(l,s)}$ from (AIWT) are the same as the nodal value $u(s2^{-l})$. Using this interpretation ($u^{(l,s)} = u(s2^{-l})$) we can consider the adaptive grid

$$\Omega_{\mathcal{T}} := \bigcup_{l} \bigcup_{s \in S(l)} s2^{-l}$$

as the minimal grid required to run the adaptive transforms, if we interpret (AWT) and (AIWT) in terms of nodal values and grid points. However, in general

$$\#\Omega_{\mathcal{T}} > \#\mathcal{T} \qquad \text{where} \qquad \mathcal{T} := \bigcup_{l=l_0}^{L} \bigcup_{t \in \mathcal{T}(l)} (l,t) .$$

The point is that under some mild conditions on $\mathcal{T}$ there holds $\#\Omega_{\mathcal{T}} = \#\mathcal{T}$ and $\Omega_{\mathcal{T}}$ has those nodes just at the right places which are required to run (AWT) and (AIWT).

Lemma 1. *Let*

$$\sigma(l,t) := s \qquad \text{if} \qquad \phi^{(l,s)} = \psi_{(l,t)}$$

$$x(l,t) := \sigma(l,t)2^{-l}$$

be the index of the unique scaling function which coincides with $\psi_{(l,t)}$ and the corresponding unique grid point. For $l = l_0$ there holds $x(l_0,t) = t2^{-l_0}$ and for $l > l_0$ there holds $x(l,t) = (2t+1)2^{-l}$ except for homogeneous Neumann b.c., see (27), (28). Since the grid point is unique one can define the inverse mapping

$$x^{-1} : \bigcup_{l=l_0}^{L} 2^{-l}S(l) \rightarrow \bigcup_{l=l_0}^{L} (l,\mathcal{T}(l)) , \qquad x^{-1}(x(l,t)) = (l,t) .$$

Now, if

$$(l,t) \in \mathcal{T} \quad \Rightarrow \quad x^{-1}\big(2^{-l}C(\tilde{G}^l,\{t\}) \big) \subseteq \mathcal{T} \tag{40}$$

then there holds

$$\mathcal{T} = x^{-1}(\Omega_{\mathcal{T}}) . \tag{41}$$

Proof. We will first show that $\mathcal{T} \subseteq x^{-1}(\Omega_{\mathcal{T}})$. For Interpolets one has

$$\sigma(l,t) \in C(\tilde{G}^l,\{t\}) . \tag{42}$$

In the periodic or shift invariant case there holds

$$\tilde{G}^l_{t,2t+1} = \tilde{g}_{(2t+1)-2t} = \tilde{g}_1 = h_0 = 1 \quad \Rightarrow \quad 2t+1 = \sigma(l,t) \in C(\tilde{G}^l,\{t\}) .$$

Now (42) yields

$$x(l,t) \in 2^{-l}C(\tilde{G}^l,\{t\}) \quad \Rightarrow \quad (l,t) \in x^{-1}\big(2^{-l}C(\tilde{G}^l,\{t\}) \big) \subset \Omega_{\mathcal{T}} .$$

The proof for $x^{-1}(\Omega_{\mathcal{T}}) \subseteq \mathcal{T}$ goes by induction for l of $(l,t) \in x^{-1}(\Omega_{\mathcal{T}})$. Let $(L,t) \in x^{-1}(\Omega_{\mathcal{T}})$. We use the following technical property of the matrices $\tilde{G}^l$

$$C(\tilde{G}^l,\{t\}) \cap \sigma(T(l)) = \{\sigma(l,t)\} . \tag{43}$$

For the inner scaling functions or for the periodic case this means that there is exactly one non-zero entry $\tilde{G}^l_{t,s}$ with odd s. This s is exactly $s = 2t+1$, see (17) and (2). For the boundary Interpolets, (43) follows from their special construction and a from longer technical calculation which we omit here. From (43) one concludes

$$(L,t) \in x^{-1}(\Omega_{\mathcal{T}}) \quad \Rightarrow \quad \sigma(L,t) \in \mathcal{S}(L) = C(\tilde{G}^L,\mathcal{T}(L)) \overset{(43)}{\Rightarrow} t \in \mathcal{T}(L) .$$

Assume now that we have shown

$$\forall (k,r) \in x^{-1}(\Omega_{\mathcal{T}}) \quad k > l \quad \Rightarrow \quad (k,r) \in \mathcal{T} .$$

Let $(l,t) \in x^{-1}(\Omega_{\mathcal{T}})$ and define

$$l' := \max\{k \mid x(l,t) \in 2^{-k}\mathcal{S}(k)\} , \quad s := 2^{l'}x(l,t) . \tag{44}$$

Two cases are possible:

(i) $s \in C(\tilde{G}^{l'},\{t'\})$ for some $t' \in \mathcal{T}(l')$. Then

$$(l,t) \in x^{-1}\big(2^{-l'}C(\tilde{G}^{l'},\{t'\}) \big) \overset{(40)}{\subseteq} \mathcal{T} .$$

(ii) $s \in R(H^{l'},\mathcal{S}(l'+1))\backslash C(\tilde{G}^{l'},\mathcal{T}(l'))$, i.e.

$$\exists\, s' \in \mathcal{S}(l'+1) : s \in R(H^{l'+1},\{s'\}) .$$

s' must be odd. Because of $\tilde{H}^k = \{\delta(n-2m)\}_{m \in S(k-1), n \in S(k)}$ and $(\tilde{H}^k)^T \cdot H^k \overset{(20)}{=} I$, each even row of H^k contains exactly one non-zero: $H^k_{2m,m} = 1$. Thus, if s' is even, then $R(H^{l'+1},\{s'\}) = \{s\}$ and $s' = 2s$ which contradicts the definition (44) of l', s. But, if s' is odd, then there is $t' \in \mathcal{T}(l'+1)$ with $\sigma(l'+1,t') = s'$. Since $l'+1 > l$ the assumption holds and $(l'+1,t') \in \mathcal{T}$. A close look on the matrices $\tilde{G}^k$ shows that $(l'+1,2s) \in C(\tilde{G}^{l'+1},\{t'\})$. This proves the assertion. $\square$

We have now shown that the technical condition (40) results in the relation (41) which allows to carry out the algorithms **(AWT)** and **(AIWT)** *in place* by using the same data structure and the same storage for $\mathbf{u}_\mathcal{T}$ and $\{\mathbf{u}^{S(l)}\}_{l=l_0}^{L}$ in a numerical code. Furthermore, (40) and (41) imply that not only the exact values $\langle \tilde{\psi}_{(l,t)} , f \circ u \rangle$ are calculated by the algorithm **(NL)**, but that also the result $\mathbf{\Psi}_\mathcal{T} \cdot \mathbf{v}_\mathcal{T}$ coincides with $f \circ u$ on $\Omega_\mathcal{T}$:

$$\left(\mathbf{\Psi}_\mathcal{T} \cdot \mathbf{v}_\mathcal{T}\right)(\Omega_\mathcal{T}) = (f \circ u)(\Omega_\mathcal{T}) . \tag{45}$$

If (40) is not satisfied (45) does not hold in general. Nevertheless, the question how to fulfill (40) remains. The idea is now to enlarge slightly a given index set $\mathcal{T}^0$ which does not fulfill (40) a priori. To this end, the following recursive algorithm calculates from $\mathcal{T}^0$ an index set $\mathcal{T}$ which satisfies (40).

---

algorithm $(\mathcal{T})$
$\mathcal{T}(L) = \mathcal{T}^0(L)$
for $l = L$ downto $l_0 + 1$
    $\mathcal{T}(l-1) = \mathcal{T}^0(l-1) \bigcup \{t \in T(l-1) \mid 2\sigma(l-1,t) \in C(\tilde{G}^l, \mathcal{T}(l))\}$
end

Remark: In [64] another algorithm was introduced which does not need (40). E.g. for the adaptive wavelet transform one starts with the nodal values $u(\Omega_\mathcal{T})$ and tries to find the interpolant $\sum_l \mathbf{\Psi}_{\mathcal{T}(l)} \cdot \mathbf{u}_{\mathcal{T}(l)}$ such that

$$\left(\sum_{l=l_0}^{L} \mathbf{\Psi}_{\mathcal{T}(l)} \cdot \mathbf{u}_{\mathcal{T}(l)} \right)(\Omega_\mathcal{T}) = u(\Omega_\mathcal{T}) .$$

To this end one exploits the fact that except for $\psi_{(l_0,t)}$ all other wavelets vanish at $x(l_0, t)$. Therefore $u_{(l_0,t)}$ and thus, $\mathbf{u}_{\mathcal{T}(l_0)}$ are easy to obtain. Then, in a second step one considers the interpolation problem

$$\left(\sum_{l=l_0+1}^{L} \mathbf{\Psi}_{\mathcal{T}(l)} \cdot \mathbf{u}_{\mathcal{T}(l)} \right)(\Omega_\mathcal{T}) = \left(u - \mathbf{\Psi}_{\mathcal{T}(l_0)} \cdot \mathbf{u}_{\mathcal{T}(l_0)}\right)(\Omega_\mathcal{T}) ,$$

which allows to determine $\mathbf{u}_{\mathcal{T}(l_0+1)}$ in the same fashion as $\mathbf{u}_{\mathcal{T}(l_0)}$. This scheme is repeated for all levels from $l = l_0 + 1$ to L. One drawback of this approach is that we have to know and handle the values of all Interpolets on all grid points of the finest level employed. Furthermore, at least in the non-adaptive case this algorithm has a complexity of $\mathcal{O}(L2^L)$ whereas our above approach is of order $\mathcal{O}(2^L)$ only. On the other hand, there is no artificial fill-in needed to ensure (40). But anyway, our experience is that this fill-in is relatively small in practical situations.

3.3   Adaptive Evaluation of Differential Operators

We will now derive an algorithm for the fast evaluation of constant coefficient differential operators. As an example consider the operator ∂_{xx}. For the discretization a Petrov-Galerkin scheme is used where the dual wavelets are chosen as test functions. We then obtain

$$\langle \tilde{\psi}_{(l,t)} \, , \, \partial_{xx} u \rangle \qquad \forall (l,t) \in \mathcal{T} \, . \tag{46}$$

Of course the trial functions $\psi_{(l,t)}$ must be at least two times continuously differentiable such that the above expression makes sense. The smoothness of the Interpolets grows with increasing N. E.g. for $N = 4$ we have $\psi_{(l,t)} \in C^{1.617}$ and for $N = 6$ we have $\psi_{(l,t)} \in C^{2.255}$. Now, we test $\partial_{xx} u$ not by just one function $\tilde{\psi}_{(l,t)}$ as in (46), but by all test functions of one level $\mathcal{T}(l)$ of the adaptive basis. We then obtain

$$\langle \tilde{\boldsymbol{\Psi}}_{\mathcal{T}(l)} \, , \, \partial_{xx} u \rangle = \langle \tilde{\boldsymbol{\Psi}}_{\mathcal{T}(l)} \, , \, \partial_{xx} \sum_{k=l_0}^{L} \boldsymbol{\Psi}_{\mathcal{T}(k)} \cdot \mathbf{u}_{\mathcal{T}(k)} \rangle$$

$$= \langle \tilde{\boldsymbol{\Psi}}_{\mathcal{T}(l)} \, , \, \partial_{xx} \sum_{k \geq l} \boldsymbol{\Psi}_{\mathcal{T}(k)} \cdot \mathbf{u}_{\mathcal{T}(k)} \rangle$$

$$+ \langle \tilde{\boldsymbol{\Psi}}_{\mathcal{T}(l)} \, , \, \partial_{xx} \sum_{k < l} \boldsymbol{\Psi}_{\mathcal{T}(k)} \cdot \mathbf{u}_{\mathcal{T}(k)} \rangle$$

$$=: \langle \tilde{\boldsymbol{\Psi}}_{\mathcal{T}(l)} \, , \, \partial_{xx} a_l \rangle \; + \; \langle \tilde{\boldsymbol{\Psi}}_{\mathcal{T}(l)} \, , \, \partial_{xx} b_{l-1} \rangle \, . \tag{47}$$

Now, the idea is to use generating systems for the calculations of the two terms in (47). For the second term in (47) we use an incomplete representation of b_{l-1} :

$$\Phi^{\mathcal{S}^b(l-1)} \cdot \mathbf{b}^{\mathcal{S}^b(l-1)} \, .$$

Here, $\mathcal{S}^b(l-1)$ must be large enough to ensure that

$$\langle \tilde{\boldsymbol{\Psi}}_{\mathcal{T}(l)} \, , \, \partial_{xx} \Phi^{\mathcal{S}^b(l-1)} \rangle \cdot \mathbf{b}^{\mathcal{S}^b(l-1)} \stackrel{!}{=} \langle \tilde{\boldsymbol{\Psi}}_{\mathcal{T}(l)} \, , \, \partial_{xx} b_{l-1} \rangle \, .$$

If we define $WS^l := \langle \tilde{\boldsymbol{\Psi}}_l \, , \, \partial_{xx} \Phi^l \rangle$ this leads to the condition

$$R(WS^l, \mathcal{T}(l)) \subseteq \mathcal{S}^b(l-1)$$

and finally to the following algorithm:

---
algorithm ($\mathcal{S}^b$)
given $\{\mathcal{T}(l)\}_{l=l_0}^{L}$
$\mathcal{S}^b(L) = \emptyset$
for $l = L - 1$ downto l_0
    $\mathcal{S}^b(l) \; = \; R(WS^l, \mathcal{T}(l)) \; \bigcup \; R(H^{l+1}, \mathcal{S}^b(l+1))$
end

The values $\mathbf{b}^{\mathcal{S}^b(l-1)}$ are calculated by algorithm (**AIWT**) with $\mathcal{S}(k)$ replaced by $\mathcal{S}^b(k)$.

A similar trick is applied to the first term in (47). Instead of a direct calculation of $\langle \tilde{\boldsymbol{\Psi}}_{\mathcal{T}(l)} \, , \, \partial_{xx} a_l \rangle$ we there compute the auxiliary values

$$\mathbf{c}^{\mathcal{S}^a(l)} := \langle \tilde{\boldsymbol{\Psi}}^{\mathcal{S}^a(l)}, \partial_{xx} a_l \rangle \, ,$$

by means of

---

algorithm (AUX)
given $\{\mathbf{u}_{\mathcal{T}(l)}\}_{l=l_0}^{L}$ and $\{\mathcal{S}^a(l)\}_{l=l_0}^{L}$
$\mathbf{c}^{\mathcal{S}^a(L)} = SW_{\mathcal{S}^a(L),\mathcal{T}(L)}^{L} \cdot \mathbf{u}_{\mathcal{T}(L)}$
for $l = L - 1$ downto l_0
$\qquad \mathbf{c}^{\mathcal{S}^a(l)} = \left(\tilde{H}_{\mathcal{S}^a(l+1),\mathcal{S}^a(l)}^{l+1}\right)^T \cdot \mathbf{c}^{\mathcal{S}^a(l+1)} + SW_{\mathcal{T}(l),\mathcal{S}^a(l)}^{l} \cdot \mathbf{u}_{\mathcal{T}(l)}$
end

---

where $SW^l := \langle \tilde{\Phi}^l, \partial_{xx}\Psi_l \rangle$. Here, the index sets $\mathcal{S}^a$ should be large enough to ensure that

$$\langle \tilde{\Psi}_{\mathcal{T}(l)}, \partial_{xx}a_l \rangle \stackrel{!}{=} \left(\tilde{G}_{\mathcal{S}^a(l),\mathcal{T}(l)}\right)^T \cdot \mathbf{c}^{\mathcal{S}^a(l)}.$$

Because of the regularity condition (40) on $\mathcal{T}$, the supports of $\Psi_{\mathcal{T}(l)}$ fulfill supp $\Psi_{\mathcal{T}(l+1)} \subseteq$ supp $\Psi_{\mathcal{T}(l)}$. Therefore

$$\mathcal{S}^a(l) = R(SW^l, \mathcal{T}(l)). \tag{48}$$

3.4   Adaptive Evaluation of Finite Difference Operators

Note that there is also another way for the adaptive evaluation of differential operators which is based on finite differences (FD). An advantage of FD is that there is no smoothness requirement for the Interpolets. Even the usual piecewise linear hat function ($N = 2$) which is only continuous can be used. The FD algorithm is similar to the algorithm (**NL**) for non-linear terms:

---

algorithm (FD)
given $\{\mathbf{u}_{\mathcal{T}(l)}\}$
Run (**AIWT**) to obtain the generating system representation $\{\mathbf{u}^{\mathcal{S}(l)}\}$
Apply the FD stencil (e.g. $2^{-l}[1 \ -2 \ \ 1]$) to $\mathbf{u}^{\mathcal{S}(l)}$ and obtain $\mathbf{v}^{\mathcal{S}(l)}$
Run (**AWT**) of $\{\mathbf{v}^{\mathcal{S}(l)}\}$ to obtain the final result

---

Here, one has to take into account that not for all points on a level l we can apply the FD stencil. For example, if $\mathcal{S}(l) = \{4, ..., 10\}$ a three point stencil like $[1 \ -2 \ 1]$ can only be used for $\{5, ..., 9\}$. Therefore, the nodal values $\{4, 10\}$ must not be employed in the Δ-correction in algorithm (**AWT**). However, note that the condition (35) ensures that the index set $\mathcal{S}(l)$ is large enough to allow for the application of the FD stencil to sufficiently many points of level l.

A similar approach has been used in [33] and [56]. In connection with the classical MRA-approach and higher order Interpolets FD have been employed in [40].

4   Multivariate Wavelets

So far, we are able to deal with the adaptive wavelet transform, the non-linear function evaluation and the evaluation of a differential operator in the 1D case on the interval. Now, we generalize the wavelet construction to the multivariate setting. To this end we consider two different tensor product-type multivariate wavelet approaches.

4.1   The MRA-Approach

This approach is used almost exclusively in the wavelet literature, see e.g. [2,5,10,18,29,48,59,64] and the references cited therein. The key idea is to establish an ordered sequence of approximation spaces like in the one-dimensional case, see (10). The simplest construction of such multivariate approximation spaces are the tensor products of the univariate V^l, i.e. $V^l \otimes ... \otimes V^l$. The complementary spaces between $V^{l-1} \otimes ... \otimes V^{l-1}$ and $V^l \otimes ... \otimes V^l$ are defined by tensor products of the univariate spaces V^{l-1} and W_l. For example in the two-dimensional case we obtain

$$V^l \otimes V^l = \left(V^{l-1} \oplus W_l\right) \otimes \left(V^{l-1} \oplus W_l\right)$$
$$= V^{l-1} \otimes V^{l-1}$$
$$\oplus \left(V^{l-1} \otimes W_l\right) \oplus \left(W_l \otimes V^{l-1}\right) \oplus \left(W_l \otimes W_l\right) .$$

Thus, the complementary space is the direct sum of three different tensor product spaces. Now, it is simple to define bases of the complementary spaces by tensor products of different combinations of univariate wavelets and/or scaling functions.

In the d-dimensional case this reads

$$\psi_{(\mathbf{e},l,\mathbf{t})}(\mathbf{x}) := \psi_{(e_1,l,t_1)}(x_1) \cdot ... \cdot \psi_{(e_d,l,t_d)}(x_d) , \qquad (49)$$

where $\psi_{(0,l,t)}(x) := \phi^{(l-1,t)}(x)$ and $\psi_{(1,l,t)}(x) := \psi_{(l,t)}(x)$. Here $\mathbf{e} = (e_1,..,e_d) \in \{0,1\} \times .. \times \{0,1\}\backslash \mathbf{0} \subset \mathbb{N}^d$ defines the type of the wavelet, $l \geq l_0$ denotes the level and $\mathbf{t} = (t_1,..,t_d) \in \mathbb{Z}^d$ denotes the spatial translation. Because of the linear ordering of the approximation spaces $V^{l-1} \otimes .. \otimes V^{l-1} \subset V^l \otimes .. \otimes V^l$ all the algorithms of the previous section can be used here as well. Of course the matrices $H^l, G^l, ...$ have to be replaced by their counterparts for the multivariate case. Due to the tensor product construction all these matrices are essentially Kronecker products [41] (sometimes called tensor products) of the matrices $H^l, G^l, ...$ We will not employ this approach in the remainder of this paper, but we will make use of the following pure tensor product technique.

4.2   The MRA-d-Approach

For this approach simple tensor products of univariate wavelets are employed as trial and test functions, i.e.

$$\psi_{(\mathbf{l},\mathbf{t})}(\mathbf{x}) := \psi_{(l_1,t_1)}(x_1) \cdot ... \cdot \psi_{(l_d,t_d)}(x_d) , \qquad (50)$$

where $\mathbf{x} = (x_1, .., x_d)$, $\mathbf{l}, \mathbf{t}$ analogous. In the same way the multivariate dual wavelets $\tilde{\psi}_{(\mathbf{l},\mathbf{t})}$ are defined. In the rest of this paper we exclusively use the MRA-d-approach. The reason for this choice is its algorithmical simplicity (as we will see below) and its superior approximation properties [34,62].

Analogously to (30) and (32) the following notation for the adaptive multivariate case is used:

$$\mathcal{T}(\mathbf{l}) \subseteq T(l_1) \otimes ... \otimes T(l_d) , \quad \mathcal{T} = \bigcup_{\mathbf{l}} \{\mathbf{l}\} \times \mathcal{T}(\mathbf{l}) ,$$

$$\mathbf{u}_{\mathcal{T}(\mathbf{l})} = \{u_{(\mathbf{l},\mathbf{t})}\}_{\mathbf{t} \in \mathcal{T}(\mathbf{l})} , \quad \mathbf{u}_{\mathcal{T}} = \{\mathbf{u}_{\mathcal{T}(\mathbf{l})}\}_{\mathbf{l}} . \tag{51}$$

Furthermore, we will denote by

$$V_{\mathcal{T}} = \operatorname{span}\{\psi_{(\mathbf{l},\mathbf{t})}\}_{(\mathbf{l},\mathbf{t}) \in \mathcal{T}}, \quad \tilde{V}_{\mathcal{T}} = \operatorname{span}\{\tilde{\psi}_{(\mathbf{l},\mathbf{t})}\}_{(\mathbf{l},\mathbf{t}) \in \mathcal{T}}$$

the adaptive multivariate trial and test spaces.

The tensor product approach has the advantage that the algorithms for the adaptive wavelet transforms, the application of discrete differential operators and, to some extent, the non-linear function evaluation just boil down to the 1D algorithms of the previous section, compare also [14]. Consider as an example the evaluation of the partial differential operator $\partial_{x_1 x_1}$. We employ a Petrov-Galerkin scheme with the dual wavelets as test functions. Therefore, we have to evaluate the terms

$$\langle \tilde{\psi}_{(\mathbf{l},\mathbf{t})} , \partial_{x_1 x_1} \sum_{(\mathbf{k},\mathbf{s}) \in \mathcal{T}} u_{(\mathbf{k},\mathbf{s})} \psi_{(\mathbf{k},\mathbf{s})} \rangle =$$

$$= \sum_{(\mathbf{k},\mathbf{s}) \in \mathcal{T}} u_{(\mathbf{k},\mathbf{s})} \langle \tilde{\psi}_{(l_1,t_1)} , \partial_{x_1 x_1} \psi_{(k_1,s_1)} \rangle \cdot \prod_{i=2}^{d} \langle \tilde{\psi}_{(l_i,t_i)} , \psi_{(k_i,s_i)} \rangle$$

$$\overset{(18)}{=} \sum_{(\mathbf{k},\mathbf{s}) \in \mathcal{T}} u_{(\mathbf{k},\mathbf{s})} \langle \tilde{\psi}_{(l_1,t_1)} , \partial_{x_1 x_1} \psi_{(k_1,s_1)} \rangle \cdot \prod_{i=2}^{d} \delta(k_i - l_i) \delta(s_i - t_i)$$

$$= \sum_{k_1,s_1} u_{((k_1,\hat{\mathbf{l}}),(s_1,\hat{\mathbf{t}}))} \langle \tilde{\psi}_{(l_1,t_1)} , \partial_{x_1 x_1} \psi_{(k_1,s_1)} \rangle$$

$$= \langle \tilde{\psi}_{(l_1,t_1)} , \partial_{x_1 x_1} \sum_{k_1,s_1} u_{((k_1,\hat{\mathbf{l}}),(s_1,\hat{\mathbf{t}}))} \psi_{(k_1,s_1)} \rangle ,$$

where $\hat{\mathbf{l}} = (l_2, .., l_d)$ and $\hat{\mathbf{t}} = (t_2, .., t_d)$. Thus, unidirectional operators like $\partial_{x_1 x_1}$ are evaluated simply by an application of the 1D algorithms of subsection 3.3 to all rows $(\hat{\mathbf{l}}, \hat{\mathbf{t}})$ of the adaptive basis. This also holds for the finite difference technique. Here the 1D algorithm (FD) is simply applied to all rows $(\hat{\mathbf{l}}, \hat{\mathbf{t}})$ of the adaptive basis.

In the same fashion the adaptive wavelet transform and its inverse can be carried over to the multivariate case. We simply apply (AWT) or (AIWT) first with respect to all rows in x_1-direction, then to all rows with respect to the x_2-direction and so on. It is important to note that only because of the special properties of Interpolets this 'tensor product type' application of e.g. the univariate algorithm **(AWT)** yields the exact solution of the interpolation problem with respect to the adaptive wavelet basis and the associated adaptive grid. This does not hold for other wavelets like the orthogonal Daubechies wavelets or biorthogonal spline wavelets. However, the multivariate version of condition (40) must be satisfied, i.e. (40) must hold for all rows with respect to the x_i-direction $(1 \le i \le d)$.

In summary, Interpolet wavelets in connection with the MRA-d-approach lead to particularly simple algorithms for the adaptive wavelet transform, its inverse, the non-linear function evaluation and the evaluation of unidirectional differential operators. All these algorithms boil down to the univariate adaptive algorithms of the previous section. In the same way as the non-linear function evaluation one can also find the approximate wavelet expansion of the product $u \cdot v$ of $u, v \in V_\mathcal{T}$. Furthermore, mixed derivatives like $\partial_{x_1 x_2} u$ are calculated simply by $\partial_{x_1}(\partial_{x_2} u)$. Altogether we have quite simple and efficient algorithms for the evaluation of all terms which may occur in the discretization of general PDEs. This is the first and most important step towards our final adaptive wavelet solver for PDEs or the Navier-Stokes equations, respectively.

Of course, the present approach is restricted to simple boxes, e.g. to $[0,1]^d$ as the computational domain. In order to deal with more general domains one can employ the usual parametric mappings where the computational domain is mapped to the physical domain. Another technique is to imbed the physical domain Ω in an enclosing box. Then, boundary conditions on $\partial\Omega$ are enforced by e.g. penalization techniques [47] or, in case of the Navier-Stokes equations, by special forcing terms on the right hand side [1].

5   Consistency and Stability

Now we will consider the convergence and the rate of convergence of our scheme. For the sake of simplicity we consider the model problem

$$\Delta u = f , \quad \text{in } \Omega = [0,1]^d \tag{52}$$

with appropriate boundary conditions. Our collocation scheme produces numerical solutions $u_\mathcal{T} = \sum_{\mathbf{l},\mathbf{t}} u_{(\mathbf{l},\mathbf{t})} \psi_{(\mathbf{l},\mathbf{t})}$ such that

$$\langle \tilde{\psi}_{(\mathbf{l},\mathbf{t})} , \Delta u_\mathcal{T} \rangle = \langle \tilde{\psi}_{(\mathbf{l},\mathbf{t})} , f \rangle \qquad \forall (\mathbf{l},\mathbf{t}) \in \mathcal{T} .$$

Therefore, there holds

$$L_\mathcal{T} u_\mathcal{T} = I_\mathcal{T} f , \tag{53}$$

where

$$I_\mathcal{T} f := \sum_{(\mathbf{l},\mathbf{t}) \in \mathcal{T}} \psi_{(\mathbf{l},\mathbf{t})} \langle \tilde{\psi}_{(\mathbf{l},\mathbf{t})} , f \rangle \quad \text{and} \quad L_\mathcal{T} u_\mathcal{T} := I_\mathcal{T} \Delta u .$$

Assume now that the discrete Laplacian L_T satisfies the following stability condition

$$\sup_{\tilde{v}\in \tilde{V}_T} \frac{\langle \tilde{v} \,,\, L_T u\rangle}{\|\tilde{v}\|_{T'}} \geq \alpha \|u\|_T \,. \tag{54}$$

Here, $\tilde{V}_T$ is the analogue of V_T for the dual wavelets, $\|.\|_T$ is the maximum norm on the adaptive grid associated to T and

$$\|\tilde{v}\|_{T'} = \sup_{u\in V_T} \frac{\langle \tilde{v} \,,\, u\rangle}{\|u\|_T} \,.$$

Then, the standard estimates yield

$$\|u - u_T\|_T = \|I_T u - u_T\|_T \leq \frac{1}{\alpha} \sup_{\tilde{v}\in \tilde{V}_T} \frac{\langle \tilde{v} \,,\, L_T(I_T u - u_T)\rangle}{\|\tilde{v}\|_{T'}}$$

$$\leq \frac{1}{\alpha}\|L_T(I_T u - u_T)\|_T$$

$$= \frac{1}{\alpha}\|(L_T I_T u - I_T \Delta u) - \underbrace{(L_T u_T - I_T \Delta u)}_{=0 \,,\, \text{see (53)}}\|_T$$

$$= \frac{1}{\alpha}\|(L_T I_T u - I_T \Delta u)\|_T \,.$$

Therefore, the error can be estimated by the consistency error

$$\|L_T I_T u - I_T \Delta u\|_T \,.$$

For the special case of so-called regular sparse grids, where

$$T = \{(1, t) \mid l_1 + .. + l_d \leq n\} \tag{55}$$

for some $n \in \mathbb{N}$, we have shown[1] in [44] that for sufficiently smooth u there holds

$$\|L_T I_T u - I_T \Delta u\|_T \leq C(u)n^{d-1}2^{-(N-2)n} \,.$$

$C(u)$ depends on Hölder-like norms of higher order mixed derivatives of u. collocation method is a special higher order FD method. Now, the interesting point of sparse grids is that we have essentially the usual error estimates as for conventional grid discretizations. But the number of degrees of freedom we have to spend is only $O(n^{d-1}2^n)$ - compared to $O(2^{dn})$ for the usual FD or FEM methods on full grids. However, for the general case of an arbitrary adaptive basis and a solution which may have singularities we know of no satisfying consistency proof yet.

The same holds for the stability (54). We are convinced that stability holds, because we never experienced any problems. Furthermore, in [56] systematic experiments were made for the sparse grid FD technique that also indicate stability.

---

[1] In [44] the consistency of general FD methods on regular sparse grids was shown. Since for regular sparse grids the collocation method is a special higher order FD method this result applies here as well.

6 Preconditioning of Linear Systems

Consider the model problem

$$-\Delta u = f \quad \text{in } [0,1]^2 \tag{56}$$

with, for example, homogeneous Dirichlet or Neumann boundary conditions. We later have to deal with such problems in the projection step for the solution of the Navier-Stokes equations. For the numerical solution of (56) we have to solve a linear system of equations

$$\mathbf{A} \cdot \mathbf{u}_{\mathcal{T}} = \mathbf{f}_{\mathcal{T}} \tag{57}$$

with the coefficient matrix

$$\mathbf{A}_{(l,t),(k,s)} = -\langle \tilde{\psi}_{(l,t)} \, , \, \Delta \psi_{(k,s)} \rangle \, , \quad (k,s),(l,t) \in \mathcal{T} \, .$$

Alternatively we also consider the FD discretization of subsections 3.4 and 4.2 for this model problem. We then denote the resulting matrix by $\mathbf{A}$ as well.

From section 3 we have fast algorithms for the matrix vector multiplication with $\mathbf{A}$, but no such algorithm for the transpose of $\mathbf{A}$. Furthermore, $\mathbf{A}$ is non-symmetric and in our experiments we found that $\mathbf{A}$ is indefinite. For all these reasons we employ a GMRES, BiCGStab or BiCGStab(l) solver [25]. As one might expect, the performance of these solvers is not very good, since the condition number grows at least like 4^L if L is the finest level of the adaptive basis. Therefore, some kind of preconditioning is required.

The first idea is to use preconditioning from the left [56], i.e. to solve

$$\mathbf{A}\mathbf{C}^{-1} \cdot \bar{\mathbf{u}}_{\mathcal{T}} = \mathbf{f}_{\mathcal{T}} \, , \tag{58}$$

where $\mathbf{C}$ is the following diagonal matrix

$$\mathbf{C}_{(k,s),(l,t)} = \delta(k-l)\delta(s-t) \sum_{i=1}^{d} 4^{l_i} \, , \quad (k,s),(l,t) \in \mathcal{T} \, . \tag{59}$$

This strategy has its origins in the preconditioning of Galerkin stiffness matrices for certain tensor product wavelets [35] which however are different from ours. We applied this preconditioner to the model problem for a non-adaptive basis with finest level $(L,..,L)$, i.e.

$$\mathcal{T} = \{ \, (l,t) \mid t \in T(l_1) \otimes .. \otimes T(l_d) \text{ and } l_0 \le l_i \le L \, , \, (1 \le i \le d) \, \} \, . \tag{60}$$

The grid associated to $\mathcal{T}$ is simply the cartesian mesh with mesh size 2^{-L} in both directions. The behaviour of the residual generated by the BiCGStab(2) solver is depicted in Figure 3. The right hand side of the model problem was chosen such that the solution is

$$u(x,y) = (x^2 + y^2)^{1/4} \cos\left(\frac{\arctan(x/y)}{2}\right)(1 - x(1-x)y(1-y)) \, . \tag{61}$$

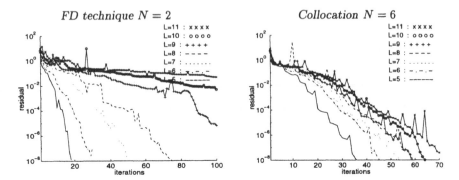

Fig. 3. BiCG(2) residuals for the simple diagonal preconditioner (58), the 2D model problem (61) and non-adaptive bases (60).

Obviously the performance of the preconditioned BiCG solver is quite different for the FD discretization with Interpolets ($N = 2$) and for the collocation discretization ($N = 6$). One explanation for this may be the following:

Although there is no satisfying convergence theory for the BiCGStab scheme, it seems to be reasonable that the performance depends on the condition number of the preconditioned coefficient matrix with respect to the l_2-norm:

$$\|\{u_l\}_{l\in\mathbb{N}}\|_{l_2}^2 := \sum_{l\in\mathbb{N}} |u_l|^2 \ .$$

The point is that the condition number scales like 2^L (or $2^{dL/2}$ in general) for the Interpolets ($N = 2$) with the FD discretization and like $\sqrt{2}^L$ (or $2^{(d-1)L/2}$ in general) for Interpolets ($N \geq 4$) with the FD or the collocation discretization.

In any case, we need a more efficient preconditioner for higher dimensional problems. In the following we will present the essentials of the Lifting Interpolet preconditioner which works for Finite Differences, the collocation method and (in slightly modified form) also for a Galerkin discretization with tensor product type trial functions (50). A more detailed description of this preconditioner will be given in a forthcoming paper [46].

As before, we apply diagonal preconditioning from the left, but we use it together with another multiscale basis $\{\hat{\psi}_{(\mathbf{l},\mathbf{t})}\}$ which is different form the Interpolets. This means that we solve the linear system

$$\mathbf{B}\mathbf{A}\mathbf{B}^{-1}\mathbf{C}^{-1} \cdot \bar{\mathbf{u}}_{\mathcal{T}} = \mathbf{B} \cdot \mathbf{f}_{\mathcal{T}} \ . \tag{62}$$

Here, $\mathbf{B}$ is the matrix which represents the basis transform

$$\{\psi_{(\mathbf{l},\mathbf{t})}\}_{(\mathbf{l},\mathbf{t})\in\mathcal{T}} \ \to \ \{\hat{\psi}_{(\mathbf{l},\mathbf{t})}\}_{(\mathbf{l},\mathbf{t})\in\mathcal{T}} \ . \tag{63}$$

The new wavelets $\hat{\psi}_{(\mathbf{l},\mathbf{t})}$ are again tensor products

$$\hat{\psi}_{(\mathbf{l},\mathbf{t})}(\mathbf{x}) = \hat{\psi}_{(l_1,t_1)}(x_1) \cdot \ldots \cdot \hat{\psi}_{(l_d,t_d)}(x_d) \ ,$$

where the univariate wavelets $\hat{\psi}$ are constructed by the lifting scheme [13], [60]. The idea is to modify the given univariate wavelets Ψ_l by coarse scaling functions:

$$\hat{\Psi}_l = 2^{l/2}\left(\Psi_l + \Phi^{l-1} \cdot Q^l\right) \tag{64}$$

such that the elements of $\hat{\Psi}_l$ have M vanishing moments, i.e.

$$\int \hat{\Psi}_l(x)x^i \, dx = 0 \quad \text{for } 0 \leq i < M .$$

The parameter M controls the stabilization effect of the lifting scheme. As we will see below $M = 2$ is a good choice for our purposes. The additional scaling by $2^{l/2}$ in (64) normalizes $\hat{\psi}_{(l,t)}$ with respect to the L_2-norm

$$\|\hat{\psi}_{(l,t)}\|_0 \sim 1 .$$

The matrix Q^l is a sparse $\#S(l-1) \times \#T(l)$ matrix. E.g. for the non-boundary-modified wavelets we have

$$\hat{\psi}_{(l,t)} = 2^{l/2}\left(\psi_{(l,t)} + \sum_{s=\lfloor \frac{M}{4}-1 \rfloor}^{\lfloor \frac{M}{2} \rfloor} q_s \phi^{(l-1,t+s)}\right) ,$$

with certain properly chosen weights q_s. In case of $M = 2$ these weights are $q_0 = q_1 = -\frac{1}{4}$. Furthermore, note that the scaling functions for both wavelet bases are the same. Therefore, the change of basis (63) is possible. However, in the fully adaptive case additional measures [46] are required to ensure that

$$\text{span}\{\hat{\psi}_{(l,t)} \mid (l,t) \in \mathcal{T}\} = \text{span}\{\psi_{(l,t)} \mid (l,t) \in \mathcal{T}\} .$$

The effect of the lifting modification (64) has to do with so-called norm equivalences between Sobolev spaces and (coefficient) sequence spaces which later result in optimal $O(1)$ iterative solvers.

Theorem 2. *There exist $\gamma^* < \gamma$ such that for $s \in]\gamma^*, \gamma[$ there are constants $\underline{\alpha}(s) < \bar{\alpha}(s)$ such that*

$$\underline{\alpha}(s)\|\hat{\Psi}_{\mathcal{T}} \cdot \mathbf{u}_{\mathcal{T}}\|_s \leq \left(\sum_{(l,t)\in\mathcal{T}} |u_{(l,t)}|^2 4^{sl}\right)^{1/2} \leq \bar{\alpha}(s)\|\hat{\Psi}_{\mathcal{T}} \cdot \mathbf{u}_{\mathcal{T}}\|_s \quad \forall \mathcal{T}, \mathbf{u}_{\mathcal{T}} . \tag{65}$$

The constants $\underline{\alpha}(s)$, $\bar{\alpha}(s)$ are independent on the adaptive univariate basis $\hat{\Psi}_{\mathcal{T}}$ and the finest level $L = \max\{l \mid (l,t) \in \mathcal{T}\}$ and the coefficient vector $\mathbf{u}_{\mathcal{T}}$.

Proof. See [59].

The lifting modification significantly enlarges the range $]\gamma^*, \gamma[$ compared to that of the original Interpolets. The reason is that the lower bound γ^* mainly depends on the smoothness and the degree of polynomial exactness of the dual scaling functions. By modifying the primal wavelets we implicitly modify the dual scaling functions. The new dual scaling functions $\tilde{\tilde{\phi}}^{(\cdot,\cdot)}$ have a degree of polynomial exactness of M and the smoothness parameter

$$-\gamma^* := \sup\{t \in \mathbb{R} \mid \tilde{\tilde{\phi}}^{(\cdot,\cdot)} \in H^t\} . \tag{66}$$

For example, with $M = 2$ the new dual scaling functions $\tilde{\tilde{\phi}}^{(\cdot,\cdot)}$ are in H^t for some $t > 0$. Table 1 shows the associated values of γ^* and γ which were numerically estimated using Lemma 7.1.2 of [21].

$M \setminus N$	2	4	6
$-$	] $0.500, 1.500[$	] $0.500, 2.171[$	] $0.500, 2.755[$
1	] $0.354, 1.500[$	] $0.401, 2.171[$	] $0.431, 2.755[$
2	] $-0.171, 1.500[$	] $-0.290, 2.171[$	] $-0.358, 2.755[$

Table 1. Ranges $]\gamma^*, \gamma[$ of norm equivalences for univariate wavelets

We will now use these norm equivalences for $s = 0, 1$ and 2 to estimate the l_2-condition number of the preconditioned operator $\mathbf{B}\mathbf{A}\mathbf{B}^{-1}\mathbf{C}^{-1}$ for both, the finite difference technique and the collocation method. Therefore we will consider the case $M = 2$ in the following. In the remainder of this subsection we denote by $\mathcal{T}$ the index set

$$\mathcal{T} = \bigcup_{l=l_0}^{L} \{l\} \times T(l) \tag{67}$$

which corresponds to a *non-adaptive* univariate basis. Furthermore, we introduce in Table 2 the $\#S(L) \times \#S(L)$ matrices which form the univariate building blocks of the preconditioned operator.

T	transformation matrix defined by $\Phi^L(x) \cdot T \overset{!}{=} \Psi_{\mathcal{T}}$
$\hat{T}$	transformation matrix defined by $2^{L/2}\Phi^L(x) \cdot \hat{T} \overset{!}{=} \hat{\Psi}_{\mathcal{T}}$
	with the scaling $2^{L/2}$ because of $\|\phi_{L,s}\|_0 \sim 2^{-L/2}$
D	$= -\langle \tilde{\Phi}^L , \partial_{xx}\Phi^L \rangle$ (collocation)
	$= -4^L[1 \ -2 \ 1]$ (finite differences)
C	diagonal scaling matrix with level dependent entries 4^l, i.e.
	$C_{(k,s),(l,t)} = \delta(l - k)\delta(s - t)4^l$
E	identity matrix

Table 2. Building blocks of operators

In the remainder of this paper it will be useful to denote by $X \sim Y$ that there are two constants $\underline{c}$ and $\bar{c}$ independent of X and Y and the various parameters X and Y may depend on such that

$$\underline{c}X \leq Y \leq \bar{c}X . \tag{68}$$

However, sometimes it is necessary to keep the constants explicitly.
From the norm equivalency (65) we deduce three important relations:

Lemma 3.

(i) There holds

$$(\mathbf{u}_T)^T \cdot \mathbf{u}_T \sim (\mathbf{u}_T)^T \cdot \hat{T}^T \hat{T} \cdot \mathbf{u}_T , \quad \forall \mathbf{u}_T . \tag{69}$$

(ii) There holds

$$(\mathbf{u}_T)^T \cdot C \cdot \mathbf{u}_T \sim (\mathbf{u}_T)^T \cdot D\hat{T}^T \hat{T} \cdot \mathbf{u}_T , \quad \forall \mathbf{u}_T . \tag{70}$$

(ii) There are $\underline{\alpha}_2$ and $\bar{\alpha}_2$ such that for all $\mathbf{u}_T$

$$\underline{\alpha}_2 (\mathbf{u}_T)^T \cdot C^2 \cdot \mathbf{u}_T \leq (\mathbf{u}_T)^T \cdot \hat{T}^T D^T D\hat{T} \cdot \mathbf{u}_T \leq \bar{\alpha}_2 (\mathbf{u}_T)^T \cdot C^2 \cdot \mathbf{u}_T . \tag{71}$$

In case of Interpolets with $N \geq 4$ $\underline{\alpha}_2$ and $\bar{\alpha}_2$ are independent of the finest level L. In case of Interpolets with $N = 2$ there holds

$$\underline{\alpha}_2(L) \sim 1 \quad and \quad \bar{\alpha}_2(L) \sim 2^L .$$

Proof. We will only sketch the main ideas of the proof for the simple case of periodic boundary conditions.
(i) follows from

$$(\mathbf{u}_T)^T \cdot \mathbf{u}_T \sim \|\hat{\Psi}_T \cdot \mathbf{u}_T\|_0^2 = 2^L (\mathbf{u}_T)^T \cdot \hat{T}^T \langle \Phi^L , \Phi^L \rangle \hat{T} \cdot \mathbf{u}_T \sim (\mathbf{u}_T)^T \cdot \hat{T}^T \hat{T} \cdot \mathbf{u}_T ,$$

since the scaled mass matrix $2^L \langle \Phi^L , \Phi^L \rangle$ is equivalent to E.
(ii) D and the stiffness matrix for $2^{L/2} \Phi^L$ are circulant. The eigenvalues λ_k of both matrices behave like $\sim k^2$. Therefore,

$$(\mathbf{u}^L)^T \cdot D \cdot \mathbf{u}^L \sim 2^L (\mathbf{u}^L)^T \cdot \langle \partial_x \Phi^L , \partial_x \Phi^L \rangle \cdot \mathbf{u}^L , \quad \forall \mathbf{u}^L .$$

This yields

$$(\mathbf{u}_T)^T \cdot \hat{T}^T D\hat{T} \cdot \mathbf{u}_T \sim \|\hat{\Psi}_T \cdot \mathbf{u}_T\|_1^2 \sim (\mathbf{u}_T)^T \cdot C \cdot \mathbf{u}_T .$$

(iii) For Interpolets ($N \geq 4$) we can proceed as above using

$$(\mathbf{u}^L)^T \cdot D^T D \cdot \mathbf{u}^L \sim 2^L (\mathbf{u}^L)^T \cdot \langle \partial_{xx} \Phi^L , \partial_{xx} \Phi^L \rangle \cdot \mathbf{u}^L .$$

For the case $N = 2$ we refer to the paper [46]. $\square$

Now, we consider the matrices $\mathbf{A}$, $\mathbf{B}$ and $\mathbf{BAB}^{-1}$ of (62) in case of our simple non-adaptive setting (67). For example, in 2D they read

$$\mathbf{A} = \left(T^{-1} \otimes T^{-1}\right)\left(D \otimes E + E \otimes D\right)\left(T \otimes T\right) ,$$
$$\mathbf{B} = 2^{-L}\left(\hat{T}^{-1} \otimes \hat{T}^{-1}\right)\left(T \otimes T\right) ,$$
$$\mathbf{BAB}^{-1} = \left(\hat{T}^{-1} \otimes \hat{T}^{-1}\right)\left(D \otimes E + E \otimes D\right)\left(\hat{T} \otimes \hat{T}\right) ,$$
$$\mathbf{C} = C \otimes E + E \otimes C . \tag{72}$$

Here, $X \otimes Y$ denotes the Kronecker product of two matrices [41]. Using all the prerequisites from above we can state the main Theorem of this section:

Theorem 4. *For the preconditioned operator* $\mathbf{BAB}^{-1}\mathbf{C}^{-1}$ *of the finite difference technique or the collocation method there holds*

$$\|\mathbf{BAB}^{-1}\mathbf{C}^{-1} \cdot \mathbf{u}_{\mathcal{T}}\|_{l_2}^2 =$$
$$= (\mathbf{u}_{\mathcal{T}})^T \cdot \mathbf{C}^{-1}\mathbf{B}^{-T}\mathbf{A}^T\mathbf{B}^T\mathbf{BAB}^{-1}\mathbf{C}^{-1} \cdot \mathbf{u}_{\mathcal{T}} \sim (\mathbf{u}_{\mathcal{T}})^T \cdot \mathbf{u}_{\mathcal{T}} = \|\mathbf{u}_{\mathcal{T}}\|_{l_2}^2 . \tag{73}$$

In case of Interpolets $N = 2$ the constants in the above equivalency relation are dependent on the finest level L and their ratio scales like 2^L. In all other cases $(N > 2)$ we obtain by (62) an optimal preconditioner with condition numbers independent on L.

Proof. Relation (73) follows immediately from

$$(\mathbf{u}_{\mathcal{T}})^T \cdot \mathbf{B}^{-T}\mathbf{A}^T\mathbf{B}^T\mathbf{BAB}^{-1} \cdot \mathbf{u}_{\mathcal{T}} \sim (\mathbf{u}_{\mathcal{T}})^T \cdot \mathbf{C}^2 \cdot \mathbf{u}_{\mathcal{T}} \quad \forall \, \mathbf{u}_{\mathcal{T}} \tag{74}$$

with the same equivalency constants. Now, for the proof of relation (74) we will use the following two general rules (see [35]):
Let X_i, Y_i $(1 \leq i \leq d)$ be $R \times R$ matrices and $\underline{\alpha}_i$, $\bar{\alpha}_i$ such that for all vectors $\mathbf{u} \in \mathbb{R}^R$

$$\underline{\alpha}_i \mathbf{u}^T \cdot X_i \cdot \mathbf{u} \ \leq \ \mathbf{u}^T \cdot Y_i \cdot \mathbf{u} \ \leq \ \bar{\alpha}_i \mathbf{u}^T \cdot X_i \cdot \mathbf{u} .$$

Then there holds for the tensor product matrices $X_1 \otimes .. \otimes X_d$ and $Y_1 \otimes .. \otimes Y_d$

$$\left(\prod_{i=1}^d \underline{\alpha}_i\right)\mathbf{v}^T \cdot \left(X_1 \otimes .. \otimes X_d\right) \cdot \mathbf{v} \ \leq \ \mathbf{v}^T \cdot \left(Y_1 \otimes .. \otimes Y_d\right) \cdot \mathbf{v}$$

$$\leq \ \left(\prod_{i=1}^d \bar{\alpha}_i\right)\mathbf{v}^T \cdot \left(X_1 \otimes .. \otimes X_d\right) \cdot \mathbf{v} .$$

Furthermore, there holds

$$\min_i(\underline{\alpha}_i)\mathbf{u}^T \cdot \left(X_1 + .. + X_d\right) \cdot \mathbf{u} \ \leq \ \mathbf{u}^T \cdot \left(Y_1 + .. + Y_d\right) \cdot \mathbf{u}$$

$$\leq \ \max_i(\bar{\alpha}_i)\mathbf{u}^T \cdot \left(X_1 + .. + X_d\right) \cdot \mathbf{u} .$$

With these rules and the results of Lemma 3 it is straightforward to show that (74) holds. $\square$

The above analysis can easily be carried over to the original preconditioned operator $\mathbf{AC}^{-1}$. There, we have to take into account that the condition number of the mass matrix and the condition number of T^TT scale like 2^L. This result cannot be improved by suitable diagonal scaling such that $\|\psi_{(l,t)}\|_0 \sim 1$. Therefore, the final results are the scaling laws for the condition numbers

$$\kappa_1 = \sup_{\mathbf{u}_T} \frac{\|\mathbf{BAB}^{-1}\mathbf{C}^{-1} \cdot \mathbf{u}_T\|_{l_2}}{\|\mathbf{u}_T\|_{l_2}} \quad \text{and} \quad \kappa_2 = \sup_{\mathbf{u}_T} \frac{\|\mathbf{AC}^{-1} \cdot \mathbf{u}_T\|_{l_2}}{\|\mathbf{u}_T\|_{l_2}}$$

of the preconditioned operators as given in Table 3.

	$\mathbf{BAB}^{-1}\mathbf{C}^{-1}$	$\mathbf{AC}^{-1}$
FD ($N = 2$)	$\sim 2^{L/2}$	$\sim 2^{dL/2}$
FD ($N \geq 4$)	~ 1	$\sim 2^{(d-1)L/2}$
Col ($N \geq 6$)	~ 1	$\sim 2^{(d-1)L/2}$

Table 3. Condition numbers of preconditioned operators on full grids (60).

It is noteworthy that for the preconditioning of the collocation / finite difference matrices one needs a larger range of norm equivalences (up to $\gamma > 2$ in Theorem 2) than for the preconditioning of the Galerkin stiffness matrix. This was also observed in [59] for the collocation method based on the MRA-approach. But in [59] the situation is even worse since norm equivalences for $s \geq 1 + d/2$ are required to obtain a $O(1)$ condition number. For our scheme there is no such dependency on the dimension.

Up to now, our analysis only covers the non-adaptive case. In Figure 5 we present the results of a few numerical experiments for sparse grids (55) which indicate the efficiency of our preconditioner also in more general adaptive cases. In summary, the use of MRA-d Interpolet wavelets leads to quite simple algorithms for the non-linear function evaluation and the discretization of differential operators on one hand, but on the other hand the condition numbers of the resulting operator matrices are strongly dependent on the finest mesh size 2^{-L}. The lifting preconditioner presented in this section solves this problem and leads to condition numbers which are not or only slightly dependent on the finest mesh size. Therefore, we can solve elliptic problems like (56) fast.

7   Time Dependent Convection Diffusion Problems

Consider now the model problem

$$\partial_t u + \nabla \cdot (\mathbf{a}u) = \nu \Delta u \quad , \quad (\mathbf{x}, t) \in \Omega \times [0, T] \tag{75}$$
$$u(\mathbf{x}, 0) = u^0(\mathbf{x})$$

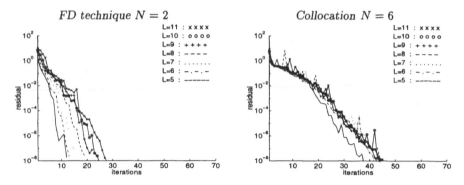

Fig. 4. BiCG(2) residuals for the lifting preconditioner (62), the 2D model problem (61) and non-adaptive bases (60).

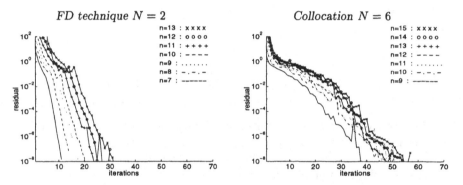

Fig. 5. BiCG(2) residuals for the lifting preconditioner (62), the 2D model problem (61) and so-called regular sparse grids (55).

with appropriate boundary conditions. Here, u is a certain quantity which is subject to diffusion and convection. The given velocity field $\mathbf{a}$ is supposed to be divergence free, i.e. we have $\nabla \cdot \mathbf{a} = 0$.

Equation (75) is of the same type as the momentum part of the Navier-Stokes equations. It is common to use the discretizations developed and studied for (75) also for the Navier-Stokes equations. In this section we will combine simple and efficient time stepping methods with our adaptive wavelet collocation scheme. More precisely we will show how the *selection of the adaptive basis* can be incorporated into such methods. Special emphasis will be put on the mass/energy-conservation properties of the adaptive scheme.

7.1   Time Discretization

For the discretization of (75) there are essentially two possibilities: Either a simultaneous space time discretization (SPTD) by e.g. our wavelet Petrov-Galerkin scheme or, simpler, a time stepping method (TSM). SPTD in connection with adaptivity would allow to exploit smoothness in space and time (see e.g. [36]) while a TSM can only exploit spatial smoothness. In this sense adaptive SPTDs are a generalization of local time stepping methods or a nonlinear Galerkin scheme [6,52]. Of course, SPTDs also have drawbacks: One has to store and handle the complete solution on $\Omega \times [0, T]$ at once, while for TSM one only needs a few time slabs of the solution. Thus, SPTD may require significantly more memory. Furthermore, SPTDs are not well understood and the arising linear systems are quite nasty and not easy to solve. Hence, we will only consider a time stepping method. For reasons of simplicity we stick to the following simple approach:

$$\frac{u^{n+1} - u^n}{k} + C(u^n, n^{n-1}, ..) = \nu \Delta u^{n+1} \tag{76}$$

Here, k is the size of the discrete time steps and $C(..)$ is an explicit approximation of the convective term, e.g. of Adams-Bashforth (ABx) or Runge-Kutta (RKx) type (see e.g. [12]). The *implicit* treatment of the diffusive term is relatively inexpensive since the arising linear systems $(id - k\nu\Delta)u = r.h.s.$ are easy to solve. Furthermore one overcomes the CFL type condition $k\nu 4^L \leq c$ (L being the finest level) which is required to maintain the numerical stability of (76). The remaining stability constraint is the usual CFL condition $k2^L \leq C$. In principle we could also overcome this constraint by an implicit treatment of the convective term (e.g. $C = \nabla \cdot (\mathbf{a} u^{n+1})$), but usually the additional effort to solve the more difficult linear system $(id + k\nabla \cdot (\mathbf{a}.) - k\nu\Delta)u = r.h.s.$ (compared to $(id - k\nu\Delta)u = r.h.s.$) does not pay off. The reason is that k and the finest mesh size 2^{-L} should be of the same order of magnitude for the sake of accuracy. Because of their simplicity, mixed implicit-explicit discretizations like (76) are used in many CFD codes [43,55].

7.2   Adaptive Basis Selection in Time Stepping Methods

An important question for adaptive methods is how to select the adaptive basis to achieve a maximum of accuracy with a given number of degrees of freedom, i.e. with a given amount of numerical work. This depends on the problem at hand. For elliptic problems an algorithm for the basis selection was developed in [15] which yields the optimum order of convergence $\alpha(s, u)$ in terms of the number of degrees of freedom (DOF):

$$\|u - u_\mathcal{T}\|_s \leq c (\#DOF)^{-\alpha(s,u)}$$

Here, the error is measured in a Sobolev norm $\|.\|_s$. In engineering applications one is usually not so much interested in minimizing the error in a Sobolev norm, but instead in minimizing the error of derived quantities, like for example the drag or lift coefficient of an airfoil or the mean rate of heat transfer at the surface of a turbo-engine. Based on this observation, an approach was presented in [4,5] which states the problem of adaptive basis selection as an optimization problem. The algorithm to solve this optimization problem involves local error estimators ([9,15,65]) and some properties of Green's function which are required to determine the source of the local errors. However, this algorithm is quite complicated to use in a rigorous way for strongly time-dependent problems like turbulent flows. Instead, we use the following simpler strategy, compare also [29]:

---

algorithm (Adaptive Time Stepping Method (ATSM))

given are the coefficients $\mathbf{u}_{\mathcal{T}^0}^0$ of u^0 and the initial adaptive basis $\mathcal{T}^0$

for $n = 0$ to T/k do

// Time Step:

Calculate the convective term $\mathbf{c}_{\mathcal{T}^n}^n := \langle \tilde{\Psi}_{\mathcal{T}^n} , \nabla \cdot (\mathbf{a}(\Psi_{\mathcal{T}^n} \cdot \mathbf{u}_{\mathcal{T}^n}^n)) \rangle$

Solve $\left(id - k\nu \langle \tilde{\Psi}_{\mathcal{T}^n} , \Delta\Psi_{\mathcal{T}^n} \rangle \right) \cdot \mathbf{u}_{\mathcal{T}^n}^{n+1} = \mathbf{u}_{\mathcal{T}^n}^n - k\mathbf{c}_{\mathcal{T}^n}^n$

// Adaptivity: new adaptive basis

Determine $\overline{\mathcal{T}}^{n+1} := \{ (\mathbf{l}, \mathbf{t}) \,\big|\, |(u_{\mathcal{T}^n}^{n+1})_{(\mathbf{l},\mathbf{t})}| > \epsilon \}$

Determine $\mathcal{T}^{n+1} := \{ (\mathbf{k}, \mathbf{s}) \big| \exists (\mathbf{l}, \mathbf{t}) \in \overline{\mathcal{T}}^{n+1} \text{with } |(\mathbf{l}, \mathbf{t}) - (\mathbf{k}, \mathbf{s})|_\infty \le K \}$

// Adaptivity: prolongation to new basis

Calculate $\mathbf{u}_{\mathcal{T}^{n+1}}^{n+1}$ by

$$(u_{\mathcal{T}^{n+1}}^{n+1})_{(\mathbf{l},\mathbf{t})} := \begin{cases} (u_{\mathcal{T}^n}^{n+1})_{(\mathbf{l},\mathbf{t})} & \text{if } (\mathbf{l}, \mathbf{t}) \in \mathcal{T}^{n+1} \bigcap \mathcal{T}^n \\ 0 & \text{if } (\mathbf{l}, \mathbf{t}) \in \mathcal{T}^{n+1} \backslash \mathcal{T}^n \end{cases}$$

end

The first part of this algorithm is the time stepping method (76). Here, we use an Euler scheme for the sake of simplicity. Note however that other higher order time discretization methods can be plugged in straightforwardly.

In the second part, we first determine the set $\overline{\mathcal{T}}^{n+1}$ of active basis functions. The analog of this step in adaptive FE/FV methods would be the selection of elements which should be refined. The refinement criterion we use is whether the magnitude of a particular wavelet coefficient is larger than a given threshold ϵ, i.e. $|u_{(\mathbf{l},\mathbf{t})}| > \epsilon$. [2] The use of the magnitude of the wavelet coefficients can also be interpreted as a traditional method of local error estimation. If one takes (3) and (4) into account it is easy to see that the magnitude of a wavelet coefficient $u_{(\mathbf{l},\mathbf{t})}$ is a measure of the magnitude of a local finite difference approximation of some higher order derivative of u. Thus, we work with a gradient type error indicator.

Now consider the second step of the adaptivity strategy, i.e. the determination of $\mathcal{T}^{n+1}$ by insertion of basis functions which are close in space and scale to the active basis functions of $\overline{\mathcal{T}}^{n+1}$. We assume that the current solution u^n is quite accurate and we want to preserve this property also for the next time slab u^{n+1}. For the sake of simplicity, let us assume that we calculate u^{n+1} by the following explicit Petrov–Galerkin scheme:

$$\langle \tilde{\psi}_{(\mathbf{l},\mathbf{t})} , u^{n+1} \rangle = \langle \tilde{\psi}_{(\mathbf{l},\mathbf{t})} , (I - \Delta t \nabla \cdot (\mathbf{v}.) + \Delta t \nu \Delta) u^n \rangle .$$

---

[2] Another possible strategy would be to choose the M largest wavelet coefficients, where M is a fixed given number. In this case the work count would be almost constant for all time steps.

The best index set $\mathcal{T}^{n+1}$ with a given number N of degrees of freedom is that with the N largest entries $\langle \tilde{\psi}_{(\mathbf{l},\mathbf{t})} , u^{n+1} \rangle$. Now, the a priori known locality properties of the differential operator $(I - \Delta t \nabla \cdot (\mathbf{v}.) + \Delta t \nu \Delta)$ show that $\langle \tilde{\psi}_{(\mathbf{l},\mathbf{t})} , u^{n+1} \rangle$ can only be significant if $(\mathbf{l},\mathbf{t})$ is near to one of the significant $(\mathbf{k},\mathbf{s}) \in \mathcal{T}^n$. In this sense, the above method is closely related to [4,5] or [15].

7.3 Conservation Properties

Consider the numerical simulation of a high Reynolds number turbulent flow. In this setting we can never expect that the numerical solution is close to the real one with respect to a Sobolev norm. However, we don't really need this. Instead, it is sufficient that the numerical solution exhibits the same statistics as the real solution. Obviously one of the prerequisites for a correct prediction of the statistics is that the global mass/energy budget of the numerical and the real solution agree, i.e.

$$\int_{\Omega} u_{\mathcal{T}}(\mathbf{x}, t) \, d\mathbf{x} \approx \int_{\Omega} u(\mathbf{x}, t) \, d\mathbf{x} \quad \text{and} \quad \int_{\Omega} \left(u_{\mathcal{T}}(\mathbf{x}, t) \right)^2 d\mathbf{x} \approx \int_{\Omega} \left(u(\mathbf{x}, t) \right)^2 d\mathbf{x} \; .$$

In adaptive schemes there are two main sources of mass or energy defects. The first is the remeshing step for each time step and the second is the discretization of the convective term. For the simple cases of periodic or homogeneous Dirichlet boundary conditions there holds

$$\int_{\Omega} \nabla \cdot (\mathbf{a}u) \, d\mathbf{x} = 0 \; , \quad \int_{\Omega} \nabla \cdot (\mathbf{a}u) u \, d\mathbf{x} = 0 \; . \tag{77}$$

In case of Galerkin or non-adaptive collocation discretizations the relations (77) usually also hold for the discretized convective term. But this is not the case for adaptive collocation methods, since here the usual partial summation arguments do not apply. Nevertheless it is relatively simple to enforce at least the first relation in (77) by a L_2-projection of $\Psi_{\mathcal{T}} \cdot c_{\mathcal{T}}^n$ to the space of functions with vanishing mean

$$\Psi_{\mathcal{T}} \cdot \hat{c}_{\mathcal{T}}^n = \Psi_{\mathcal{T}} \cdot c_{\mathcal{T}}^n - \frac{1}{Vol(\Omega)} \int_{\Omega} \Psi_{\mathcal{T}} \cdot c_{\mathcal{T}}^n \, d\mathbf{x} \; .$$

Note that this projection even improves the accuracy of the discrete convective term with respect to the L_2-norm. Also, the conservation of mass can be easily maintained in the remeshing step. To this end, one switches to the Lifting-Interpolet basis in the second part of algorithm (ATSM). Except for the wavelets of the coarsest scale $(l_0, .., l_0)$ which are always kept in the adaptive basis, all Lifting Interpolet wavelets have a vanishing mean. Therefore the removal or insertion of basis functions has no effect on the mass balance.

Unfortunately such tricks are not available for the energy conservation for both, the convective term and the remeshing step. However, in our numerical experiments we observed that the energy defect introduced by the prolongation in (**ATSM**) and the non-energy-conservative discretization of the convective term is quite small in the temporal mean. Thus, the correct rate of energy dissipation is predicted very well.

As a numerical example we consider the standard test problem of a hill rotating around the origin [66], i.e.

$$\partial_t u + \nabla \cdot \begin{pmatrix} -yu \\ xu \end{pmatrix} = \nu \Delta u \quad \text{in} \quad [-1,1]^2 \times [0, 2\pi]$$

$$u(\mathbf{x}, 0) = \max(0, -3 + 4e^{-30|\mathbf{x}-(.25,.25)|^2}),$$

see Figure 7 (right). Compared to [66] we increased the difficulty of the problem by using a non-smooth initial condition. Furthermore, we set $\nu = 10^{-5}$ and $\nu = 10^{-6}$ to show that even for such high Reynolds numbers the true energy dissipation $\nu \int |\nabla u|^2 \, d\mathbf{x}$ dominates the energy defects produced by our scheme - at least if the resolution is sufficiently fine. Because of the small dissipative effects the solution is small at the boundaries. Therefore, periodic boundary conditions are used.

In Figure 6 the energy for three different adaptive solutions is compared to the energy of a reference solution which was calculated by a mass- and energy-conservative Fourier spectral code. Obviously the errors of the energy budget are relatively small and remain bounded. Nevertheless, the perturbations which appear for the coarsest adaptive solutions seem to indicate an unsufficient spatial resolution. Such effects are also known from higher order finite difference methods.

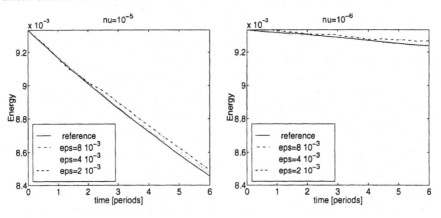

Fig. 6. Evolution of energy. Left: $\nu = 10^{-5}$. Right: $\nu = 10^{-6}$.

The evolution of the number of degrees of freedom for the adaptive solutions is shown in Figure 7 (left). At the beginning the number of degrees of freedom (DOF) is high, because of the non-smooth initial condition. Then, the smoothing by the diffusive term allows for a reduction of the number of DOF. However, in the adaptive case a significant part of the DOF is spent

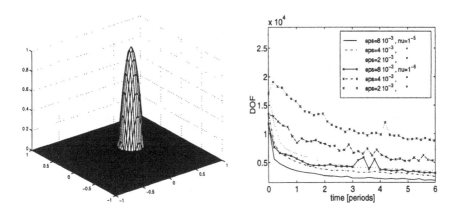

Fig. 7. Left: initial condition. Right: evolution of number of degrees of freedom.

to represent small spurious oscillations (wiggles) which are typical for higher order finite difference schemes. These wiggles are suppressed very well in case of $\nu = 10^{-5}$, but not so well for $\nu = 10^{-6}$. This, together with the weaker smoothing of the hill, explains why the number of DOF is larger for $\nu = 10^{-6}$ than for $\nu = 10^{-5}$.

8 Incompressible Navier-Stokes Equations

Now, we consider the incompressible Navier-Stokes equations. There, momentum and continuity equation are coupled. This makes them numerically more complicated to treat than the advection-diffusion problem of the previous section. A first idea to overcome this difficulty is to use (weakly) divergence-free trial functions. Then the Navier-Stokes equation boil down to an advection-diffusion problem. In the FE world this approach has been advocated by several authors (see [31] for a recent overview). In the wavelet world the analogue are divergence-free wavelets. They have been introduced in [3] and in [63]. Of course these wavelets are no longer scalar functions but vector valued functions where each component is the tensor product (either of type (49) or of type (50)) of univariate wavelets. One difference to our setting in section 4 is that different types of univariate wavelets are used in each coordinate direction. However the construction of these different univariate wavelets is not practical if the divergence operator contains variable coefficient terms which is the case if the domain Ω is a parametric image of e.g. $[0,1]^3$.

Another solution to the coupling problem is to switch to the velocity-vorticity formulation of the Navier-Stokes equations (see [38,39,43])

$$\partial_t \omega + \nabla \cdot (\omega \times \mathbf{u}) = -\nu \nabla \times \nabla \times \omega , \quad \text{in } \Omega$$
$$\Delta \mathbf{u} = -\nabla \times \omega , \tag{78}$$
$$\omega = \nabla \times \mathbf{u} , \quad \text{at } t = 0 .$$

The most expensive part of a discretization based on the velocity-vorticity formulation is the evaluation of the nonlinear term $\nabla \cdot (\omega \times \mathbf{u})$ which requires the calculation of the velocity $\mathbf{u}$ for a given vorticity field. To this end, three Poisson equations must be solved in (78) in the three-dimensional case. Besides of rather analytical difficulties such as the lack of boundary conditions for ω or the treatment of irrotational forcing terms, the costs for the treatment of these three Poisson equations are the main drawback of the velocity-vorticity formulation. In contrast to that only one Poisson equation must be solved in splitting methods which we will describe below.

8.1   Splitting Methods and Stabilization

The Navier-Stokes equations in primitive variable formulation are given by

$$\partial_t \mathbf{u} \, + \, \nabla \cdot (\mathbf{u} \otimes \mathbf{u}) = \mathbf{f} - \nabla p + \nu \Delta \mathbf{u} \, , \quad (\mathbf{x}, t) \in \Omega \times [0, T] \, ,$$
$$\nabla \cdot \mathbf{u} = 0 \, ,$$
$$\mathbf{u}(\mathbf{x}, 0) = u^0(\mathbf{x}) \tag{79}$$

with appropriate boundary conditions. Now, the discretization of the Navier-Stokes equations is split into two subproblems:

(Transport step)   Calculate the auxiliary velocity $\hat{\mathbf{u}}^{n+1}$ by

$$\frac{\hat{\mathbf{u}}^{n+1} - \mathbf{u}^n}{k} \, + \, \mathbf{C}(\mathbf{u}^n, \mathbf{u}^{n-1}, ..) = \mathbf{f}^{n+1} - \nabla p^n + \nu \Delta \hat{\mathbf{u}}^{n+1} \, . \tag{80}$$

For this problem we apply the **(ATSM)** algorithm of the last section.

(Projection step)   Calculate $\mathbf{u}^{n+1}$, p^{n+1} as the solution of

$$\frac{\mathbf{u}^{n+1} - \hat{\mathbf{u}}^{n+1}}{k} = -\nabla (p^{n+1} - p^n) \, ,$$
$$\nabla \cdot \mathbf{u}^{n+1} = 0 \, . \tag{81}$$

Of course both subproblems have to be augmented by boundary conditions; not only for the $\mathbf{u}$ and $\hat{\mathbf{u}}$ but also for the pressure p. Especially the (in general non-physical) pressure boundary conditions are the subject of a controversy (cf. [31,43,54]) which is beyond the scope of the paper. Therefore we will assume for the rest of this work that periodic boundary conditions hold for $\mathbf{u}$ and p.

The saddle point problem (81) is treated by solving the Schur complement equation

$$\nabla \cdot \nabla (p^{n+1} - p^n) = \frac{1}{k} \nabla \cdot \hat{\mathbf{u}}^{n+1} \, , \tag{82}$$

followed by the correction of the velocity

$$\mathbf{u}^{n+1} = \hat{\mathbf{u}}^{n+1} - k \nabla (p^{n+1} - p^n) \, . \tag{83}$$

Since the equations of the saddle point problem (81) must hold only in the weak sense (i.e. tested by $\tilde{\Psi}_T$), the discrete Schur complement operator $\nabla \cdot \nabla$ is not the usual discrete Laplacian $\langle \tilde{\Psi}_T, \Delta\Psi_T \rangle$, but the product of a discrete gradient and a discrete divergence operator. Thus, the matrix representation of the discrete $\nabla \cdot \nabla$ operator reads

$$\mathbf{A} = \sum_{i=1}^{d} \langle \tilde{\Psi}_T, \partial_{x_i}\Psi_T \rangle \langle \tilde{\Psi}_T, \Psi_T \rangle^{-1} \langle \tilde{\Psi}_T, \partial_{x_i}\Psi_T \rangle$$

$$= \sum_{i=1}^{d} \langle \tilde{\Psi}_T, \partial_{x_i}\Psi_T \rangle \langle \tilde{\Psi}_T, \partial_{x_i}\Psi_T \rangle . \tag{84}$$

Obviously the matrix-vector multiplication with $\mathbf{A}$ is approximately twice as expensive as that with $\langle \tilde{\Psi}_T, \Delta\Psi_T \rangle$. However as it was mentioned in [31] for FEM the replacement of $\nabla \cdot \nabla$ by Δ in (82) is no good advise since non-physical solutions may occur [30]. Furthermore in our own experiments with unsteady problems and the naive discretization of Δ we observed that the projection step constantly injects an amount of divergence into the velocity field $\mathbf{u}^{n+1}$ which is of the order of the discretization error. This leads to a relatively large number of BiCG iterations required for the accurate solution of the pressure Poisson equation.

However, the use of the operator $\nabla \cdot \nabla$ also causes some problems. Consider the matrix representation $\mathbf{A}$ for the 2D non-adaptive case:

$$\mathbf{A} = \left(T^{-1} \otimes T^{-1} \right) \left(G^2 \otimes E + E \otimes G^2 \right) \left(T \otimes T \right) ,$$

where $G = \langle \tilde{\Phi}^L, \partial_x \Phi^L \rangle$. Here, $-G^2$ is not spectrally equivalent to the stiffness matrix $\langle \partial_x \Phi^L, \partial_x \Phi^L \rangle$ (unlike $-\langle \tilde{\Phi}^L, \partial_{xx}\Phi^L \rangle$). See Figure 8 for a comparison of the scaled eigenvalues of the one-dimensional operators for Interpolets ($N = 6$). As usual the eigenvalues λ_j are normalized by the factor 4^{-L} and they are plotted against the so-called wavenumber $2^{-L} \cdot 2\pi j$. Hence, the resulting plots are almost independent on the level L.

Now, the point is that spectral equivalence was one of the prerequisites for the efficiency of both the $\mathbf{AC}^{-1}$ and the $\mathbf{BAB}^{-1}\mathbf{C}^{-1}$ preconditioning technique of section 6. Hence, even the preconditioned version of (82) is very ill conditioned. Furthermore, G^2 has an additional zero eigenvalue for the even-odd mode. Thus, spurious checker-board modes occur in the solution of (82). The FE analog of the above problems is the lack of a LBB condition for the saddle point problem (81).

Our solution to this problem is to replace G in the gradient operator by $G + 2^{-6L}F$ where F is a central FD discretization of the 6th order derivative. Conversely, in the divergence operator we replace G by $G - 2^{-6L}F$. Then, F and G commute in the periodic case and

$$(G - 2^{-6L}F)(G + 2^{-6L}F) = (G^2 - 2^{-12L}F^2) . \tag{85}$$

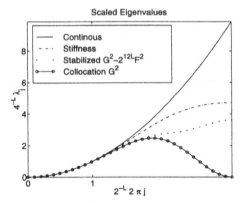

Fig. 8. Eigenvalues of 1D operators

Hence, the matrix representation of the stabilized Schur complement operator reads

$$\mathbf{A} = \left(T^{-1} \otimes T^{-1}\right)\left((G^2 - 2^{-12L}F^2) \otimes E + E \otimes (G^2 - 2^{-12L}F^2)\right)\left(T \otimes T\right) .$$

Figure 8 shows that the eigenvalues of the stabilized one-dimensional operator are equivalent (in the sense of (68)) to the stiffness matrix. Therefore this operator does not lead to spurious pressure oscillations and it ensures the efficiency of the preconditioners of section 6.

In contrast to the usual stabilization methods known from the FE literature [8,42,67] where the original operator

$$\begin{pmatrix} \Delta & \nabla \\ \nabla \cdot & 0 \end{pmatrix} \quad \text{or} \quad \begin{pmatrix} I & \nabla \\ \nabla \cdot & 0 \end{pmatrix} , \text{respectively},$$

is stabilized by

$$\begin{pmatrix} \Delta & \nabla \\ \nabla \cdot & S \end{pmatrix} \quad \text{or} \quad \begin{pmatrix} I & \nabla \\ \nabla \cdot & S \end{pmatrix} , \text{respectively}, \tag{86}$$

the above method replaces the gradient and the divergence operator by up- and down-winded operators ∇^+, $\nabla^- \cdot$ leading to the stabilized operator

$$\begin{pmatrix} I & \nabla^+ \\ \nabla^- \cdot & 0 \end{pmatrix} .$$

Hence the Schur complement operator $\nabla^- \cdot \nabla^+$ has the same structure as the original operator $\nabla \cdot \nabla$. This is not the case for $\nabla \cdot \nabla - S$. Therefore our velocity field is always (weakly) divergence-free with respect to $\nabla^- \cdot$ and we do not run into all that trouble which comes with non-divergence-free velocity fields.

In the adaptive case the stabilization terms are evaluated by the adaptive FD algorithm (**FD**). In all our experiments this method performed very well.

8.2   Numerical Experiment: Three Vortices Interaction

Now, we will apply the adaptive wavelet solver developed in the previous sections to the model problem of the interaction of three vortices in a two-dimensional flow, see [58]. We use MRA-d Interpolet wavelets where $N = 6$.

The initial velocity is induced by three vortices each with a Gaussian vorticity profile

$$\omega(\mathbf{x}, 0) = \omega_0 + \sum_{i=1}^{3} \omega_i \exp\left(\frac{\|\mathbf{x} - \mathbf{x}_i\|^2}{\sigma_i^2}\right), \quad \mathbf{x} \in [0, 1]^2 .$$

The first two vortices have the same positive sign ω_1, $\omega_2 > 0$ and the third has a negative sign, see Figure 9. The different parameters ω_0, ω_i and σ_i are

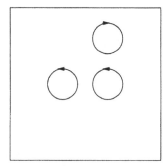

Fig. 9. Initial configuration for the three vortices interaction.

chosen such that the mean value of $\omega(., 0)$ vanishes and that $\omega(., 0)\,|_{\partial[0,1]^2}$ is almost ω_0 to allow for periodic boundary conditions.

Due to the induced velocity field the three vortices start to rotate around each other. In a later stage the two same-sign vortices merge which leads to a configuration of two counter-rotating vortices. This process is the basic mechanism in 2D turbulence and it takes place e.g. in the shear layer problem of the next subsection or during the convergence of ω to the solution of the Joyce-Montgomery equation [17] for random initial vorticity fields.

In our numerical experiments the viscosity was set to $\nu = 3.8 \cdot 10^{-6}$. The maximum velocity of the initial condition is $7 \cdot 10^{-2}$ which corresponds to a Reynolds number of ≈ 55200. A reference solution was obtained with a Fourier spectral code (512×512 modes) which is based on the splitting method with a third order Adams-Bashforth scheme for the convective term. This third order scheme was also used for the adaptive wavelet solver. Here MRA-d Interpolet wavelets of 6th order (i.e. $N = 6$) were employed in a collocation method for the spatial discretization. The values of the threshold parameter ϵ were set to $\{8, 4, 1\} \cdot 10^{-4}$ and the finest level l was limited to $(10, 10)$ to avoid very fine time steps. The time step used in both codes was $dt = 10^{-2}$.

In Figure 12 the contour lines of the vorticity of the spectral and the adaptive solutions are shown. The contour levels are equally spaced from -1.5 to 3 which is approximately the minimum/maximum value of $\omega(.,0)$.

Although the reference solutions and the adaptive solutions are qualitatively the same, the adaptive solution seems to overestimate the rotation of the cores of the vortices. Also in the time behaviour of the energy $E(t)$ and the enstrophy $Z(t)$

$$E(t) = \int_\Omega |\mathbf{u}|^2 \, d\mathbf{x} \, , \quad Z(t) = \frac{1}{2} \int_\Omega |\omega|^2 \, d\mathbf{x}$$

there are differences, see Figure 10. However, these differences are very small. Further experiments have shown that this error stems from the spatial discretization and may be reduced by relaxing the level limitation and by decreasing the threshold parameter ϵ.

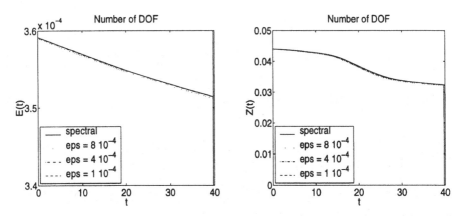

Fig. 10. Left: Evolution of energy. Right: Evolution of enstrophy.

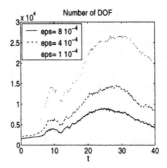

Fig. 11. Evolution of the number of DOF.

The evolution of the number of degrees of freedom is shown in Figure 11. Obviously our method recognizes the arising complicated flow patterns in space and time (see Figure 12) and spends there much more degrees of

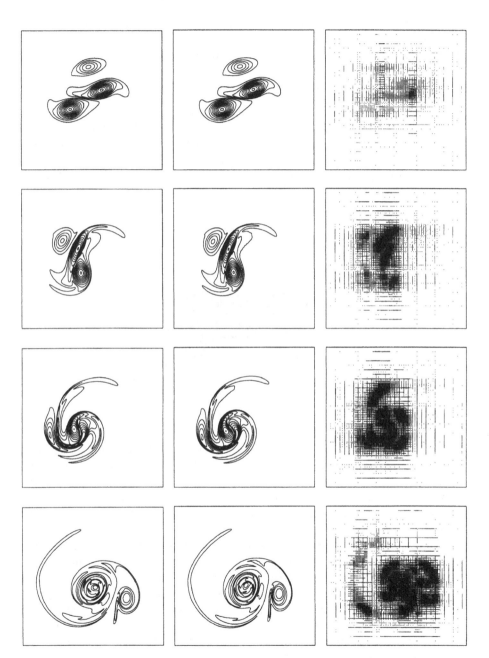

Fig. 12. Isolines of ω at $t = 5$, 10, 15, 35 (top–bottom), left: spectral code, middle: adaptive wavelet solver ($\epsilon = 4e - 4$), right: adaptive grid.

freedom than in smooth regions. Note that the left and middle pictures of Figure 12 are enhanced by 20 %.

Despite of small discrepancies the results of the adaptive wavelet solver are quite accurate. To achieve this accuracy only a small number of degrees of freedom was needed – less than 1% in comparison to the reference solution.

8.3   Numerical Experiment: the Shear Layer

Now we apply the adaptive wavelet solver to a shear layer model problem. The initial configuration of the temporally developing shear layer is a velocity field with a hyperbolic–tangent profile $u(y) = U \tanh(2y/\delta)$, where δ is the vorticity shear layer thickness $\delta = 2U/\max(du/dy)$, see Figure 13. From

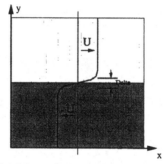

Fig. 13. Initial configuration for the mixing layer.

linear stability analysis this initial configuration is known to be inviscidly unstable. Slight perturbations are amplified by Kelvin-Helmholtz instabilities, where the most amplified mode corresponds to a wavelength $\lambda = 7\delta_0$ [50].

In our numerical experiments the vorticity thickness δ was chosen such that 10 vortices should develop in the computational domain $[0, 1]^2$. The maximum velocity was $U = 1.67 \cdot 10^{-2}$ and the viscosity was $\nu = 3.8 \cdot 10^{-6}$. The instabilities were triggered by a superimposed weak white noise in the shear layer, compare [57].

The numerical simulations were applied to the periodized version of the problem with two shear layers on $[0, 1] \times [0, 2]$. The reference simulation was carried out with a Fourier spectral code (splitting method, AB3) with 512×1024 modes. We used the same wavelet solver as in the previous experiment. The values of the threshold parameter ϵ were set to $\{12, 8, 4\} \cdot 10^{-4}$ and the finest level $\mathbf{l}$ was limited to $(10, 10)$. Both, the spectral and the adaptive solver used a time step of $5 \cdot 10^{-3}$.

In Figure 16 the results of the run for $\epsilon = 8 \cdot 10^{-4}$ are compared with the reference solution. Both results agree quite well. For $t = 36$ we observe a phase shift for the right-most two vortices in the adaptive simulation. As in the first experiment, the adaptive code seems to overestimate the rotation of the vortices. Despite of these slight differences in the instantaneous vorticity fields, the statistical quantities energy and enstrophy agree very well, see Figure 14.

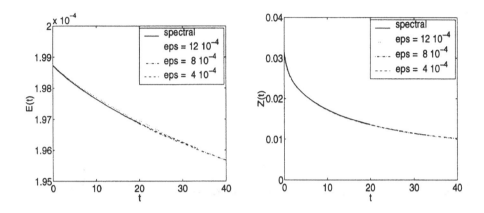

Fig. 14. Left: Evolution of energy. Right: Evolution of enstrophy.

The evolution of the number of degrees of freedom (DOF) is shown in Figure 15. The initial velocity is not very smooth due to the white noise added for triggering the instability. Therefore, a large number of DOF is required to resolve the initial velocity in the shear layer sufficiently well. Then, in a first phase, the diffusion smooths the velocity very fast which

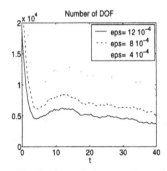

Fig. 15. Evolution of the number of DOF.

leads to the strong decay of the number of DOF. This process is stopped by the development of the Kelvin-Helmholtz instabilities leading to an increase of the number of DOF ($4 < t < 10$). In the last phase ($t > 10$) the number of coherent vortices constantly decreases by successive merging. Therefore, in this stage the number of DOF decreases almost constantly.

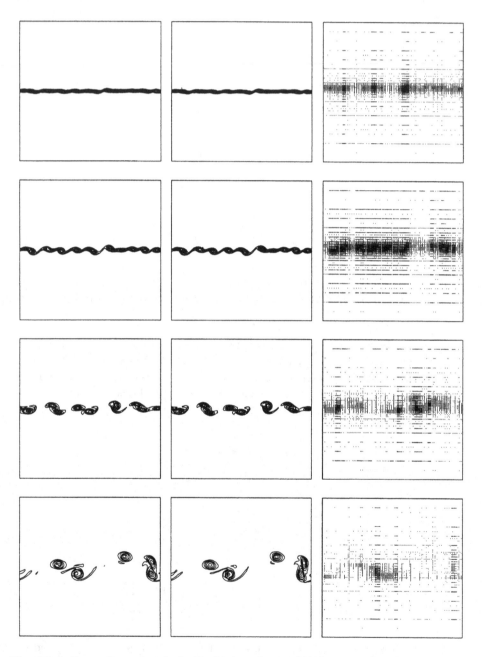

Fig. 16. Isolines of ω at $t = 4$, 8, 16, 36 (top–bottom), left: spectral code, middle: adaptive wavelet solver ($\epsilon = 8e - 4$), right: adaptive grid.

The numerical experiments of subsections 8.2 and 8.3 show that our adaptive wavelet scheme is a quite efficient and accurate solution method. The adaptive time stepping method predicts very well the locations where adaptive refinement is needed. Furthermore, the consistent discretization of the Schur complement operator in connection with our stabilization technique and the preconditioners of section 6 lead to an efficient solution of the projection problem which is usually the most time consuming part of a Navier-Stokes solver.

9   Concluding Remarks

In this paper we presented a new method for the numerical treatment of the Navier-Stokes equations which uses simple tensor products of univariate Interpolets as trial functions in an adaptive Petrov-Galerkin scheme. The tensor product approach allows to carry the univariate adaptive algorithms (and codes) directly over to the multivariate case for arbitrary dimensions. Furthermore this MRA-d-approach leads to superior approximation properties of the resulting multivariate basis. We also presented a simple preconditioning technique for the linear systems of equations which stem from the discretization of elliptic PDEs. This technique is based on the lifting scheme and results in optimal $O(1)$ condition numbers. Then, a simple time-stepping method is employed for the discretization of time-dependent problems. There, also the algorithm for the adaptive basis selection is incorporated and the adaptive basis is now evolved from time slab to time slab. In our numerical experiments this method proved to be robust and efficient.

In summary, we always used the simplest and most efficient parts of the wavelet approaches to build up our method. The Interpolets lead to simple algorithms for evaluation of non-linear terms. The MRA-d tensor product approach and the Petrov-Galerkin discretization with the dual wavelets as test functions form the most efficient and easiest way to carry the univariate adaptive algorithms over to the case of arbitrary dimensions. We could even use the complete C++ code developed for the univariate case in the multivariate code as well. Another implication of the MRA-d-approach is the pronounced locality of data within the univariate algorithms. This is important for modern microprocessor architectures. Finally the lifting preconditioner solved in a simple way the remaining problem of our approach: the badly conditioned linear systems.

All these techniques (tensor product wavelets, lifting preconditioner, time stepping method and adaptive basis selection) form the building blocks of our adaptive wavelet solver for the Navier-Stokes equations. For the discretization of these equations we used a Chorin-type projection method. In comparison to simple advection-diffusion problems the additional difficulty is the stable discretization of the pressure Poisson equation. To this end we employed an upwind-downwind scheme for the pressure gradient and the divergence operator, respectively. This approach mimics the structure the continuous operator which is required to avoid spurious oscillations and to maintain the efficiency of our preconditioners. Finally we demonstrated the efficiency and accuracy of our method for two examples of high Reynolds-number flows in 2D.

However, there is still a lot to do. On the theoretical side this is mainly the proof of the consistency and stability of our Petrov-Galerkin scheme and the proof of the efficiency of our lifting preconditioner in the fully adaptive case. On the practical side the main improvements will concern the parallelization of the method. Here we intend to use the space-filling curve technique [37]. Also a generalization of the method to more complicated geometries by means of parametric mappings and block structuring has to be accomplished in the next future.

Acknowledgements. The authors would like to thank Marie Farge and Kai Schneider for proposing the numerical experiments of section 8 and for kindly providing the initial conditions of the mixing layer simulation.

Notations

c	generic constant
C	diagonal matrix, see Table (2)
$\mathbf{C}$	diagonal matrix, (59)
$C(.,.)$	$C(< \text{matrix} >, < \text{index set} >)$, (33)
$C(.,.)$	$C(u^n, u^{n-1}, ..)$, convective term, (76)
$\mathbf{C}(.,.)$	$\mathbf{C}(u^n, u^{n-1}, ..)$, convective term, (80)
d	dimension of computational domain, e.g. $[0,1]^d$
$\mathbf{e}$	$= (e_1, .., e_d)$, $e_i \in \{0,1\}$, (49)
g_s	scaling coefficient for ψ, similar to (16)
$\tilde{g}_s$	" , for $\tilde{\psi}$, (16)
G^l	scaling matrix for level l, (19)
$\tilde{G}^l$	" , for dual wavelets, (19)
h_s	scaling coefficient for ϕ, (1)
$\tilde{h}_s$	" , for $\tilde{\phi}$, (7)
$\mathbf{h}$	$= \{h_s\}_{s \in \mathbb{Z}}$, sequence/vector of scaling coefficients
H^l	scaling matrix for level l, (19,29)
$\tilde{H}^l$	" , for dual scaling functions, (19)
k	level of refinement, see l
$\mathbf{k}$	$= (k_1, .., k_d)$, see $\mathbf{l}$
l	refinement level of scaling function $\phi^{(l,s)}$ or wavelet $\psi_{(l,t)}$, (9,11)
$\mathbf{l}$	$= (l_1, .., l_d)$, multivariate refinement level, (50)
L	a certain finest level in e.g. an adaptive basis
n	number of time step
N	degree of polynomial exactness of ϕ, (4)
M	" of $\tilde{\phi}$, (4,64)

p	the pressure, (79)
$R(.,.)$	$R(< \text{matrix} >, < \text{index set} >)$, (33)
s	(in most cases) translation index for scaling functions, (9)
$\mathbf{s}$	$= (s_1, .., s_d)$
$S(l)$	index set of all scaling functions on level l, $S(l) = \{0, .., 2^l\}$ or $\{1, .., 2^l - 1\}$, $\{0, 2, .., 2^l - 2, 2^l\}$,... depending on the b.c., (25)
S	pressure stabilization term, (86)
$\mathcal{S}(l)$	$\subseteq S(l)$, (31)
$\mathcal{S}^a(l)$	$\subseteq S(l)$, (48)
$\mathcal{S}^b(l)$	$\subseteq S(l)$, see algorithm $(\mathcal{S}^b)$
t	translation index for wavelets, (11,12,13)
$\mathbf{t}$	$= (t_1, .., t_d)$, " for multivariate wavelets, (50)
$T(l)$	index set of all wavelets on level l, $T(l_0) = \{0, .., 2^{l_0}\}$ and $T(l) = \{0, .., 2^{l-1} - 1\}$ for $l > l_0$, (26)
T	see Table 2
$\hat{T}$	see Table 2
$\mathcal{T}(l)$	$\subseteq T(l)$, (30)
$\boldsymbol{\mathcal{T}}(\mathbf{l})$	$\subseteq T(l_1) \otimes ... \otimes T(l_d)$, (51)
$\boldsymbol{\mathcal{T}}$	see (51)
u	a function $[0,1]^d \to \mathbb{R}$
$\mathbf{u}$	the velocity, (79)
$u^{(l,s)}$	a single scaling function coefficient, (13,32)
$u_{(l,t)}$	a single wavelet coefficient, (13,32)
$u_{(\mathbf{l},\mathbf{t})}$	" , multivariate case, (51)
$\mathbf{u}^l$	sequence/vector of scaling function coefficients, (13)
$\mathbf{u}^{\mathcal{S}(l)}$	" , adaptive case, (32)
$\mathbf{u}_l$	sequence/vector of wavelet coefficients, (13)
$\mathbf{u}_{\mathcal{T}(l)}$	" , adaptive case, (32)
$\mathbf{u}_{\boldsymbol{\mathcal{T}}(\mathbf{l})}$	" , adaptive, multivariate case, (51)
$\mathbf{u}_{\mathcal{T}}$	" , for complete adaptive basis, (32)
$\mathbf{u}_{\boldsymbol{\mathcal{T}}}$	" , for complete, adaptive, multivariate basis, (51)
ϕ	mother function for primal scaling functions, (1)
$\tilde{\phi}$	" , for dual scaling functions, (6)
$\phi^{(l,t)}$	a single scaling function, (9,21)
Φ^l	row vector of all scaling functions $\phi^{(l,s)}$ on level l, (12,32)
$\Phi^{\mathcal{S}(l)}$	" , adaptive case, (51)
ψ	mother function for primal wavelets
$\tilde{\psi}$	" , for dual wavelets
$\hat{\psi}$	" , for Lifting-wavelets
$\tilde{\hat{\psi}}$	" , for dual Lifting-wavelets
$\psi_{(l,t)}$	a single wavelet, (11)
$\hat{\psi}_{(l,t)}$	a single Lifting-wavelet, (11)
Ψ_l	row vector of all wavelets $\psi_{(l,t)}$ on level l, (26)
$\Psi_{\mathcal{T}(l)}$	" , adaptive case, (32)

References

1. Angot P., Bruneau C., Fabrie P.: *A penalization method to take into account obstacles in incompressible flows*; Preprint No. 97017, Mathematiques Appliquees, University Bordeaux (1997).
2. Barinka A., Barsch T., Charton P., Cohen A., Dahlke S., Dahmen W., Urban K.: *Adaptive wavelet schemes for elliptic problems – implementation and numerical experiments*; Institut für Geometrie und praktische Math., RWTH Aachen (1999), preprint.
3. Battle G., Federbush P.: *Divergence-free vector wavelets*; Michigan Math. J., 40 (1993) pp. 181–195.
4. Becker R., Rannacher R.: *Weighted a posteriori error control in FE methods*; in ENUMATH-95 in Proc. ENUMATH-97, World Scientific Publ., Singapore (1998).
5. Becker R.: *Weighted a posteriori error estimators for FE approximations of the incompressible Navier-Stokes equations*; Universität Heidelberg, IWR, SFB 359, Preprint 48 (1999).
6. Burie J.M, Marion M.: *Multilevel Methods in Space and Time for the Navier-Stokes Equations*; SIAM J. Num. Anal. 34 No. 4 (1997), pp. 1574–1599.
7. Bramble J.H., Pasciak J.E., Xu J.: *Parallel multilevel preconditioners*; Math. Comp. 55 (1990), pp. 1–22.
8. Brezzi F., Pitkäranta J.: *On the Stabilization of Finite Element Approximations of the Stokes Problem*; in Efficient Solutions of the Stokes Problem, Hackbusch W. (Eds.), Vieweg (1984).
9. Bank R., Weiser A.: *Some a posteriori error estimators for elliptic PDEs*; Math. Comp. 44 (1985), pp. 283–301.
10. Bertoluzza S.: *An adaptive collocation method based on interpolating wavelets*; Istituto di analisi numerica del consiglio nazionale delle ricerche, Pavia (1997)
11. Beylkin G., Keiser J.M.: *An Adaptive Pseudo-Wavelet Approach for Solving Nonlinear Partial Differential Equations*; in Multi Scale Wavelet Methods for Partial Differential Equations, Vol. 6 in the Wavelet Analysis and Applications series, Academic Press.
12. Canuto C., Hussaini M., Quarteroni A., Zang T.: *Spectral Methods in Fluid Dynamics*; Springer Verlag (1987).
13. Carnicer J.M., Dahmen W., Pena J.M.: *Local decomposition of refinable spaces*; Appl. Comp. Harm. Anal. 3 (1996), pp. 127–153.
14. Charton P., Perrier V.: *A Pseudo-Wavelet Scheme for the two-dimensional Navier-Stokes Equations*; Matematica Aplicada e Computacional (1996)
15. Cohen A., Dahmen W., DeVore R.: *Adaptive wavelet methods for elliptic operator equations – convergence rates*; IGPM-Report, (1998), RWTH Aachen.
16. Cohen A., Daubechies I., Feauveau J.-C.: *Biorthogonal bases of compactly supported wavelets*; Comm. pure and Appl. Math. 45 (1992), pp.485–560.
17. Chorin A.: *Vorticity and Turbulence*; Springer Verlag (1998).
18. Dahmen W., Schneider R., Xu Y.: *Nonlinear Functionals of Wavelet Expansions – Adaptive Reconstruction and Fast Evaluation*; IGPM-Report 160, (1998), RWTH Aachen, to appear in Numerische Mathematik.
19. Daubechies I.: *Orthonormal Bases of compactly supported Wavelets I*; Comm. Pure Appl. Math. (1988).
20. Daubechies I.: *Orthonormal Bases of compactly supported Wavelets II*; SIAM J. Math. Anal. 24 (1993).

21. Daubechies I.: *Ten lectures on wavelets*; CBMS-NSF Regional Conference Series in Appl. Math. 61 (1992).
22. Deslauriers G., Dubuc S.: *Symmetric iterative interpolation processes*; Constr. Approx. 5 (1989), pp. 49–68.
23. DeVore R.: *Nonlinear Approximation*; Acta Numerica, 7 (1998), pp. 51–150.
24. DeVore R., Konyagin S.V., Temlyakov V.N.: *Hyperbolic wavelet approximation*; Constr. Approx. 14 (1998), pp. 1–26.
25. Dongarra J., Duff I., Sorensen D., van der Vorst H.: *Numerical Linear Algebra for High-Performance Computers*; SIAM (1998).
26. Donoho D.: *Interpolating wavelet transform*; Stanford University (1992), preprint.
27. Donoho D., Yu P.: *Deslauriers-Dubuc: Ten Years After*; in Deslauriers G., Dubuc S. (Eds), CRM Proceedings and Lecture Notes Vol. 18 (1999).
28. Frisch U.: *Turbulence*; Cambridge University press (1995).
29. Fröhlich J., Schneider K.: *An Adaptive Wavelet-Vaguelette Algorithm for the Solution of PDEs*; J. Comput. Physics 130 (1997), pp. 174–190.
30. Gresho P., Chan S.: *On the theory of semi-implicit projection methods*; Int. J. Numer. Meth. Fluids 11(5) (1990).
31. Gresho P., Sani R.: *Incompressible Flow and the Finite Element Method*; Wiley (1998).
32. Griebel M.: *Multilevelmethoden als Iterationsverfahren ueber Erzeugendensystemen*; Teubner Verlag (1994).
33. Griebel M.: *Adaptive sparse grid multilevel methods for elliptic PDEs based on finite differences*; Computing 61 No. 2 (1998), pp. 151–180.
34. Griebel M., Knapek M.: *Optimized tensor-product approximation spaces*; Constr. Approx., in print.
35. Griebel M., Oswald P.: *Tensor product type subspace splitting and multilevel iterative methods for anisotropic problems*; Adv. Comp. Math. 4 (1995), pp. 171–206.
36. Griebel M., Zumbusch G.: *Adaptive Sparse Grids for Hyperbolic Conservation Laws.*; Proceedings of Seventh International Conference on Hyperbolic Problems: Theory, Numerics, Applications, ISNM 129, Birkh"user (1999), Vol. 1, pp. 411–421.
37. Griebel M., Zumbusch G.: *Parallel Adaptive Subspace Correction Schemes with Applications to Elasticity*; accepted for Computer Methods in Applied Mechanics and Engineering, Elsevier, (1999).
38. Gunzburger M., Mundt M., Peterson P.: *Experiences with FEM for the velocity-vorticity formulation of 3D viscous incompressible flows*; in Computational methods for viscous flows Vol 4. Brebbia C. (Eds.) (1990).
39. Hafez M., Dacles J., Soliman M.: *A velocity/vorticity method for viscous incompressible flow calculations*; in Lecture Notes in Physics Vol. 323 Dwoyer D., Hussaini M. (Eds.) (1989).
40. Holmstroem M.: *Solving Hyperbolic PDEs Using Interpolating Wavelets*; SIAM J. Sc. Comp. 21/2 (1999) pp. 405–420.
41. Horn R., Johnson C.: *Topics in Matrix Analysis*; Cambridge Univ. press (1989).
42. Hughes T., Franca L.: *A new finite element formulation for computational fluid dynamics: VII*; Comput. Meth. Appl. Mech. Eng. Vol 65 (1987) pp. 85–96.
43. Karniadakis G., Sherwin S.: *Spectral/hp Element Methods for CFD*; Oxford University Press (1999).

44. Koster F.: *A Proof of the Consistency of the Finite Difference Technique on Sparse Grids*; University Bonn, (2000) SFB-Report No. 642.
45. Koster F., Griebel M.: *Orthogonal wavelets on the interval*; University Bonn, (1998) preprint No. 576.
46. Koster F., Griebel M.: *Efficient Preconditioning of Linear Systems for the Finite Difference and the Collocation Method on Sparse Grids*; University Bonn, (2000), in preparation.
47. Kunoth A.: *Fast Iterative Solution of Saddle Point Problems in Optimal Control Based on Wavelets*; University Bonn, (2000) SFB-Report.
48. Lippert R., Arias T., Edelman A.: *Multi Scale Computation with Interpolating Wavelets*; J. Comp. Physics 140 (1998), pp. 278–310.
49. Metais O., Lesieur M.: *Turbulence and Coherent Structures*; Kluwer (1991).
50. Michalke A.: *On the inviscid instability of the hyperbolic tangent velocity profile*; J. Fluid Mech. 19 (1964) 543–556.
51. Monzon L., Beylkin G., Hereman W.: *Compactly supported wavelets based on almost interpolating and nearly linear phase filters (Coiflets)*; Appl. Comp. Harm. Anal., 7, (1999), pp. 184–210.
52. Marion M., Temam R.: *Nonlinear Galerkin Methods*; SIAM J. Num. Anal. 26 (1989), pp. 1139–1157.
53. Oswald P.: *Multilevel Finite Element Approximation*; Teubner Verlag (1994).
54. Prohl A.: *Projection and Quasi-Compressibility Methods for Solving the Incompressible Navier-Stokes Equations*; Teubner Verlag (1997).
55. Peyret R., Taylor T.: *Computational Methods for Fluid Flow*; Springer (1983).
56. Schiekofer T.: *Die Methode der finiten Differenzen auf dünnen Gittern zur Lösung elliptischer und parabolischer PDEs*; PhD thesis, Bonn (1998).
57. Schneider K., Farge M.: *Numerical simulation of a temporally growing mixing layer in an adaptive wavelet basis*; C.R.A.S. Paris, Ser. IIb, (2000)
58. Schneider K., Kevlahan N., Farge M.: *Comparison of an Adaptive Wavelet Method and Nonlinearly Filtered Pseudospectral Methods for Two-Dimensional Turbulence*; Theor. Comput. Fluid Dyn. 9, (1997), pp. 191-206
59. Schneider R.: *Multiskalen- und Wavelet-Matrixkompression: Analysisbasierte Methoden zur effizienten Lösung großer vollbesetzter Gleichungssysteme*; Advances in Numerical Mathematics, Teubner Stuttgart (1998).
60. Sweldens W.: *The lifting scheme: a construction of second generation wavelets*; SIAM J. Math. Anal. 29,2 (1997), pp. 511–546.
61. Sweldens W., Piessens R.: *Quadrature formulae and asymptotic error expansions for wavelet approximations of smooth functions*; SIAM J. Num. Anal. 36,3 (1994), pp. 377–412.
62. Temlyakov V.: *Approximation of periodic functions*; Nova Science Publ. New York (1994).
63. Urban K.: *On Divergence-free Wavelets*; Adv. Comp. Math. Vol. 4 (1995).
64. Vasilyev O., Paolucci S.: *Fast Adaptive Wavelet Collocation Algorithm for Multidimensional PDEs* ; J. Comp. Phys. 138/1 (1997) pp. 16–56.
65. Verfürth R.: *A Review of a Posteriori Error Estimation and Adaptive Mesh-Refinement Techniques*; Teubner Verlag (1996).
66. Vreugdenhil C.B., Koren B.: *Numerical methods for Advection-Diffusion Problems*; Notes on Numerical Fluid Mechanics Vol. 45, Vieweg Verlag (1993).
67. Zienkiewicz O., Wu J.: *Incompressibility without tears – how to avoid restrictions of mixed formulations*; Int. J. Numer. Meth. Eng. Vol 32 (1991), pp. 1189–1203.

Asymptotic Problems and Compressible-Incompressible Limit

Nader Masmoudi[*]

CEREMADE-UMR CNRS 7534, Université Paris Dauphine,
Place de Lattre de Tassigny, 75775 Paris cedex 16, France
e-mail: masmoudi@dmi.ens.fr

Abstract. This series of lectures is devoted to the study of some asymptotic problems in fluid mechanics and mainly to the study of the compressible-incompressible limit when the Mach number goes to zero or when the adiabatic constant γ goes to infinity.

1 Introduction

Asymptotic problems arise when a dimensionless parameter ε goes to zero in an equation describing the motion of some type of fluid (or any other physical system). Physically, this allows a better knowledge of the system in this limit case by describing (usually by a simpler equation) the prevailing phenomenon when this parameter is small. Indeed, this small parameter, usually describes a physical reality. For instance, a slightly compressible fluid is characterized by a low Mach number, whereas a slightly viscous fluid is characterized by a high Reynolds number (a low viscosity). In many cases, we have different small parameters (we can be in presence of a slightly compressible and slightly viscous fluid at the same time). Depending on the way these small parameters go to zero, we can recover different systems at the limit. For instance, if $\varepsilon, \delta, \nu, \eta \ll 1$, the limit system can depend on the magnitude of the ratio of ε/δ.

These asymptotic problems allow us to get simpler models at the limit, due to the fact that we usually have fewer variables or (and) fewer unknowns. This simplifies the numerical simulations, in fact, instead of solving the initial system, we can solve the limit system and then add a corrector.

Many mathematical problems are encountered when we try to justify the passage to the limit, which are mainly due to the change of the type of the equations, the presence of many spatial and temporal scales, the presence of boundary layers (we can no longer impose the same boundary conditions for the initial system and the limit one), the presence of oscillations in time at high frequency, etc. Different type of questions can be asked:

1) What do the solutions of the initial system (S_ε) converge to? Is the convergence strong or weak?

---

[*] Current address: Courant Institute, New York University 251 Mercer St, New York, NY 10012, USA.

2) In the case of weak convergence, can we give a more detailed description of the sequences of solutions? (Can we describe the time oscillations for instance?)

In Section 2 and Section 3, we will state some results concerning respectively the convergence from the Navier-Stokes system towards the Euler system and the study of rotating fluid at high frequency. Next in Section 4, we will study in more detail the compressible incompressible limit when the Mach number goes to 0. Finally, Section 5 is devoted to the limit γ (the adiabatic constant) going to ∞.

2   The Navier-Stokes Euler Limit

The zero-viscosity limit for the incompressible Navier-Stokes equation in a bounded domain, with Dirichlet boundary conditions is a challenging open problem due to the formation of a boundary layer satisfying the Prandtl equations, which seem to be ill-posed. The whole space case was performed by several authors, we can refer for instance to Swann [33] and Kato [16]. In [3], Bardos treats the case of a bounded domain with a boundary condition on the vorticity, which does not engender any boundary layer. In the sequel, we will study this limit in the case we consider different vertical and horizontal viscosities. We consider the following system of equations $(NS_{\nu,\eta})$

$$\partial_t u^n + \operatorname{div}(u^n \otimes u^n) - \nu\partial_z^2 u^n - \eta\Delta_{x,y} u^n = -\nabla p \quad \text{in } \Omega \tag{1}$$
$$\operatorname{div} u^n = 0 \quad \text{in } \Omega \tag{2}$$
$$u^n = 0, \quad \text{on } \partial\Omega \tag{3}$$
$$u^n(0) = u_0^{\eta,\nu} \quad \text{with } \operatorname{div} u_0^n = 0 \tag{4}$$

where $\Omega = \omega \times (0,h)$, or $\Omega = \omega \times (0,\infty)$, and $\omega = \mathbb{T}^2$, or $\mathbb{R}^2$, $\nu = \nu_n, \eta = \eta_n$. When η, ν go to 0, we expect that u^n converges to the solution of the Euler system

$$\begin{cases} \partial_t w + \operatorname{div}(w \otimes w) = -\nabla p & \text{in } \Omega, \\ \operatorname{div} w = 0 & \text{in } \Omega, \\ w.n = \pm w_3 = 0 & \text{on } \partial\Omega, \\ w(t=0) = w^0 . \end{cases} \tag{5}$$

It turns out that we are able to justify this formal derivation under an additional condition on the ratio of the vertical and horizontal viscosities.

Theorem 1. *Let $s > 5/2$, and*

$$w^0 \in H^s(\Omega)^3, \quad \operatorname{div} w^0 = 0, \quad w^0.n = 0 \text{ on } \partial\Omega .$$

We assume that $u^n(0)$ converges to w^0 in $L^2(\Omega)$, and $\nu, \eta, \nu/\eta$ go to 0. Then any sequence of global weak solutions (à la Leray) u^n of (1)–(4) satisfying the energy inequality satisfies

$$u^n - w \to 0 \quad \text{in} \quad L^\infty(0,T^*,L^2(\Omega)),$$

$$\sqrt{\eta}\nabla_{x,y}u^n, \sqrt{\nu}\partial_z u^n \to 0 \text{ in } L^2(0,T^*,L^2(\Omega)),$$

where w is the unique solution of (5) in $L^\infty(0,T^*;H^s(\Omega)^3)$.

We give here a sketch of the proof and refer to [28] for a complete proof. The existence of global weak solutions for $(NS_{\nu,\eta})$, satisfying the energy inequality is due to J. Leray

$$\frac{1}{2}\|u^n(t)\|_{L^2}^2 + \nu \int_0^t \|\partial_z u^n\|_{L^2}^2 + \eta \int_0^t \|\partial_x u^n\|_{L^2}^2 + \|\partial_y u^n\|_{L^2}^2 \le \frac{1}{2}\|u_0^n\|_{L^2}^2 . \quad (6)$$

This estimate does not show that u^n is bounded in $L^2(0,T;H^1)$ and hence if we extract a subsequence still denoted by u^n converging weakly to u in $L^\infty(0,T;L^2)$, we cannot deduce that $u^n \otimes u^n$ converges weakly to $w \otimes w$. If we try to use energy estimates to show that $u^n - w$ remains small we see that the integrations by parts introduce terms that we cannot control, since $u^n - w$ does not vanish at the boundary. Hence, we must construct a boundary layer which allows us to recover the Dirichlet boundary conditions: $\mathcal{B}^n$ will be correctors of small L^2 norm, and localized near $\partial\Omega$ (we take here the case where $\Omega = \omega \times (0,\infty)$ not to deal with boundary conditions near $z = h$)

$$\begin{cases} \mathcal{B}^n(z=0) + w(z=0) = 0, \quad \mathcal{B}^n(z=\infty) = 0, \\ \\ \operatorname{div}\mathcal{B}^n = 0, \quad \mathcal{B}^n \to 0 \text{ in } L^\infty(0,T^*;L^2), \end{cases}$$

a possible choice is to take $\mathcal{B}^n$ of the form

$$\mathcal{B}^n = -w(z=0)\exp\left(-\frac{z}{\sqrt{\nu\zeta}}\right) + \dots$$

where ζ is a free parameter to be chosen later. We want to explain now the idea of the proof. Instead of using energy estimates on $u^n - w$, we will work with $v^n = u^n - (w + \mathcal{B}^n)$. Next we write the following equation satisfied by $w^\mathcal{B} = w + \mathcal{B}^n$ (in what follows, we will write $\mathcal{B}$ instead of $\mathcal{B}^n$)

$$\partial_t w^\mathcal{B} + w^\mathcal{B}.\nabla w^\mathcal{B} - \nu\partial_z^2 w^\mathcal{B} - \eta\Delta_{x,y}w^\mathcal{B}$$
$$= \partial_t\mathcal{B} + \mathcal{B}.\nabla w^\mathcal{B} + w.\nabla\mathcal{B} - \nu\partial_z^2 w^\mathcal{B} - \eta\Delta_{x,y}w^\mathcal{B} - \nabla p \quad (7)$$

which yields the following energy equality

$$\frac{1}{2}\|w^\mathcal{B}(t)\|_{L^2}^2 + \nu \int_0^t \|\partial_z w^\mathcal{B}\|_{L^2}^2 + \eta \int_0^t \|\partial_x w^\mathcal{B}\|_{L^2}^2 + \|\partial_y w^\mathcal{B}\|_{L^2}^2$$
$$= \frac{1}{2}\|w^\mathcal{B}(0)\|_{L^2}^2 + \int_0^t w^\mathcal{B}.[\partial_t\mathcal{B} + w.\nabla\mathcal{B} - \nu\partial_z^2 w^\mathcal{B} - \eta\Delta_{x,y}w^\mathcal{B}] \quad (8)$$

Next, using the weak formulation of (1), we get for all t

$$\int_\Omega (u^n.w^B)(t) + \nu \int_0^t \int_\Omega \partial_z w^B u^n + \eta \int_0^t \int_\Omega \partial_x w^B \partial_x u^n + \partial_y w^B \partial_y u^n$$

$$= \int_\Omega (u^n.w^B)(0) + \int_0^t \int_\Omega u^n.\nabla w^B u^n \qquad (9)$$

$$+ u^n.[\partial_t B - w.\nabla w - \nu \partial_z^2 w^B - \eta \Delta_{x,y} w^B] .$$

Then adding up (6), (8) and subtracting (9), we get

$$\frac{1}{2}\|v(t)\|_{L^2}^2 + \nu \int_0^t \|\partial_z v\|_{L^2}^2 + \eta \int_0^t \|\partial_x v\|_{L^2}^2 + \|\partial_y u^n\|_{L^2}^2$$

$$\leq \frac{1}{2}\|v_0\|_{L^2}^2 + \int_0^t \int_\Omega v.[\partial_t B - \nu \partial_z^2 w^B - \eta \Delta_{x,y} w^B] \qquad (10)$$

$$+ w.\nabla B w^B - u^n.\nabla w^B u^n + w.\nabla w u^n .$$

Finally, using that $\int (u.\nabla q)q = 0$, we get

$$\int_\Omega w.\nabla B w^B - u^n.\nabla w^B u^n + w.\nabla w u^n = \int_\Omega -w^B.\nabla B v - B.\nabla w v - v.\nabla w^B v$$

Now, we want to use a Gronwall lemma to deduce that $\|v(t)\|_{L^2}^2$ remains small. By studying two terms among those occurring in the left hand side of the energy estimate (10), we want to show why we need the condition $\nu/\eta \to 0$. In fact

$$\left| \int_\Omega v_3 \partial_z B v \right| \leq \int_\Omega \frac{v_3}{z} z^2 \partial_z B \frac{v}{z}$$

$$\leq C\|\partial_z v_3\|_{L^2} \sqrt{\nu\zeta}\|w\|_{L^\infty} \|\partial_z v\|_{L^2}$$

$$\leq C\zeta\|\partial_z v_3\|_{L^2}^2\|w\|_{L^\infty}^2 + \frac{\nu}{4}\|\partial_z v\|_{L^2}^2$$

where, we have used the divergence-free condition $\partial_z v_3 = -\partial_x v_1 - \partial_y v_2$. We see from this term that we need the following condition to absorb the first term by the viscosity in (10): $C\zeta\|w\|_{L^\infty}^2 \leq \eta$. On the other hand, the second term can be treated as follows

$$\left| \nu \int_\Omega \partial_z^2 B v \right| \leq \nu\|\partial_z v\|_{L^2}\|\partial_z B\|_{L^2}$$

$$\leq \frac{\nu}{4}\|\partial_z v\|_{L^2}^2 + \nu\|\partial_z B\|_{L^2}^2$$

$$\leq \frac{\nu}{4}\|\partial_z v\|_{L^2}^2 + \nu\|w\|_{L^\infty}^2 \frac{1}{\sqrt{\nu\zeta}} .$$

The second term of the left hand side must go to zero, this is the case if we have $\nu/\zeta \to 0$. Finally, we see that

$$\text{if } \frac{\nu}{\eta} \to 0 \text{ then } \zeta = \frac{\eta}{C\|w\|_{L^\infty}^2}$$

is a possible choice.

We conclude this section by recalling that the method used here is an energy method based on the regularity of the limiting system and the introduction of some suitable corrector. It turned out that some stability condition, namely the fact that $\frac{\nu}{\eta} \to 0$ was necessary for the convergence result to hold.

3   Study of Rotating Fluids at High Frequency

We consider the following system of equations

$$\partial_t u^n + \operatorname{div}(u^n \otimes u^n)$$

$$-\nu \partial_z^2 u^n - \eta \Delta_{x,y} u^n + \frac{e_3 \times u^n}{\varepsilon} = -\frac{\nabla p}{\varepsilon} \qquad \text{in } \Omega \qquad (11)$$

$$\operatorname{div} u^n = 0 \qquad \text{in } \Omega \qquad (12)$$

$$u^n(0) = u_0^n \qquad \text{with } \operatorname{div} u_0^n = 0 \quad (13)$$

$$u^n = 0 \qquad \text{on } \partial\Omega \qquad (14)$$

where $\Omega = \mathbb{T}^2 \times]0, h[$, ν and η are respectively the vertical and horizontal viscosities, whereas ε is the Rossby number. This system describes the motion of a rotating fluid as the Ekman and Rossby numbers go to zero (see Pedlovsky [31], and Greenspan [13]). It can model the ocean, the atmosphere, or a rotating fluid in a container. When there is no boundary ($\Omega = \mathbb{T}^3$ for instance) and when $\nu = \eta = 1$, the problem was studied by several authors ([14], [5], [1], [2], [9], [11], ...) by using the group method [32]. The method introduced in [32] fails when Ω has a boundary (excepted in very particular cases where there is no boundary layer, or where boundary layers can be eliminated by symmetry [4]). In domains with boundaries, only results in the "well-prepared" case (which means that there are no oscillations in time) were known ([6], [15], [28]). In our case, namely $\Omega = \mathbb{T}^2 \times]0, h[$, we have a domain with boundary and we want to study the oscillations in time and also show that they do not affect the averaged flow.

To study this system, we introduce the spaces V_{sym}^s consisting of functions from H^s with some extra conditions on the boundary (see [29]). We also set $Lu = -P(e_3 \times u)$, where P is the projection onto divergence-free vector fields such that the third component vanishes on the boundary and $\mathcal{L}(\tau) = e^{\tau L}$. Let us denote w the solution in $L^\infty(0, T^*, V_{\text{sym}}^s)$ of the following system

$$\begin{cases} \partial_t w + \overline{Q}(w, w) - \Delta_{x,y} w + \gamma \overline{S}(w) = -\nabla p & \text{in } \Omega, \\ \operatorname{div} w = 0 & \text{in } \Omega, \\ w.n = \pm w_3 = 0 & \text{on } \partial\Omega, \\ w(t = 0) = w^0 \end{cases} \qquad (15)$$

where $\overline{Q}(w,w)$, $\overline{S}(w)$ are respectively a bilinear and a linear operators of w, given by

$$\overline{Q}(w,w) = \sum_{\substack{l,m,k \\ k\in A(l,m) \\ \lambda(l)+\lambda(m)=\lambda(k)}} b(t,l)\, b(t,m)\, \alpha_{lmk} N^k(X)\,, \tag{16}$$

$$\overline{S}(w) = \sum_k \frac{1}{h}(D(k) + iI(k))b(t,k)N^k(X)$$

where the N^k are the eigenfunctions of L and $i\lambda(k)$ are the associated eigenvalues, α_{lmk} are constants which depends on (l,m,k) and $A(l,m) = \{l+m,\ Sl+m,\ l+Sm,\ Sl+Sm\}$, $(Sl = (l_1,l_2,-l_3))$ is the set of possible resonances. Finally,

$$D(k) = \sqrt{2}\left\{(1-\lambda(k)^2)^{\frac{1}{2}}\right\}, \quad I(k) = \sqrt{2}\left\{\lambda(k)(1-\lambda(k)^2)^{\frac{1}{2}}\right\}\,.$$

The bilinear term $\overline{Q}$ is due to the fact that only resonant modes in the advective term $w.\nabla w$ are present in the limit equation, and $\overline{S}(w)$ is a damping term that depends on the frequencies $\lambda(k)$, since $D(k) \geq 0$. It is due to the presence of a boundary layer which creates a second flow of order ε responsible of this damping (called damping of Ekman).

Theorem 2. *Let $s > 5/2$, and $w^0 \in V_{\text{sym}}^s(\Omega)^3$, div $w^0 = 0$. We assume that u_0^n converges in $L^2(\Omega)$ to w^0, $\eta = 1$ and ε, ν go to 0 such that $\sqrt{\frac{\nu}{\varepsilon}} \to \gamma$. Then any sequence of global weak solutions (à la Leray) u^n of (11)–(14) satisfying the energy inequality satisfies*

$$u^n - \mathcal{L}\left(\frac{t}{\varepsilon}\right)w \to 0 \quad \text{in } L^\infty(0,T^*,L^2(\Omega)),$$

$$\nabla_{x,y}\left(u^n - \mathcal{L}\left(\frac{t}{\varepsilon}\right)w\right),\ \sqrt{\nu}\,\partial_z u^n \to 0 \qquad \text{in } L^2(0,T^*,L^2(\Omega))$$

where w is the solution in $L^\infty(0,T^*,V_{\text{sym}}^s)$ of (15).

The above theorem gives a precise description of the oscillations in the sequence u^n. Next, we also show that the oscillations do not affect the averaged flow (also called the quasi-geostrophic flow). We see then that $\overline{w}$ (the weak limit of w) satisfies a 2-D Navier-Stokes equation with a damping term, namely

Theorem 3. *$\overline{w}$ satisfies the following system*

$$\begin{cases} \partial_t \overline{w} + \overline{w}.\nabla\overline{w} - \Delta_{x,y}\overline{w} + \sqrt{\frac{\nu}{\varepsilon}}\frac{\sqrt{2}}{h}\overline{w} = -\nabla p & \text{in } \mathbb{T}^2, \\ \text{div }\overline{w} = 0 & \text{in } \mathbb{T}^2, \\ \overline{w}(t=0) = S(w^0) = \overline{w}^0, \end{cases} \tag{17}$$

where S is the projection onto the slow modes, namely that do not depend on z, $\overline{w}(t,x,y) = S(w) = (1/h)\int_0^h w(t,x,y,z)\,dz$.

This theorem is proved by studying the operator Q and showing that if $k \in \mathcal{A}(l, m)$ with $k_3 = 0$, $l_3 m_3 \neq 0$, then $\alpha_{lmk} + \alpha_{mlk} = 0$.

In [29], we also deal with other boundary conditions, and construct Ekman layers near a non flat bottom

$$\Omega_\delta = \{(x, y, z) \text{ where } (x, y) \in \mathbb{T}^2, \text{ and } \delta f(x, y) < z < h\},$$

with the following boundary conditions

$$u(x, y, \delta f(x, y)) = 0 . \tag{18}$$

We also treat the case of a free surface,

$$u_3^n(z = h) = 0, \qquad \partial_z \begin{pmatrix} u_1^n \\ u_2^n \end{pmatrix}\bigg|_{z=h} = \frac{1}{\beta} \sigma \left(\frac{t}{\varepsilon}, t, x, y \right) \tag{19}$$

where σ describes the wind (see [31]). Next, we have the following theorem

Theorem 4. *Let u^n be global weak solutions of (11)–(13), (18), (19). If $(\varepsilon, \nu, \beta, \delta) \to (0, 0, 0, 0)$, then*

$$u^{\eta, \nu} - \mathcal{L}\left(\frac{t}{\varepsilon} \right) w \to 0 \qquad in \; L^\infty(0, T^*; L^2(\Omega)),$$

$$\nabla_{x,y} \left(u^{\eta, \nu} - \mathcal{L}\left(\frac{t}{\varepsilon} \right) w \right), \; \sqrt{\nu} \, \partial_z u^{\eta, \nu} \to 0 \qquad in \; L^2(0, T^*; L^2(\Omega))$$

where w is the solution of the following system ($\sqrt{\frac{\nu}{\varepsilon}}, \frac{\nu}{\beta}, \frac{\delta}{\varepsilon}$ stand for the limit of these quantities when n goes to infinity)

$$\begin{cases} \partial_t w + \overline{Q}(w, w) - \Delta_{x,y} w + \frac{1}{2}\sqrt{\frac{\nu}{\varepsilon}}\overline{S}(w) + \frac{\nu}{\beta}\overline{S}_1(\sigma) + \frac{\delta}{\varepsilon}\overline{S}_2(f, w) = -\nabla p + \overline{F} \\ \qquad\qquad\qquad \mathrm{div}\, w = 0 \qquad in \; \Omega, \\ \qquad\qquad\qquad w.n = \pm w_3 = 0 \qquad on \; \partial\Omega, \\ \qquad\qquad\qquad w(t = 0) = w^0 \end{cases}$$
$$\tag{20}$$

where $\overline{S}_1(\sigma)$, and $\overline{S}_2(f, w)$ are source terms that are due respectively to the wind, and to the non-flat bottom.

We conclude this section by noticing that the proofs of the above theorems are based (as in the previous section) on energy estimates and use more complicated correctors due to the presence of oscillations in time as well as the presence of different types of boundary layers. For more details about the proof, we refer to the original paper [29].

4   Compressible-Incompressible Limit (Mach $\to$ 0)

We now concentrate on the third example which is taken from joint works with P.-L. Lions. For this example, we will use weak convergence techniques to show our convergence results. Indeed, in this case we will not impose any regularity on the limiting system. We begin by some physical motivations.

As is well-known from a Fluid Mechanics viewpoint, one can derive formally incompressible models such as the Navier-Stokes equations or the Euler equations from compressible ones namely compressible Navier-Stokes equations (CNS) when the Mach number goes to 0. In these section, we will give a complete justification (from the mathematical point of view) of these formal derivations and, more precisely, pass to the limit in the global weak solutions of compressible Navier-Stokes equations (which were recently proven to exist by P.-L. Lions [23]) when the Mach number ε goes to 0. In the limit, we shall recover in particular the global weak solutions obtained in 1933 by J. Leray [19], [21], [20] as well as the local (or global in dimension 2) strong solutions of the Euler system (we refer to the book [22] of P.-L.Lions for an extensive bibliography concerning incompressible models). The next section will be devoted to another compressible-incompressible limit, namely the case where γ goes to ∞ and we will recover other type of incompressible models at the limit.

We first wish to recall a general setup for such asymptotic problems, which is a straightforward adaptation of the one introduced by S. Klainerman and A. Majda [17], [18] in the inviscid case (Euler equations) and for local strong solutions. We recall the compressible Navier-Stokes equations in the so-called isentropic regime (even though our results are trivially adapted to the case of general barotropic flows i.e. when the pressure is a function of the density only). The unknowns $(\tilde{\rho}, v)$ are respectively the density and the velocity of the fluid (gas) and solve on $\mathbb{R}^N \times (0, \infty)$

$$\frac{\partial \tilde{\rho}}{\partial t} + \operatorname{div}(\tilde{\rho} v) = 0 , \qquad \tilde{\rho} \geq 0, \tag{21}$$

$$\frac{\partial \tilde{\rho} v}{\partial t} + \operatorname{div}(\tilde{\rho} v \otimes v) - \tilde{\mu} \Delta v - \tilde{\xi} \nabla \operatorname{div} v + \nabla \tilde{p} = 0 , \tag{22}$$

and

$$\tilde{p} = a \tilde{\rho}^\gamma \tag{23}$$

where $N \geq 2$, $\tilde{\mu} > 0$, $\tilde{\mu} + \tilde{\xi} > 0$, $a > 0$ and $\gamma > 1$ are given. At this stage, we want to avoid confusing the issue and ignore all questions about boundary conditions and force terms.

From a physics view-point, the fluid should behave (asymptotically) like an incompressible one when the density is almost constant, the velocity is small and we look at large time scales. More precisely, we scale ρ and v (and thus p) in the following way

$$\tilde{\rho} = \rho(x, \varepsilon t), \quad v = \varepsilon u(x, \varepsilon t) \tag{24}$$

and we assume that the viscosity coefficients μ, ξ are also small and scale like

$$\tilde{\mu} = \varepsilon \mu_\varepsilon, \quad \tilde{\xi} = \varepsilon \xi_\varepsilon \tag{25}$$

where $\varepsilon \in (0,1)$ is a "small parameter " and the normalized coefficient $\mu_\varepsilon, \xi_\varepsilon$ satisfy

$$\mu_\varepsilon \to \mu, \quad \mu_\varepsilon \to \xi \quad \text{as } \varepsilon \text{ goes to } 0_+ . \tag{26}$$

We shall always assume that we have either $\mu > 0$ and $\mu + \xi > 0$ or $\mu = 0$.

With the preceding scalings, the system (21)–(23) yields

$$\begin{cases} \dfrac{\partial \rho}{\partial t} + \operatorname{div}(\rho u) = 0 , \quad \rho \geq 0, \\[4mm] \dfrac{\partial \rho u}{\partial t} + \operatorname{div}(\rho u \otimes u) - \mu_\varepsilon \Delta u - \xi_\varepsilon \nabla \operatorname{div} u + \dfrac{a}{\varepsilon^2} \nabla \rho^\gamma = 0 . \end{cases} \tag{27}$$

We may now explain the heuristics which lead to incompressible models. First of all, the second equation (for the momentum ρu) indicates that ρ should be like $\bar{\rho} + O(\varepsilon^2)$ where $\bar{\rho}$ is a constant. Of course, $\bar{\rho} \geq 0$ and we always assume that $\bar{\rho} > 0$ (in order to avoid the trivial case $\bar{\rho} = 0$). Obviously, we need to assume this property holds initially (at $t = 0$). And, let us also remark that by a simple (multiplicative) scaling, we may always assume without loss of generality that $\bar{\rho} = 1$.

Since ρ goes to 1, we expect that the first equation in (27) yields at the limit: $\operatorname{div} u = 0$. And writing $\nabla \rho^\gamma = \nabla(\rho^\gamma - 1)$, we deduce from the second equation in (27) that we have in the case when $\mu > 0$

$$\frac{\partial u}{\partial t} + \operatorname{div}(u \otimes u) - \mu \Delta u + \nabla \pi = 0 \tag{28}$$

or when $\mu = 0$

$$\frac{\partial u}{\partial t} + \operatorname{div}(u \otimes u) + \nabla \pi = 0 \tag{29}$$

where π is the "limit" of $\frac{\rho^\gamma - 1}{\varepsilon^2}$. In other words, we recover the incompressible Navier-Stokes equations (28) or the incompressible Euler equations (29), and the hydrostatic pressure appears as the limit of the "renormalized" thermodynamical pressure $(\frac{\rho^\gamma - 1}{\varepsilon^2})$. In fact, as we shall see later on, the derivation of (28) (or (29)) is basically correct even globally in time, for global weak solutions ; but the limiting process for the pressure is much more involved and may, depending on the initial conditions, incorporate additional terms coming from the oscillations in $\operatorname{div}(\rho_\varepsilon u_\varepsilon \otimes u_\varepsilon)$.

In the first subsection, we state the convergence result in the periodic case. In the second part, we proof some uniform estimates and give a formal (but wrong) proof. In the third subsection, we present a first rigorous method of proof using the group method introduced by Schochet [32] (see also Grenier [14]). This method will be called a global method since it is based on some type of Fourier decomposition and depends highly on the boundary conditions. In the forth subsection, we present another method based on some special properties of the wave equations. Indeed, as will be seen later on one of the major difficulties in the passage to the limit is the presence of acoustic waves. This latter method has the advantage of being a local one and holds for any type of boundary conditions. In the fifth subsection, the case of Dirichlet boundary conditions will be studied and more precise results will be stated. In the sixth subsection, we shall consider the passage to the limit towards solutions of (incompressible) Euler equations using an energy method.

4.1   Statement of the Result in the Periodic Case

We begin by introducing some notation and formulating precisely the notions of weak solutions we shall use. We assume that $\mu > 0$ and we consider only the periodic case i.e. all the functions we consider are required to be periodic in each variable x_i of period T_i (for $1 \leq i \leq N$). In fact, in order to simplify the notations, we assume that $T_i = 2\pi$ for $1 \leq i \leq N$ and it will be clear from the proofs below that any choice of the periods would be treated exactly in the same way. In particular, all functional spaces appearing below are composed of periodic functions (of period 2π in each variable) and all integrals are over $(0, 2\pi)^N$ unless explicitly mentioned.

We then consider a solution $(\rho_\varepsilon, u_\varepsilon)$ of the compressible Navier-Stokes equations (27) and we assume that $\rho_\varepsilon \in L^\infty(0, \infty; L^\gamma) \cap C([0, \infty), L^p)$ for all $1 \leq p < \gamma$, $u_\varepsilon \in L^2(0, T; H^1)$ for all $T \in (0, \infty)$, $\rho_\varepsilon |u_\varepsilon|^2 \in L^\infty(0, \infty; L^1)$ and $\rho_\varepsilon u_\varepsilon \in C([0, \infty) ; L_w^{2\gamma/(\gamma+1)})$ i.e. is continuous with respect to $t \geq 0$ with values in $L^{2\gamma/(\gamma+1)}$ endowed with the weak topology. We require (27) to hold in the sense of distributions. Finally, we prescribe initial conditions

$$\rho_\varepsilon\Big|_{t=0} = \rho_\varepsilon^0 \ , \rho_\varepsilon u_\varepsilon\Big|_{t=0} = m_\varepsilon^0 \tag{30}$$

where $\rho_\varepsilon^0 \geq 0$, $\rho_\varepsilon^0 \in L^\gamma$, $m_\varepsilon^0 \in L^{2\gamma/(\gamma+1)}$, $m_\varepsilon^0 = 0$ a.e. on $\{\rho_\varepsilon^0 = 0\}$ and $\rho_\varepsilon^0 |u_\varepsilon^0|^2 \in L^1$, denoting by $u_\varepsilon^0 = \frac{m_\varepsilon^0}{\rho_\varepsilon^0}$ on $\{\rho_\varepsilon^0 > 0\}$, $u_\varepsilon^0 = 0$ on $\{\rho_\varepsilon^0 = 0\}$. Furthermore, we assume that $\sqrt{\rho_\varepsilon^0}\, u_\varepsilon^0$ converges weakly in L^2 to some $\widetilde{u^0}$ and that we have

$$\int \rho_\varepsilon^0 |u_\varepsilon^0|^2 + \frac{1}{\varepsilon^2} \int (\rho_\varepsilon^0)^\gamma - \gamma \rho_\varepsilon^0 (\overline{\rho_\varepsilon})^{\gamma-1} + (\gamma-1)(\overline{\rho_\varepsilon})^\gamma \leq C , \tag{31}$$

with

$$\overline{\rho_\varepsilon} = (2\pi)^{-N} \int \rho_\varepsilon^0 \underset{\varepsilon}{\rightarrow} 1 ,$$

where, here and below, C denotes various positive constants independent of ε. Let us notice that (31) implies in particular that, roughly speaking, ρ_ε^0 is of order $\overline{\rho_\varepsilon} + 0(\varepsilon^2)$: indeed, we just need to rewrite

$$\left(\rho_\varepsilon^0\right)^\gamma - \gamma\rho_\varepsilon^0\left(\overline{\rho_\varepsilon}\right)^{\gamma-1} + (\gamma-1)\left(\overline{\rho_\varepsilon}\right)^\gamma = \left(\rho_\varepsilon^0\right)^\gamma - \left(\overline{\rho_\varepsilon}\right)^\gamma - \gamma(\rho_\varepsilon^0 - \overline{\rho_\varepsilon})\left(\overline{\rho_\varepsilon}\right)^{\gamma-1}$$

and recall that $(t \mapsto t^\gamma)$ is convex on $[0, \infty)$ since $\gamma \geq 1$.

Our last requirement on $(\rho_\varepsilon, u_\varepsilon)$ concerns the total energy: we assume that we have

$$E_\varepsilon(t) + \int_0^t D_\varepsilon(s)\,ds \leq E_\varepsilon^0 \quad \text{a.a. } t, \quad \frac{dE_\varepsilon}{dt} + D_\varepsilon \leq 0 \quad \text{in } \mathcal{D}'(0, \infty) \qquad (32)$$

where $E_\varepsilon(t) = \displaystyle\int \frac{1}{2}\rho_\varepsilon|u_\varepsilon|^2(t) + \frac{a}{\varepsilon^2(\gamma-1)}(\rho_\varepsilon)^\gamma(t)$, $D_\varepsilon(t) = \displaystyle\int \mu_\varepsilon|Du_\varepsilon|^2(t) +$

$\xi_\varepsilon\,(\text{div }u_\varepsilon)^2(t)$ and $E_\varepsilon^0 = \displaystyle\int \frac{1}{2}\rho_\varepsilon^0|u_\varepsilon^0|^2 + \frac{a}{\varepsilon^2(\gamma-1)}(\rho_\varepsilon^0)^\gamma$.

The inequality in (32) can be deduced (at least) formally by multiplying the momentum equation by u and by using the continuity equation. The fact that we have an inequality rather than an equality comes from the construction of weak solutions and the possible loss of compactness while passing to the limit in the approximating solutions. We also emphasize the fact that we *assume* the existence of a solution with the above properties, and we shall also assume that $\gamma > \frac{N}{2}$. And we recall the results in [23] which yield the *existence* of such a solution precisely when $\gamma > \frac{N}{2}$ and $N \geq 4$, $\gamma \geq \frac{9}{5}$ and $N = 3$, $\gamma \geq \frac{3}{2}$ and $N = 2$. However, the proof of the convergence towards the incompressible Navier-Stokes system only requires that $\gamma > \frac{N}{2}$ in all cases.

Next, we explain the notion of weak solutions of the incompressible Navier-Stokes equations (28) with the incompressibility conditions i.e.

$$\frac{\partial u}{\partial t} + \text{div}\,(u \otimes u) - \mu\Delta u + \nabla\pi = 0 \,, \quad \text{div }u = 0 \,. \qquad (33)$$

Given an initial condition $u_0 \in L^2$ such that $\text{div }u_0 = 0$, u is a solution of (33) satisfying

$$u\Big|_{t=0} = u^0 \qquad (34)$$

if $u \in C([0, \infty); L_w^2) \cap L^2(0, T; H^1)(\forall T \in (0, \infty))$ and if we have for all $\varphi \in C^\infty$ with $\text{div }\varphi = 0$ and for all $\psi \in C_0^\infty([0, \infty))$

$$\left(\int u^0.\varphi\right)\psi(0) + \int dt\psi'(t)\int u.\varphi + \int dt\psi(t)\int u_iu_j\partial_i\varphi_j$$
$$- \mu\int dt\psi(t)\int Du.D\varphi = 0 \,.$$

This is in fact equivalent to request that (33) holds in the sense of (periodic) distributions for some distribution π. As is well-known, there exists such a global solution of (33)–(34) and we have in addition: if $N = 2$,

$u \in C([0, \infty); L^2)$ is unique, $\frac{\partial u}{\partial t} \in L^2(0, T; H^{-1})$, $\pi \in L^2(0, T; L^2)$ for all $T \in (0, \infty)$, while if $N \geq 3$, $u \in C([0, \infty); L^p)$ for all $1 \leq p < 2$, $\frac{\partial u}{\partial t}$ and $\nabla \pi \in L^2(0, T; H^{-1}) + (L^s(0, T; W^{-1, Ns/(Ns-2)}) \cap L^q(0, T; L^r))$ for $1 \leq s < \infty$, $1 \leq q < 2$ and $r = \frac{Nq}{Nq+q-2}$, $\pi \in L^2(0, T; L^2) + L^s(0, T; L^{\frac{Ns}{Ns-2}})$ for $1 \leq s < \infty$, for all $T \in (0, \infty)$. Furthermore, the following energy inequality holds

$$\frac{1}{2} \int |u(t)|^2 + \mu \int_0^t ds \int |Du|^2 \leq \frac{1}{2} \int |u^0|^2 \quad \text{for all } t \geq 0 . \tag{35}$$

When $N = 2$, this inequality is in fact an equality.

Finally, we denote by P the (orthogonal) projection onto divergence-free vector fields and Q the projection onto gradient vector fields i.e. $v = Pv + Qv$ for all $v \in (L^2)^N$ where $\operatorname{div}(Pv) = 0$, and $\operatorname{curl}(Qv) = 0$ and $\int Qv = 0$.

The first result we want to state is the following theorem.

Theorem 5. *In addition to the above notations and conditions, we assume that $\gamma > \frac{N}{2}$. Then, ρ_ε converges to 1 in $C([0, T]; L^\gamma)$ and u_ε is bounded in the norm of $L^2(0, T; H^1)$ for all $T \in (0, \infty)$. In addition, for any subsequence of u_ε (still denoted by u_ε) weakly converging in $L^2(0, T; H^1)$ $(\forall T \in (0, \infty))$ to some u, u is a solution of the incompressible Navier-Stokes equation (33) (as defined above) corresponding to the initial condition $u^0 = P\widetilde{u^0}$.*

We want to point out that the solution u of the incompressible Navier-Stokes system does not satisfy exactly the energy inequality (35) but only the following one

$$\frac{1}{2} \int |u(t)|^2 + \mu \int_0^t ds \int |Du|^2 \leq \frac{1}{2} C_0 \quad \text{for all } t \geq 0 , \tag{36}$$

where $C_0 = \liminf_\varepsilon E_\varepsilon^0$. Moreover, in view of the uniqueness (of u) recalled above, when $N = 2$, the whole sequence u_ε weakly converges to u in $L^2(0, T; H^1)$ for all $T \in (0, \infty)$ and one can recover the energy inequality (35) and even the equality in (35).

4.2   Uniform Bounds and Formal Proof

In this subsection, we derive some uniform a priori bounds and then deduce various straightforward convergences and finally sketch the (method of) proof of the main step namely the passage to the limit in the term $\operatorname{div}(\rho_\varepsilon u_\varepsilon \otimes u_\varepsilon)$.

We first deduce from (32) and from the conservation of mass (first equation of (27)) that we have for (almost) all $t \geq 0$

$$\int \frac{1}{2} \rho_\varepsilon |u_\varepsilon|^2 + \frac{a}{\varepsilon^2(\gamma - 1)} \left(\rho_\varepsilon^\gamma - \gamma \rho_\varepsilon (\overline{\rho_\varepsilon})^{\gamma-1} + (\gamma - 1)(\overline{\rho_\varepsilon})^\gamma \right)(t)$$

$$+ \int_0^t ds \int \mu_\varepsilon |Du_\varepsilon|^2 + \xi_\varepsilon |\text{div } u_\varepsilon|^2$$

$$\leq \int \frac{1}{2} \rho_\varepsilon^0 |u_\varepsilon^0|^2 + \frac{a}{\varepsilon^2(\gamma-1)} \left((\rho_0^\varepsilon)^\gamma - \gamma \rho_0^\varepsilon (\overline{\rho_\varepsilon})^{\gamma-1} + (\gamma-1)(\overline{\rho_\varepsilon})^\gamma \right) \leq C$$

in view of (31). We deduce from this inequality that $\rho_\varepsilon |u_\varepsilon|^2$ and $\frac{1}{\varepsilon^2} \left(\rho_\varepsilon^\gamma - \right.$
$\left. \gamma \rho_\varepsilon (\overline{\rho_\varepsilon})^{\gamma-1} + (\gamma-1)(\overline{\rho_\varepsilon})^\gamma \right)$ are bounded in $L^\infty(0, \infty; L^1)$ and that Du_ε is
bounded in $L^2(0, \infty; L^2)$. In particular, ρ_ε is bounded in $L^\infty(0, \infty; L^\gamma)$ and
we deduce as in [23] a bound on u_ε in $L^2(0, T; H^1)$ for all $T \in (0, \infty)$: we
briefly recall the argument for the sake of completeness. We deduce from
Sobolev and Poincaré's inequalities that we have for all $T \in (0, \infty)$

$$\int_0^T dt \int \rho_\varepsilon \left| u_\varepsilon - (2\pi)^{-N} \int u_\varepsilon \right|^2 \leq C \|\rho_\varepsilon\|_{L^\infty(0,T;L^\gamma)} \|Du_\varepsilon\|^2_{L^2(0,T;L^2)} \leq C$$

hence, in view of the above bound on $\rho_\varepsilon |u_\varepsilon|^2$,

$$C \geq \int_0^T dt \int \rho_\varepsilon \left| \left(\int u_\varepsilon \right) \right|^2 = \left(\int \rho_\varepsilon^0 \right) \int_0^T dt \left(\int u_\varepsilon \right)^2$$

Since (31) implies that ρ_ε^0 converges to 1 in measure and thus in L^1 (in fact
in L^γ as we shall show below), we deduce a bound on u_ε in $L^2(0, T; L^2)$
and our claim is shown. From now on, we assume, extracting a subsequence
(still denoted by u_ε) if necessary, that u_ε converges weakly to some u in
$L^2(0, T; H^1)$ for all $T \in (0, \infty)$.

Convergence of ρ_ε to 1

We next claim that ρ_ε converges to 1 in $C([0, \infty); L^\gamma)$: indeed, for ε small
enough $\bar{\rho}_\varepsilon \in \left(\frac{1}{2}, \frac{3}{2} \right)$ and thus for all $\delta > 0$, there exists some $\nu_\delta > 0$ such that

$$x^\gamma + (\gamma-1)(\bar{\rho}_\varepsilon)^\gamma - \gamma x (\bar{\rho}_\varepsilon)^{\gamma-1} \geq \nu_\delta |x - \bar{\rho}_\varepsilon|^\gamma \text{ if } |x - \bar{\rho}_\varepsilon| \geq \delta, \, x \geq 0 .$$

Hence,

$$\sup_{t \geq 0} \int |\rho_\varepsilon - 1|^\gamma \leq \delta^\gamma (2\pi)^N + \sup_{t \geq 0} \left[\int 1_{(|\rho_\varepsilon - 1| \geq \delta)} |\rho_\varepsilon - \bar{\rho}_\varepsilon|^\gamma \right] + C |\bar{\rho}_\varepsilon - 1|^\gamma$$

$$\leq (2\pi)^N \delta^\gamma + \frac{C\varepsilon^2}{\nu_\delta} + C |\bar{\rho}_\varepsilon - 1|^\gamma$$

and we conclude upon letting first ε go to 0 and then δ go to 0. In the next
subsection, we will need more information about this convergence and more
precisely the following bounds valid for all $R \in (1, \infty)$ will be used to control

the oscillating terms

$$\begin{cases} \|\varphi_\varepsilon\|_{L^\infty(0,\infty;L^2)} \leq C & \text{if } \gamma \geq 2 \\[2mm] \|\varphi_\varepsilon 1_{(\rho_\varepsilon < R)}\|_{L^\infty(0,\infty;L^2)} \leq C & \text{if } \gamma < 2 \\[2mm] \|\varphi_\varepsilon 1_{(\rho_\varepsilon \geq R)}\|_{L^\infty(0,\infty;L^\gamma)} \leq C\,\varepsilon^{(\frac{2}{\gamma}-1)} & \text{if } \gamma < 2 \end{cases} \tag{37}$$

where we denote by $\varphi_\varepsilon = \frac{1}{\varepsilon}(\rho_\varepsilon - \overline{\rho_\varepsilon})$. These bounds are deduced immediately (choosing $x = \frac{\rho_\varepsilon}{\overline{\rho_\varepsilon}}$) from the following straightforward inequalities: we have, for some $\nu > 0$ and for all $x \geq 0$

$$x^\gamma + \gamma - 1 - \gamma(x-1) \geq \nu|x-1|^2 \quad \text{if } \gamma \geq 2 \,,$$
$$x^\gamma + \gamma - 1 - \gamma(x-1) \geq \nu|x-1|^2 \quad \text{if } x \leq R \quad \text{and } \gamma < 2 \,,$$
$$x^\gamma + \gamma - 1 - \gamma(x-1) \geq \nu|x-1|^\gamma \quad \text{if } x \geq R \quad \text{and } \gamma < 2 \,.$$

We want also to point out that, in the case we take $\overline{\rho_\varepsilon} = 1$ (or in the case $\overline{\rho_\varepsilon} - 1$ converges to 0 rapidly enough with respect to ε), we can deduce a rate of convergence of $\rho_\varepsilon - 1$, namely

$$\begin{cases} \|\rho_\varepsilon - 1\|_{L^\infty(0,\infty;L^p)} \leq C\varepsilon^{\frac{2}{p}} & \text{if } 2 \leq p \leq \gamma \\[2mm] \|\rho_\varepsilon - 1\|_{L^\infty(0,\infty;L^p)} \leq C\varepsilon & \text{if } 1 \leq p \leq \min(2,\gamma) \,. \end{cases} \tag{38}$$

Convergence of u_ε

We next deduce from the previous bounds that $\operatorname{div} u_\varepsilon$ converges weakly to 0 in $L^2(0,T;L^2)$ and that Pu_ε converges to $u(= Pu)$ *strongly* in $L^2(0,T;L^2)$ and thus by Sobolev embeddings in $L^2(0,T;L^q)$ for all $T \in (0,\infty)$ and for all $2 \leq q < \frac{2N}{N-2}$. Of course, these facts also imply that Qu_ε converges weakly to 0 in $L^2(0,T;H^1)$ for all $T \in (0,\infty)$. Indeed, since ρ_ε converges to 1 in $C([0,\infty);L^\gamma)$ and $\gamma > \frac{N}{2}$ ($\frac{2N}{N+2}$ would in fact suffice...), we deduce from (27) the first fact. The second fact is proven observing first that we have projecting (27) onto divergence-free vector-fields:

$$\frac{\partial}{\partial t}P(\rho_\varepsilon u_\varepsilon) + P\Big[\operatorname{div}(\rho_\varepsilon u_\varepsilon \otimes u_\varepsilon)\Big] - \mu_\varepsilon \Delta P u_\varepsilon = 0 \,. \tag{39}$$

Using the fact that P is a bounded linear mapping in all Sobolev space $W^{s,p}$ (for all $s \in \mathbb{R}$, $1 < p < \infty$) and the preceding bounds, (39) yields a bound on $\frac{\partial}{\partial t}P(\rho_\varepsilon u_\varepsilon)$ in $L^2(0,T;H^{-1}) + L^\infty(0,\infty;W^{-1-\delta,1})$ for all $\delta > 0$ since $P(L^1) \subset W^{-\delta,1}$ for all $\delta > 0$. In addition, $P(\rho_\varepsilon u_\varepsilon)$ is bounded in $L^\infty\Big((0,\infty);L^{\frac{2\gamma}{\gamma+1}}\Big) \cap L^2\Big(0,T;L^r\Big)$ with $\frac{1}{r} = \frac{1}{\gamma} + \frac{N-2}{2N}$ if $N \geq 3$, $1 \leq r < \gamma$ if $N = 2$. Finally, we recall that Pu_ε is bounded in $L^2(0,T;H^1)$. All these bounds allow to apply the following Lemma taken from P.-L. Lions [23] (Chapter 5, Section 5.2).

Lemma 6. *Let* g^n, h^n *converge weakly to* g, h *respectively in* $L^{p_1}(0,t;L^{p_2})$, $L^{q_1}(0,t;L^{q_2})$, *where* $1 \le p_1$, $p_2 \le +\infty$, $\frac{1}{p_1} + \frac{1}{q_1} = \frac{1}{p_2} + \frac{1}{q_2} = 1$. *We assume in addition that*

$$\begin{cases} \dfrac{\partial g^n}{\partial t} & is\ bounded\ in \quad L^1(0,T;W^{-m,1}(\Omega)) \\[2mm] & for\ some \quad m \ge 0 \quad independent\ of \quad n, \end{cases} \tag{40}$$

$$\|h^n\|_{L^1(0,T;H^s)} \quad is\ bounded\ for\ some\ s > 0 . \tag{41}$$

Then $g^n h^n$ *converges to* gh *in* $\mathcal{D}'$.

With $p_1 = q_1 = 2$, $q_2 < \frac{2N}{N-2}$, $q_1 = r$ and $\frac{1}{q_1} + \frac{1}{q_2} = 1$, such a choice is possible since $\gamma > \frac{N}{2}$ and $\frac{1}{r} + \frac{N-2}{2N} = \frac{1}{\gamma} - \frac{2}{N} + 1 < 1$. We thus deduce that $P(\rho_\varepsilon u_\varepsilon)$. Pu_ε converges in the sense of distributions to $|u|^2$. We then conclude easily that Pu_ε converges in $L^2(0,T;L^2)(\forall T \in (0,\infty))$ to u upon remarking that we have

$$\left| \int_0^T dt \int |Pu_\varepsilon|^2 - P(\rho_\varepsilon u_\varepsilon).Pu_\varepsilon \right| \le C \|\rho_\varepsilon - 1\|_{C([0,\infty);L^\gamma)} \|u_\varepsilon\|^2_{L^2(0,T;L^s)}$$

where $s = \frac{\gamma}{2(\gamma-1)} < \frac{2N}{N-2}$ since $\gamma > \frac{N}{2}$.

To end this subsection, we describe a formal proof of the passage to the limit, next the main difficulty and finally the strategy of proof used in order to circumvent that difficulty.

It is not difficult to check that the main (and only) difficulty with the passage to the limit lies with the term $\operatorname{div}(\rho_\varepsilon u_\varepsilon \otimes u_\varepsilon)$ and more precisely with the term $\operatorname{div}\left(Q(\rho_\varepsilon u_\varepsilon) \otimes Qu_\varepsilon\right)$ since Pu_ε converges strongly to u in $L^2_{x,t}$. Formally, this term should not create an obstruction since we can rewrite the term $\left(\frac{\partial}{\partial t}(\rho_\varepsilon u_\varepsilon) + \operatorname{div}(\rho_\varepsilon u_\varepsilon \otimes u_\varepsilon)\right)$ as $\left(\rho_\varepsilon \frac{\partial u_\varepsilon}{\partial t} + \rho_\varepsilon(u_\varepsilon.\nabla)u_\varepsilon\right)$, itself comparable to $\frac{\partial u_\varepsilon}{\partial t} + (u_\varepsilon.\nabla)u_\varepsilon$. Next, the "dangerous" term $\left[(Qu_\varepsilon).\nabla\right]Qu_\varepsilon$ can be incorporated in the pressure π at the limit since $Qu_\varepsilon = \nabla\psi_\varepsilon$ (for some $\psi_\varepsilon\ldots$) and then

$$\left[(Qu_\varepsilon).\nabla\right] Qu_\varepsilon = \frac{1}{2}\nabla |\nabla\psi_\varepsilon|^2 .$$

However, this formal argument does not seem to be easy to justify and we shall have to make a rather different rigorous proof in order to show that $\operatorname{div}\left(Q(\rho_\varepsilon u_\varepsilon) \otimes Qu_\varepsilon\right)$ converges in the sense of distributions (say) to a distribution which is a gradient and can thus be incorporated in the pressure π. This is of course the crucial step in our proof and we will present two different methods to handle this term, namely a global method and a local one.

4.3   Global Method

We try here to make the convergence argument for $\operatorname{div}\left(Q(\rho_\varepsilon u_\varepsilon) \otimes Qu_\varepsilon\right)$ rigorous, using a global method based on the introduction of some group (see for instance [32]). We project the second equation of (27) onto the space of "gradient vector fields" and we find

$$
\begin{cases}
\dfrac{\partial}{\partial t} Q(\rho_\varepsilon u_\varepsilon) + Q\Big[\mathrm{div}\,(\rho_\varepsilon u_\varepsilon \otimes u_\varepsilon)\Big] - (\mu_\varepsilon + \xi_\varepsilon)\nabla \mathrm{div}\, u_\varepsilon \\[2mm]
\qquad + \dfrac{a}{\varepsilon^2}\nabla\Big(\rho_\varepsilon^\gamma - \gamma\rho_\varepsilon(\bar\rho_\varepsilon)^{\gamma-1} + (\gamma-1)(\overline{\rho_\varepsilon})^\gamma\Big) \\[2mm]
\qquad + \dfrac{a\gamma(\bar\rho_\varepsilon)^{\gamma-1}}{\varepsilon^2}\nabla(\rho_\varepsilon - \bar\rho_\varepsilon) = 0\ .
\end{cases}
\tag{42}
$$

Then, we write the first equation of (27) together with (42) as

$$
\varepsilon\frac{\partial \varphi_\varepsilon}{\partial t} + \mathrm{div}\, Q(\rho_\varepsilon u_\varepsilon) = 0, \quad \varepsilon\frac{\partial}{\partial t} Q(\rho_\varepsilon u_\varepsilon) + b\nabla\varphi_\varepsilon = \varepsilon F_\varepsilon
\tag{43}
$$

where $b = a\gamma(\bar\rho_\varepsilon)^{\gamma-1}$, $F_\varepsilon = (\mu_\varepsilon + \xi_\varepsilon)\nabla\,\mathrm{div}\,u_\varepsilon - Q\big[\,\mathrm{div}\,(\rho_\varepsilon u_\varepsilon \otimes u_\varepsilon)\big] - a\nabla\big[\frac{1}{\varepsilon^2}(\rho_\varepsilon^\gamma - \gamma\rho_\varepsilon(\overline{\rho_\varepsilon})^{\gamma-1} + (\gamma-1)(\overline{\rho_\varepsilon})^\gamma\big]$.

Let us observe that b, in general, depends upon ε and $b_\varepsilon \to a\gamma$ as ε goes to 0_+. We shall ignore this dependence in all that follows (in order to simplify the notation) since it creates no difficulty.

We next introduce the following group $(\mathcal{L}(\tau), \tau \in \mathbb{R})$ defined by $e^{\tau L}$ where L is the operator defined on $\mathcal{D}'_0 \times (\mathcal{D}')^N$, where $\mathcal{D}'_0 = \big\{\varphi \in \mathcal{D}', \int \varphi = 0\big\}$, by

$$
L\begin{pmatrix}\varphi \\ v\end{pmatrix} = -\begin{pmatrix}\mathrm{div}\, v \\ b\nabla\varphi\end{pmatrix}.
\tag{44}
$$

We shall check that $e^{\tau L}$ is an isometry on each $H^s \times (H^s)^N$ for all $s \in \mathbb{R}$ and for all τ. It follows that $e^{\tau L}\begin{pmatrix}\varphi \\ v\end{pmatrix} = \begin{pmatrix}\varphi(\tau) \\ v(\tau)\end{pmatrix}$ solves

$$
\frac{\partial\varphi}{\partial\tau} = -\mathrm{div}\, v\ , \quad \frac{\partial v}{\partial\tau} = -b\nabla\varphi
$$

and thus $\frac{\partial^2\varphi}{\partial\tau^2} - b\Delta\varphi = 0$. In other words, the group $e^{\tau L}$ is nothing but a reformulation of the group generated by the wave equation (whence the isometry in H^s...). Then, using (43), we shall prove that $\begin{pmatrix}\varphi_\varepsilon \\ Q(\rho_\varepsilon u_\varepsilon)\end{pmatrix}$ is equivalent (modulo compact terms) to $\mathcal{L}\big(\frac{t}{\varepsilon}\big)\begin{pmatrix}\psi \\ m\end{pmatrix}$ for some $\psi, m \in L^2(0, T; L^2)$. This in turn allows to check that $\mathrm{div}\,(Q(\rho_\varepsilon u_\varepsilon) \otimes Qu_\varepsilon)$ behaves asymptotically like $\mathrm{div}\,\Big(\mathcal{L}_2\big(\frac{t}{\varepsilon}\big)\begin{pmatrix}\psi \\ m\end{pmatrix} \otimes \mathcal{L}_2\big(\frac{t}{\varepsilon}\big)\begin{pmatrix}\varphi \\ m\end{pmatrix}\Big)$ where we denote by $\mathcal{L}_2(\tau)\begin{pmatrix}\psi \\ m\end{pmatrix} = m(\tau)$ if $\mathcal{L}(\tau)\begin{pmatrix}\psi \\ m\end{pmatrix} = \begin{pmatrix}\varphi(t) \\ m(\tau)\end{pmatrix}$. Finally, we shall check by direct verification that this term converges in the sense of distributions to a gradient-like distribution, the precise fact that we expect from the formal considerations above and that we need to prove.

Next, we claim that $\mathcal{L}(\tau) = e^{\tau L}$ is an isometry from $H^s \times (H^s)^N$ into itself endowed with the norm $\|(\varphi, v)\| = \left(\|\varphi\|_{H^s}^2 + \frac{1}{b}\|v\|_{H^s}^2\right)^{1/2}$ for all $s \in \mathbb{R}$. Since L has constant coefficients, it suffices to check this fact for $s = 0$. Then, writing $e^{\tau L}\begin{pmatrix}\varphi\\v\end{pmatrix} = \begin{pmatrix}\varphi(\tau)\\v(\tau)\end{pmatrix}$, we have

$$\frac{\partial \varphi}{\partial \tau} + \operatorname{div} v = 0, \quad \frac{\partial v}{\partial \tau} + b\nabla\varphi = 0 \tag{45}$$

and multiplying the first equation by φ and the second one by v, we obtain

$$\frac{1}{2}\frac{d}{d\tau}\int \varphi^2 + \frac{1}{b}|v|^2 = -\int (\operatorname{div} v)\,\varphi + \nabla\varphi.v = 0 \ .$$

(In other words, L is anti-selfadjoint in the appropriate scalar product of $L^2 \times (L^2)^N \dots$).

In order to understand how the group $\mathcal{L}(\tau)$ acts, it is worth making a few remarks (that are not really needed in the arguments below...). First of all, denoting by $\mathcal{L}_1(\tau)$ and $\mathcal{L}_2(\tau)$ the components of $\mathcal{L}(\tau)$, we remark that

$$\int \mathcal{L}_1(\tau)\begin{pmatrix}\varphi\\v\end{pmatrix} \quad \text{and} \quad P\,\mathcal{L}_2(\tau)\begin{pmatrix}\varphi\\v\end{pmatrix} \quad \text{are independent of } \tau \in \mathbb{R} \ ,$$

and thus, in particular, $\mathcal{L}(\tau)\begin{pmatrix}\varphi\\v\end{pmatrix}$ is independent of $\tau \in \mathbb{R}$ if v is divergence free ($Pv = 0$) and φ is constant. Next, we deduce from (22) that we have

$$\frac{\partial^2 \varphi}{\partial \tau^2} - b\Delta\varphi = 0 \ , \quad \frac{\partial^2 \Phi}{\partial \tau^2} - b\Delta\Phi = 0 \tag{46}$$

where we denote by $v(\tau) = Pv + \nabla\Phi(\tau)$ (with $\int \Phi(\tau) = 0, \ \forall \tau \in \mathbb{R}$) .

We next claim that $\mathcal{L}\left(-\dfrac{t}{\varepsilon}\right)\begin{pmatrix}\varphi_\varepsilon\\Q(\rho_\varepsilon u_\varepsilon)\end{pmatrix}$ is relatively compact in the space $L^2(0, T; H^{-m})$ for some $m \in (0, 1)$, where T is from now on a fixed but arbitrary constant in $(0, \infty)$ (and we are going to prove Theorem 5 on $(0, T)\dots$). In order to do so, we are going to prove first that $\begin{pmatrix}\varphi_\varepsilon\\Q(\rho_\varepsilon u_\varepsilon)\end{pmatrix}$ (and thus $\mathcal{L}\left(-\dfrac{t}{\varepsilon}\right)\begin{pmatrix}\varphi_\varepsilon\\Q(\rho_\varepsilon u_\varepsilon)\end{pmatrix}$ since L is an isometry) is bounded in $L^2(0, T; H^{-s})$ for some fixed $s \in (0, 1)$ and then that $\dfrac{\partial}{\partial t}\left\{\mathcal{L}\left(-\dfrac{t}{\varepsilon}\right)\begin{pmatrix}\varphi_\varepsilon\\Q(\rho_\varepsilon u_\varepsilon)\end{pmatrix}\right\}$ is bounded in $L^2(0, T; H^{-r})$ for some $r > 0$ large enough. Our claim then follows by classical compactness theorems choosing m in $(s, 1)$.

We can see easily from (37) that φ_ε is bounded in $L^\infty(0,\infty;L^p)$ where $p = \min(2,\gamma)$. Since $p > \frac{2N}{N+2}$, Sobolev embeddings imply the desired bound in $L^2(0,T;H^{-m})$ for some $m \in [0,1)$. Next, we also deduce from above that $\rho_\varepsilon u_\varepsilon$ and thus $Q(\rho_\varepsilon u_\varepsilon)$ is bounded in $L^2(0,T;L^q)$ where $\frac{1}{q} = \frac{1}{\gamma} + \frac{N-2}{2N}$ if $N \geq 3$, $1 < q < \gamma$ if $N = 2$. Once more, $q > \frac{2N}{N+2}$ since $\gamma > \frac{N}{2}$ ($\frac{1}{q} < \frac{2}{N} + \frac{N-2}{2N} = \frac{N+2}{2N}$) and the desired bounds on $\mathcal{L}\left(-\frac{t}{\varepsilon}\right)\begin{pmatrix}\varphi_\varepsilon \\ Q(\varphi_\varepsilon u_\varepsilon)\end{pmatrix}$ are established.

Next, we set $\psi_\varepsilon(t) = \mathcal{L}_1(-\frac{t}{\varepsilon})\begin{pmatrix}\varphi_\varepsilon \\ Q(\rho_\varepsilon u_\varepsilon)\end{pmatrix}$, $m_\varepsilon(t) = \mathcal{L}_2(-\frac{t}{\varepsilon})\begin{pmatrix}\varphi_\varepsilon \\ Q(\varphi_\varepsilon u_\varepsilon)\end{pmatrix}$ and we use (43) to deduce

$$\frac{\partial}{\partial t}\begin{pmatrix}\psi_\varepsilon \\ m_\varepsilon\end{pmatrix} = \mathcal{L}\left(-\frac{t}{\varepsilon}\right)\left\{\frac{\partial}{\partial t}\begin{pmatrix}\varphi_\varepsilon \\ Q(\rho_\varepsilon u_\varepsilon)\end{pmatrix} + \frac{1}{\varepsilon}\begin{pmatrix}\mathrm{div}\, Q(\rho_\varepsilon u_\varepsilon) \\ b\nabla\varphi_\varepsilon\end{pmatrix}\right\}$$

$$= \mathcal{L}\left(-\frac{t}{\varepsilon}\right)\begin{pmatrix}0 \\ F_\varepsilon\end{pmatrix}.$$

Let us recall that $F_\varepsilon = (\mu_\varepsilon + \xi_\varepsilon)\nabla\mathrm{div}\,u_\varepsilon - Q\left[\mathrm{div}\,(\rho_\varepsilon u_\varepsilon \otimes u_\varepsilon)\right] - a\nabla\left[\frac{1}{\varepsilon^2}(\rho_\varepsilon^\gamma - \gamma\rho_\varepsilon(\overline{\rho_\varepsilon})^{\gamma-1} + (\gamma-1)(\overline{\rho_\varepsilon})^\gamma)\right]$ and thus is bounded, in view of the bounds derived above, in $L^2(0,T;H^{-1}) + L^\infty(0,\infty;W^{-1-\delta,1})$ for all $\delta > 0$ since $Q(L^1) \subset W^{-\delta,1}$ for all $\delta > 0$. Hence, F_ε is bounded in $L^2(0,T;H^{-r})$ for all $r > \frac{N}{2} + 1$ and thus $\frac{\partial}{\partial t}\begin{pmatrix}\psi_\varepsilon \\ m_\varepsilon\end{pmatrix}$ is bounded in $L^2(0,T;H^{-r})$.

We deduce from the compactness of $(\psi_\varepsilon, m_\varepsilon)$ that we may assume without loss of generality (extracting a subsequence if necessary) that $(\psi_\varepsilon, m_\varepsilon)$ converges in $L^2(0,T;H^{-m})$ to some (ψ, m). Since $Pm_\varepsilon \equiv 0$, we also have $Pm \equiv 0$. Similarly, $\int \psi = 0$ a.e. $t \in (0,T)$. In conclusion, we have shown that we have on $(0,T) \times \mathbb{R}^N$

$$\begin{pmatrix}\varphi_\varepsilon \\ Q(\rho_\varepsilon u_\varepsilon)\end{pmatrix} = \mathcal{L}\left(\frac{t}{\varepsilon}\right)\begin{pmatrix}\psi \\ m\end{pmatrix} + r_\varepsilon, \quad r_\varepsilon \to 0 \text{ in } L^2(0,T;H^{-m}) \text{ as } \varepsilon \to 0_+,$$

(47)

using once more the fact that $\mathcal{L}$ is an isometry in H^s for all $s \in \mathbb{R}$.

Next, we want to show that $\psi, m \in L^2(0,T;L^2)$, which will allow use to use them in nonlinear terms. In view of (37) and of the fact that ρ_ε converges to 1 in $C([0,T];L^\gamma)$, we deduce that

$$\left\|\varphi_\varepsilon - \varphi_\varepsilon \mathbf{1}_{(\rho_\varepsilon \leq R)}\right\|_{L^\infty(0,T;L^p)} \xrightarrow[\varepsilon]{} 0, \quad \left\|\rho_\varepsilon u_\varepsilon - u_\varepsilon\right\|_{L^2(0,T;L^q)} \xrightarrow[\varepsilon]{} 0$$

where $R = +\infty$ if $\gamma \geq 2$, R is fixed in $(1,+\infty)$ if $\gamma < 2$, $p = 2$ if $\gamma \geq 2$, $p = \gamma$ if $\gamma < 2$, $\frac{1}{q} = \frac{1}{\gamma} + \frac{N-2}{2N}$ if $N \geq 3$, $1 < q < \gamma$ if $N = 2$. This implies in particular that $\mathcal{L}\left(-\frac{t}{\varepsilon}\right)\begin{pmatrix}\varphi_\varepsilon\mathbf{1}_{(\rho_\varepsilon \leq R)} \\ Q(u_\varepsilon)\end{pmatrix}$ converges to $\begin{pmatrix}\psi \\ m\end{pmatrix}$ in $L^2(0,T;H^{-s})$ for some $s > 0$ large enough. Next, we observe, using (37), that $\varphi_\varepsilon 1_{(\rho_\varepsilon \leq R)}$ and $Q(u_\varepsilon)$ are bounded in $L^2(0, T; L^2)$. Since L is an isometry, we deduce that ψ and m belong to $L^2(0, T; L^2)$.

Next, we show that

$$\text{div}\,(\rho_\varepsilon u_\varepsilon \otimes u_\varepsilon) - \text{div}(v_\varepsilon \otimes v_\varepsilon) \underset{\varepsilon}{\rightharpoonup} \text{div}\,(u \otimes u) \tag{48}$$

in the sense of distributions (for instance), where $v_\varepsilon = \mathcal{L}_2\left(\dfrac{t}{\varepsilon}\right)\begin{pmatrix} \psi \\ m \end{pmatrix}$. In order to prove this claim, we recall that Pu_ε converges strongly to u in $L^2(0, T; H^r)$ for all $0 \leq r < 1$ while Pu_ε and Qu_ε converge weakly respectively to u and 0 in $L^2(0, T; H^1)$. Hence,

$$\text{div}\,(\rho_\varepsilon u_\varepsilon \otimes Pu_\varepsilon) \underset{\varepsilon}{\rightharpoonup} \text{div}\,(u \otimes u) \quad \text{in } \mathcal{D}'$$

since $\rho_\varepsilon u_\varepsilon$ converges weakly to u in $L^2(0, T; L^p)$ where $\frac{1}{p} = \frac{1}{\gamma} + \frac{N-2}{2N}$ if $N \geq 3$, $1 < p < \gamma$ if $N = 2$, and $\gamma > \frac{N}{2}$. Next, we observe that

$$\text{div}\,(P(\rho_\varepsilon u_\varepsilon) \otimes Qu_\varepsilon) \underset{\varepsilon}{\rightharpoonup} 0 \quad \text{in } \mathcal{D}' \,,$$

since $P(\rho_\varepsilon u_\varepsilon)$ converges to u in $L^2(0, T; L^p)$ where p is arbitrary in $(1, \bar{p})$ and $\frac{1}{\bar{p}} = \frac{1}{\gamma} + \frac{N-2}{2N}$ (recall that ρ_ε converges to 1 in $C([0, T]; L^\gamma)$).

The next reduction consists of showing that we have

$$\text{div}\,\left((Q(\rho_\varepsilon u_\varepsilon) - v_\varepsilon) \otimes Qu_\varepsilon\right) \underset{\varepsilon}{\rightharpoonup} 0 \quad \text{in } \mathcal{D}' \,.$$

This is immediate in view of (47): observe indeed that r_ε converges strongly to 0 in $L^2(0, T; H^{-1})$ while Qu_ε is bounded in $L^2(0, T; H^1)$.

Finally, (48) is shown provided we check that

$$\text{div}\,\left(v_\varepsilon \otimes (Qu_\varepsilon - v_\varepsilon)\right) \underset{\varepsilon}{\rightharpoonup} 0 \quad \text{in } \mathcal{D}' \,.$$

One possible proof consists in observing that we have

$$\left\|(v_\varepsilon - v_\varepsilon^\delta) \otimes (Qu_\varepsilon - v_\varepsilon)\right\|_{L^1(0,T;L^1)} \leq \delta \left\|Qu_\varepsilon - v_\varepsilon\right\|_{L^2(0,T;L^2)} \leq C\delta \,,$$

where we used twice the fact that $\mathcal{L}$ is an isometry and where we denote by $v_\varepsilon^\delta = \mathcal{L}_2(\frac{t}{\varepsilon})\begin{pmatrix} \psi_\delta \\ m_\delta \end{pmatrix}$ with $\psi_\delta, m_\delta \in C^\infty_{x,t}$ are such that $\left\|\psi_\delta - \psi\right\|_{L^2(0,T;L^2)} + \left\|m_\delta - m\right\|_{L^2(0,T;L^2)} \leq \delta$. Next, we remark that v_ε^δ is bounded is $L^2(0, T; H^s)$ for all $s \geq 0$, for each $\delta > 0$. Therefore, our claim follows upon proving that $Qu_\varepsilon - v_\varepsilon$ converges to 0 in $L^2(0, T; H^{-s})$ for some $s > 0$. In view of (47), it suffices to show this fact for $Qu_\varepsilon - Q(\rho_\varepsilon u_\varepsilon)$ and this is immediate since $u_\varepsilon - \rho_\varepsilon u_\varepsilon = (1 - \rho_\varepsilon)u_\varepsilon$ converges to 0 in $L^2(0, T; L^p)$ with $\frac{1}{p} = \frac{1}{\gamma} + \frac{N-2}{2N}$ if $N \geq 3$ and $1 < p < \gamma$ if $N = 2$.

The last step in this global method is to show that $\operatorname{div}(v_\varepsilon \otimes v_\varepsilon)$ converges (in $\mathcal{D}'$) to a distribution which is a gradient. In order to do so, we make a rather explicit computation of $\operatorname{div}(v_\varepsilon \otimes v_\varepsilon)$ using Fourier series. We thus write, recalling that $m = \nabla\phi$,

$$\psi = \sum_{k \in \mathbb{Z}^N} \psi_k(t)e^{ik \cdot x} \,, \quad m = i \sum_{k \in \mathbb{Z}^N} k\phi_k(t)e^{ik \cdot x} \,,$$

$\psi_0 = 0$, $\displaystyle\int_0^T \sum_{k \in \mathbb{Z}^N} |\psi_k(t)|^2 + |\phi_k(t)|^2|k|^2 \, dt < \infty$, and ψ_k and ϕ_k are scalar functions for all k. An explicit computation yields the following formula

$$v_\varepsilon = i \sum_{k \in \mathbb{Z}^N} k \, e^{ik \cdot x} \left(\cos\left(\sqrt{b}\,|k|\,\frac{t}{\varepsilon}\right) \phi_k(t) - \frac{\sqrt{b}}{|k|} \sin\left(\sqrt{b}\,|k|\,\frac{t}{\varepsilon}\right) \psi_k(t) \right) \,.$$

Hence,

$$v_\varepsilon \otimes v_\varepsilon = \sum_{k,\ell} e^{i(k+\ell) \cdot x} (k \otimes \ell) \, a_k^\varepsilon(t) \, a_\ell^\varepsilon(t) \tag{49}$$

where $a_k^\varepsilon(t) = \cos\left(\sqrt{b}|k|\frac{t}{\varepsilon}\right) \phi_k(t) - \frac{\sqrt{b}}{|k|} \sin\left(\sqrt{b}|k|\frac{t}{\varepsilon}\right) \psi_k(t)$ for all $k \in \mathbb{Z}^N$.

We next observe that $a_k^\varepsilon \, a_\ell^\varepsilon$ converges weakly in $L^1(0,T)$ to 0 as ε goes to 0_+ unless $|k| = |\ell|$. We thus denote by

$$A_\varepsilon = \sum_{\substack{k,\ell \\ |k| \neq |\ell|}} e^{i(k+\ell) \cdot x} \, (k \otimes \ell) \, a_k^\varepsilon(t) \, a_\ell^\varepsilon(t)$$

$$B_\varepsilon = \sum_{\substack{k,\ell \\ |k| = |\ell|}} e^{i(k+\ell) \cdot x} \, (k \otimes \ell) \, a_k^\varepsilon(t) \, a_\ell^\varepsilon(t) \,.$$

We claim that A_ε converges to 0 in $\mathcal{D}'$ (for instance). Indeed, for any $\chi \in L^\infty(0,T)$, we have

$$\int \chi A_\varepsilon \, dt = \sum_{\substack{k,\ell \\ |k| \neq |\ell|}} e^{i(k+\ell) \cdot x} \, (k \otimes \ell) \int a_k^\varepsilon a_\ell^\varepsilon \chi \, dt$$

where $\displaystyle\int a_k^\varepsilon a_\ell^\varepsilon \chi \, dt$ converges to 0 as ε goes to 0_+ for each (k, ℓ) such that $|k| \neq |\ell|$. Then, observing that we have

$$\left| (k \otimes \ell) \int a_k^\varepsilon a_\ell^\varepsilon \chi \, dt \right| \leq C \, \Delta_k \Delta_\ell \,,$$

where $\Delta_k = \|\psi_k\|_{L^2(0,T)} + |k| \, \|\phi_k\|_{L^2(0,T)}$, our claim follows easily upon proving that for $s > N/2$

$$\sum_n \frac{1}{(t + |n|^2)^s} \left(\sum_n \Delta_k \Delta_{n-k}\right)^2 \leq \left(\sum_n \frac{1}{(1 + |n|^2)^s}\right)\left(\sum_k \Delta_k^2\right)^2 < \infty \,.$$

We complete the proof of Theorem 5 by checking that div B_ε is a gradient. Indeed, we have:

$$B_\varepsilon = \frac{1}{2} \sum e^{i(k+\ell)\cdot x} \, (k \otimes \ell + \ell \otimes k) \, a_k^\varepsilon a_\ell^\varepsilon$$

$$= \sum_{n \in \mathbb{Z}^N} e^{in\cdot x} \left[\sum_{|k|=|n-k|} (k \otimes (n-k) + (n-k) \otimes k) \, a_k^\varepsilon \, a_{n-k}^\varepsilon \right]$$

We conclude remarking that if $|k| = |n - k|$,

$$k\big[(n-k).n\big] + (n-k)\big[k-n\big]$$
$$= k\big[|n-k|^2 + k.(n-k)\big] + (n-k)\big[|k|^2 + k(n-k)\big]$$
$$= n(k.n) = \frac{1}{2} n|n|^2$$

and thus

$$\text{div } B_\varepsilon = \frac{-i\nabla}{2} \left\{ \sum_{n \in \mathbb{Z}^N} e^{in\cdot x} \, |n|^2 \sum_{|k|=|n-k|} a_k^\varepsilon \, a_{n-k}^\varepsilon \right\}.$$

Remark 7. The preceding proof shows that the hydrostatic pressure and more precisely its gradient is the limit of the *sum of two terms* namely $\frac{a}{\varepsilon^2}\nabla\rho^\gamma$ (as expected!) *and* div $v_\varepsilon \otimes v_\varepsilon$ (or even $-\text{div}B_\varepsilon$) with the notations introduced above. Furthermore, this last term in general does not vanish since the quantity $\frac{\partial}{\partial t}\left(\frac{\psi_\varepsilon}{m_\varepsilon}\right)$ is bounded in $L^2(0,T;H^{-r})$ while $\psi_\varepsilon(0) = \frac{1}{\varepsilon}(\rho_\varepsilon - \overline{\rho_\varepsilon})$ and $m_\varepsilon(0) = Q(\rho_\varepsilon^0 u_\varepsilon^0)$ (notice that $m_\varepsilon(0)$ converges to $Q(\widetilde{u_0}) = \widetilde{u}^0 - u^0$).

Remark 8. When $N = 2$, one can check that u_ε converges to u in $L^2(0,T;H^1)$ for all $T \in (0,\infty)$ if we assume that $\frac{1}{\varepsilon^2}\left[(\rho_\varepsilon^0)^\gamma - \gamma\rho_\varepsilon^0(\overline{\rho_\varepsilon})^{\gamma-1}(\overline{\rho_\varepsilon})^\gamma\right] \underset{\varepsilon}{\rightarrow} 0$ in L^1 and

$$\sqrt{\rho_\varepsilon^0}\, u_\varepsilon^0 \underset{\varepsilon}{\rightarrow} u^0 \quad \text{in } L^2,$$

(in particular $\widetilde{u}^0 = u^0$ ...).

Indeed, passing to the limit in (32), we deduce

$$\frac{1}{2}\int |u(t)|^2 + \mu \int_0^t ds \int |Du|^2 \leq \varlimsup_\varepsilon E_\varepsilon(t) + \int_0^t D_\varepsilon(s)\, ds$$
$$\leq \varlimsup_\varepsilon E_\varepsilon(t) + \int_0^t D_\varepsilon(s)\, ds$$
$$\leq \lim_\varepsilon E_\varepsilon^0 = \frac{1}{2}\int |u^0|^2,$$

while the "conservation of energy" holds since $N = 2$. The assertion follows then easily.

4.4 Local Method

Now, we present a more direct and more general proof (see [27]) based on some properties of the wave equation. We will divide the proof into two steps. In the first one, we assume that all the quantities are regular in x uniformly in ε and then show how we can deduce the general case from the regular one by some regularization. For simplicity, in all this subsection we will assume that $\bar{\rho}_\varepsilon = 1$.

The regular case:

In this first step, we assume that φ_ε, $\pi_\varepsilon = \frac{1}{\varepsilon^2}(\rho_\varepsilon^\gamma - 1 - \gamma(\rho_\varepsilon - 1))$, $m_\varepsilon = \rho_\varepsilon u_\varepsilon$, u_ε are regular in x, uniformly in ε, i.e. φ_ε, π_ε and m_ε, are bounded in $L^\infty(0,T;H^s)$ and u_ε is bounded in $L^2(0,T;H^s)$ for all $s \geq 0$. Next, we want to show that

$$\operatorname{div}(\rho_\varepsilon u_\varepsilon \otimes u_\varepsilon) \underset{\varepsilon}{\rightharpoonup} \operatorname{div}(u \otimes u) + \nabla q \tag{50}$$

for some distribution q. To this end, we will pass to the limit locally in x when ε goes to 0. Let B be a ball in our domain Ω (here taken for simplicity to be the torus $\mathbb{T}^N$ even though this local proof applies without any additional technicalities to any domain with any kind of boundary conditions). We want to prove the convergence stated in (50) locally in $B \times (0,T)$. We introduce the orthogonal projections P and Q defined on $L^2(B)$ by $I = P + Q$; $\operatorname{div}Pu = 0$, $\operatorname{curl}(Qu) = 0$ in B ; $Pu.n = 0$ on $\partial\Omega$ where n stands for the exterior normal to ∂B.

Applying P to the second equation of (27), we deduce easily that $\frac{\partial}{\partial t}Pm_\varepsilon$ is bounded in $L^\infty(0,T;H^s)(\forall\, s \geq 0)$ and hence that Pm_ε converges to Pu in $C([0,T];H^s)$ ($\forall\, s \geq 0$). Here, we have used that the injection of $H^r(B)$ in $H^s(B)$ is compact since B is bounded. We also deduce that Pu_ε converges to Pu in $L^2(0,T;H^s)$ since $P(u_\varepsilon - u) = P\big((1 - \rho_\varepsilon)u_\varepsilon\big) + P\big(\rho_\varepsilon u_\varepsilon - u\big)$.

Next, we decompose in B, m_ε in $u + P(m_\varepsilon - u) + Q(m_\varepsilon - u)$ and u_ε in $u + P(u_\varepsilon - u) + Q(u_\varepsilon - u)$. Hence, we can decompose in $\mathcal{D}'(B)$ $\operatorname{div}(\rho_\varepsilon u_\varepsilon \otimes u_\varepsilon)$ in 8 different terms and it is easy to see that it is sufficient to show that $\operatorname{div}\big(Q(m_\varepsilon - u) \otimes Q(u_\varepsilon - u)\big)$ converges to some gradient. Moreover since $Q(m_\varepsilon - u)$ and $Q(u_\varepsilon - u)$ converge weakly to 0 and that $Q(m_\varepsilon - u) - Q(u_\varepsilon - u) = Q\big(\big((1 - \rho_\varepsilon)u\big)$ converges to 0 in $L^2(0,T;H^s)$ ($\forall\, s \geq 0$), we see that it is equivalent to show the above requirement for the following term $\operatorname{div}\big(Q(m_\varepsilon - u) \otimes Q(m_\varepsilon - u)\big)$. Next, we introduce ψ_ε such that $\int_B \psi_\varepsilon\, dx = 0$, $\nabla\psi_\varepsilon = Qm_\varepsilon$. Besides, it is easy to see that ψ_ε is bounded in $L^\infty(0,T;H^s)(\forall\, s \geq 0)$. With the above notations, we deduce from the initial system (27) the following one

$$\frac{\partial\varphi_\varepsilon}{\partial t} + \frac{1}{\varepsilon}\,\Delta\psi_\varepsilon = 0 \;,\quad \frac{\partial\nabla\psi_\varepsilon}{\partial t} + \frac{a\gamma}{\varepsilon}\,\nabla\varphi_\varepsilon = F_\varepsilon \tag{51}$$

where $F_\varepsilon = \xi\nabla\operatorname{div}u_\varepsilon + \nabla\pi_\varepsilon + \mu Q\big[\Delta u_\varepsilon - \operatorname{div}(\rho_\varepsilon u_\varepsilon \otimes u_\varepsilon)\big]$ is bounded in $L^2(0,T;H^s)(\forall\, s \geq 0)$.

Next, we observe that in $\mathcal{D}'(B \times]0, T[)$, we have on one hand

$$\text{div}\,(Qu \otimes Qu) = \frac{1}{2}\nabla|Qu|^2 + (\text{div}\,Qu)Qu = \frac{1}{2}\nabla|Qu|^2$$

and on the other hand

$$
\begin{aligned}
\text{div}\left(\nabla\psi_\varepsilon \otimes \nabla\psi_\varepsilon\right) &= \frac{1}{2}\nabla|\nabla\psi_\varepsilon|^2 + \Delta\psi_\varepsilon\nabla\psi_\varepsilon \\
&= \frac{1}{2}\nabla\left(|\nabla\psi_\varepsilon|^2\right) - \frac{\partial}{\partial t}(\varepsilon\varphi_\varepsilon\nabla\psi_\varepsilon) + \varepsilon\varphi_\varepsilon F_\varepsilon - a\gamma\varphi_\varepsilon\nabla\varphi_\varepsilon \\
&= \frac{1}{2}\nabla\left(|\nabla\psi_\varepsilon|^2 - a\gamma\varphi_\varepsilon^2\right) - \frac{\partial}{\partial t}(\varepsilon\varphi_\varepsilon\nabla\psi_\varepsilon) + \varepsilon\varphi_\varepsilon F_\varepsilon \;.
\end{aligned}
$$

Using that $\varepsilon\,\varphi_\varepsilon\nabla\psi_\varepsilon$ converges strongly to 0 in $L^2(0,T;H^s)$ ($\forall s \geq 0$) and that $\varepsilon\varphi_\varepsilon F_\varepsilon$ converges strongly to 0 in $L^\infty(0,T;H^s)$ ($\forall s \geq 0$), we deduce that

$$\text{div}\left(Q(m_\varepsilon - u) \otimes Q(m_\varepsilon - u)\right) \underset{\varepsilon}{\rightharpoonup} \nabla q \tag{52}$$

and finally, we obtain that

$$\text{div}\,(\rho_\varepsilon u_\varepsilon \otimes u_\varepsilon) \underset{\varepsilon}{\rightharpoonup} \text{div}\,(u \otimes u) + \nabla q \quad \text{in } B \times (0,T) \tag{53}$$

and the theorem is proved in the regular case.

The general case:

Now, we pass to the second step where we are going to show how we can regularize in x the above quantities (uniformly in ε). To do so let $K_\delta = \frac{1}{\delta^N}K(\frac{\cdot}{\delta})$, where $K \in C_0^\infty(\mathbb{R}^N)$, $\int_{\mathbb{R}^N} K\,dz = 1$, $\delta \in (0,1)$. We can then regularize by convolution in the following way: $\varphi_\varepsilon^\delta = \varphi_\varepsilon * K_\delta$, $m_\varepsilon^\delta = m_\varepsilon * K_\delta$, $u_\varepsilon^\delta = u_\varepsilon * K_\delta$, $\pi_\varepsilon^\delta = \pi_\varepsilon * K_\delta$. We can then follow the same proof as in the regular case by replacing φ_ε, π_ε, m_ε and u_ε by their regularizations and we conclude by observing that $\|u_\varepsilon^\delta - u_\varepsilon\|_{L^2(0,T;L^2)} \leq C\delta$, $\|u_\varepsilon^\delta\|_{L^2(0,T;H^1)} \leq C$ and $\|u_\varepsilon^\delta \otimes u_\varepsilon^\delta - u_\varepsilon \otimes u_\varepsilon\|_{L^1(0,T;L^p)} \leq C\delta$, $\|u_\varepsilon^\delta \otimes u_\varepsilon^\delta\|_{L^1(0,T;L^p)} \leq C$ ($p = \frac{N}{N-2}$ if $N \geq 3$, $1 \leq p < +\infty$ if $N = 2$). Indeed, from the above uniform bounds, we deduce that

$$\sup_{\varepsilon \in]0,1]}\left\{\|\rho_\varepsilon^\delta u_\varepsilon^\delta - m_\varepsilon^\delta\|_{L^2(L^q)} + \|m_\varepsilon^\delta - u_\varepsilon^\delta\|_{L^2(L^q)} + \|\rho_\varepsilon u_\varepsilon - m_\varepsilon^\delta\|_{L^2(L^q)}\right\} \underset{\delta}{\longrightarrow} 0\;,$$

$$\sup_{\varepsilon \in]0,1]}\left\{\|\rho_\varepsilon^\delta u_\varepsilon^\delta \otimes u_\varepsilon^\delta - m_\varepsilon^\delta \otimes u_\varepsilon^\delta\|_{L^1(L^r)} + \|m_\varepsilon^\delta \otimes u_\varepsilon^\delta - m_\varepsilon \otimes u_\varepsilon\|_{L^1(L^r)}\right.$$

$$\left. + \|m_\varepsilon \otimes u_\varepsilon - u_\varepsilon^\delta \otimes u_\varepsilon^\delta\|_{L^1(L^r)}\right\} \underset{\delta}{\longrightarrow} 0\;,$$

with $\frac{1}{q} > \frac{1}{\gamma} + \frac{N-2}{2N}$, $\frac{1}{r} > \frac{1}{\gamma} + \frac{N-2}{N}$, since $\frac{1}{\gamma} + \frac{N-2}{N} < 1$. Moreover, it is easy to see that for all δ and all s, we have that $\|m_\varepsilon^\delta - u_\varepsilon^\delta\|_{L^2(H^s)}$ goes to 0 when ε goes to 0.

Remark 9. The idea we have introduced in the first step is the following classical fact for a solution ψ of the wave equation $\left[\frac{\partial^2 \psi}{\partial t^2} - c^2 \Delta \psi = 0\right]$:

$$\operatorname{div}\left(\nabla\psi \otimes \nabla\psi\right) = \nabla\left(\frac{1}{2}|\nabla\psi|^2 - \frac{1}{2c^2}\left(\frac{\partial\psi}{\partial t}\right)^2\right) + \frac{\partial}{\partial t}\left(\frac{1}{c^2}\nabla\psi\frac{\partial\psi}{\partial t}\right) .$$

Remark 10. The theorem 5 stated above was only about the case of periodic boundary conditions. However, it is easy to see from this local proof that the result also holds for any domain with any kind of boundary conditions as soon as all the uniform bounds derived in the second subsection hold.

In the next subsection, we will state a more precise result in the case of Dirichlet boundary conditions. Indeed, depending on some geometrical property of the domain, we will prove a strong convergence result towards the incompressible Navier-Stokes system, which means that all the oscillations are damped in the limit.

4.5   The Case of Dirichlet Boundary Conditions

In this subsection, Ω is taken to be a bounded domain. We consider the system (27) with the following Dirichlet boundary condition

$$u_\varepsilon = 0 \quad \text{on} \quad \partial\Omega . \tag{54}$$

The existence of global weak solutions is also known [23]. In order to state precisely our main Theorem, we need to introduce a geometrical condition on Ω. Let us consider the following overdetermined problem

$$-\Delta\phi = \lambda\phi \text{ in } \Omega, \quad \frac{\partial\phi}{\partial\mathbf{n}} = 0 \text{ on } \partial\Omega, \text{ and } \phi \text{ is constant on } \partial\Omega . \tag{55}$$

A solution of (55) is said to be trivial if $\lambda = 0$ and ϕ is a constant. We will say that Ω satisfies assumption (H) if all the solutions of (55) are trivial. Schiffer's conjecture says that every Ω satisfies (H) excepted the ball (see for instance [12]). In two dimensional space, it is proved that every bounded, simply connected open set $\Omega \subset \mathbb{R}^2$ whose boundary is Lipschitz but not real analytic satisfies (H), hence property (H) is generic in $\mathbb{R}^2$. The main result reads as follows

Theorem 11. *Under the above conditions, ρ_ε converges to 1 in the space $C([0,T]; L^\gamma(\Omega))$ and extracting a subsequence if necessary u_ε converges weakly to u in $L^2((0,T) \times \Omega)^N$ for all $T > 0$, and strongly if Ω satisfies (H). In addition, u is a global weak solution of the incompressible Navier-Stokes equations (33) with Dirichlet boundary conditions satisfying $u_{|t=0} = u_0$ in Ω.*

For the proof of this result, we refer to [8]. We only sketch below the phenomenon going on. Let $(\lambda_{k,0}^2)_{k\geq1}$, $(\lambda_{k,0} > 0)$, be the nondecreasing sequence of eigenvalues and $(\Psi_{k,0})_{k \geq 1}$ the orthonormal basis of $L^2(\Omega)$ functions with zero mean value of eigenvectors of the Laplace operator $-\Delta_N$ with homogeneous Neumann boundary conditions

$$-\Delta\Psi_{k,0} = \lambda_{k,0}^2 \Psi_{k,0} \ \text{in} \ \Omega, \qquad \frac{\partial\Psi_{k,0}}{\partial\mathbf{n}} = 0 \ \text{on} \ \mathcal{D}\Omega \ . \tag{56}$$

We can split these eigenvectors $(\Psi_{k,0})_{k \in \mathbb{N}}$ (which represent the acoustic eigenmodes in Ω) into two classes: those which are not constant on $\partial\Omega$ will generate boundary layers and will be quickly damped, thus converging strongly to 0; those which are constant on $\partial\Omega$ (non trivial solutions of (55)), for which no boundary layer forms, will remain oscillating forever, leading to only weak convergence. Indeed, if (H) is not satisfied, u^ε will in general only converge weakly and not strongly to u (like in the periodic case $\Omega = \mathbb{T}^d$ for instance). However, if at initial time $t = 0$, no modes of second type are present in the velocity, the convergence to the incompressible solution is strong in L^2.

Theorem 11 also applies to $\Omega = \mathbb{T}^{N-1} \times [0,1]$, which does not satisfy (H). Notice that according to Schiffer's conjecture the convergence is not strong for general initial data when Ω is two or three dimensional ball, but is expected to be always strong in any other domain. Another case where we can get a better result than the simple weak convergence is the whole space case. Indeed, this case was treated in [25] where only weak convergence was proved. In [7] a more precise result was given using the dispersion for the acoustic waves. We will come back to this issue in the next subsection.

4.6   Convergence towards the Euler System

In this subsection, we study the case where μ_ε converges to 0 too. We will state two results in the periodic case and in the whole space case taken from [30] (see also [25]). The case of domains with boundaries is open even in the incompressible case (see the second section).

The Whole Space Case. Let us begin with the case of the whole space. We consider a sequence of global weak solutions $(\rho_\varepsilon, u_\varepsilon)$ of the compressible Navier-Stokes equations (27) and we assume that $\rho_\varepsilon - 1 \in L^\infty(0, \infty; L_2^\gamma) \cap C([0,\infty), L_2^p)$ for all $1 \leq p < \gamma$, where we put $L_2^p = \{f \in L_{loc}^1, |f|1_{|f| \geq 1} \in L^p, |f|1_{|f| \leq 1} \in L^2\}$, $u_\varepsilon \in L^2(0,T;H^1)$ for all $T \in (0,\infty)$ (with a norm which can explode when ε goes to 0), $\rho_\varepsilon|u_\varepsilon|^2 \in L^\infty(0,\infty;L^1)$ and $\rho_\varepsilon u_\varepsilon \in C([0,\infty);L_w^{2\gamma/(\gamma+1)})$ i.e. $\rho_\varepsilon u_\varepsilon$ is continuous with respect to $t \geq 0$ with values in $L^{2\gamma/(\gamma+1)}$ endowed with its weak topology. We require (27) to hold in the sense of distributions and we impose the following conditions at infinity

$$\rho_\varepsilon \to 1 \ \text{as} \ |x| \to +\infty \ , \quad u_\varepsilon \to 0 \ \text{as} \ |x| \to +\infty \ . \tag{57}$$

Finally, we prescribe initial conditions

$$\rho_\varepsilon\big|_{t=0} = \rho_\varepsilon^0 \ , \ \rho_\varepsilon u_\varepsilon\big|_{t=0} = m_\varepsilon^0 \tag{58}$$

where $\rho_\varepsilon^0 \geq 0$, $\rho_\varepsilon^0 - 1 \in L^\gamma$, $m_\varepsilon^0 \in L^{2\gamma/(\gamma+1)}$, $m_\varepsilon^0 = 0$ a.e. on $\{\rho_\varepsilon^0 = 0\}$ and $\rho_\varepsilon^0 |u_\varepsilon^0|^2 \in L^1$, denoting by $u_\varepsilon^0 = \frac{m_\varepsilon^0}{\rho_\varepsilon^0}$ on $\{\rho_\varepsilon^0 > 0\}$, $u_\varepsilon^0 = 0$ on $\{\rho_\varepsilon^0 = 0\}$. We also introduce the following notation $\rho_\varepsilon = 1 + \varepsilon\varphi_\varepsilon$. Notice that if $\gamma < 2$, we cannot deduce any bound for φ_ε in $L^\infty(0,T;L^2)$. This is why we introduce the following approximation which belongs to L^2

$$\Phi_\varepsilon = \frac{1}{\varepsilon}\sqrt{\frac{2a}{\gamma-1}(\rho_\varepsilon^\gamma - 1 - \gamma(\rho_\varepsilon - 1))} \ .$$

Furthermore, we assume that $\sqrt{\rho_\varepsilon^0}\, u_\varepsilon^0$ converges strongly in L^2 to some $\tilde{u}^0$. Then, we denote by $u^0 = P\tilde{u}^0$, where P is the projection on divergence-free vector fields, we also define Q (the projection on gradient vector fields), hence $\tilde{u}^0 = P\tilde{u}^0 + Q\tilde{u}^0$. Moreover, we assume that Φ_ε^0 converges strongly in L^2 to some φ^0. This also implies that φ_ε^0 converges to φ^0 in L_2^γ. We also assume that $(\rho_\varepsilon, u_\varepsilon)$ satisfies the energy inequality.

When ε goes to zero and μ_ε goes to 0, we expect that u_ε converges to v, the solution of the Euler system

$$\begin{cases} \partial_t v + \text{ div}\,(v \otimes v) + \nabla\pi = 0 \\[2mm] \text{div}\,v = 0\,, \qquad v_{|t=0} = u^0 \end{cases} \tag{59}$$

in $C([0,T^*);H^s)$. We show the following theorem.

Theorem 12. *We assume that $\mu_\varepsilon \xrightarrow{\varepsilon} 0$ (such that $\mu_\varepsilon + \xi_\varepsilon > 0$ for all ε) and that $P\tilde{u}^0 \in H^s$ for some $s > N/2 + 1$, then $P(\sqrt{\rho_\varepsilon}u_\varepsilon)$ converges to v in $L^\infty(0,T;L^2)$ for all $T < T^*$, where v is the unique solution of the Euler system in $L_{\text{loc}}^\infty([0,T^*);H^s)$ and T^* is the existence time of (59). In addition $\sqrt{\rho_\varepsilon}u_\varepsilon$ converges to v in $L^p(0,T;L_{\text{loc}}^2)$ for all $1 \leq p < +\infty$ and all $T < T^*$.*

The proof is based on energy estimates, since we lose compactness in x at the limit. For this, we have to describe the oscillations in time and incorporate them in the energy estimates. It turns out that in the whole space case the acoustic waves disperses to infinity (see [34] in the framework of strong solutions and [7] in the framework of weak solutions).

The Periodic Case. Now, we take $\Omega = \mathbb{T}^N$ and consider a sequence of solutions $(\rho_\varepsilon, u_\varepsilon)$ of (27), satisfying the same conditions as in the whole space case (the functions are now periodic in space and all the integration are performed over $\mathbb{T}^N$). Of course, the conditions at infinity are removed and the spaces L_2^p can be replaced by L^p. Here, we have to impose more conditions on the oscillating part (acoustic waves), namely we have to assume that $Q\tilde{u}^0$ is more regular than L^2. In fact, in the periodic case, we do not have a dispersion phenomenon as in the case of the whole space and the acoustic waves will not go to infinity, but they are going to interact with each other.

This is way, we have to include them in the energy estimates to show our convergence result.

For the next theorem, we assume that $Q\tilde{u}^0$, $\varphi^0 \in H^{s-1}$ and that there exists a nonnegative constant ν such that $\mu_\varepsilon + \xi_\varepsilon \geq 2\nu > 0$ for all ε. For simplicity, we assume that $\mu_\varepsilon + \xi_\varepsilon$ converges to 2ν.

Theorem 13. *(The periodic case.) We assume that $\mu_\varepsilon \underset{\varepsilon}{\rightarrow} 0$ (such that $\mu_\varepsilon + \xi_\varepsilon \rightarrow 2\nu > 0$) and that $P\tilde{u}^0 \in H^s$ for some $s > N/2+1$, and $Q\tilde{u}^0$, $\varphi^0 \in H^{s-1}$ then $P(\sqrt{\rho_\varepsilon}u_\varepsilon)$ converges to v in $L^\infty(0,T;L^2)$ for all $T < T^*$, where v is the unique solution of the Euler system in $L^\infty_{loc}(0,T^*;H^s)$ and T^* is the existence time of (59). In addition $\sqrt{\rho_\varepsilon}u_\varepsilon$ converges weakly to v in $L^\infty(0,T;L^2)$*

In the above theorem, one can remove the condition $2\nu > 0$. In that case, we still have the result of Theorem 13 but only on an interval of time $(0,T^{**})$ which is the existence interval for the equation governing the oscillating part, namely

$$\begin{cases} \partial_t V + Q_1(v,V) + Q_2(V,V) - \nu\Delta V = 0 \\ V|_{t=0} = (\varphi^0, Q\tilde{u}^0) \end{cases} \tag{60}$$

where $Q_1(v,V)$ and $Q_2(V,V)$ take into account the resonances. Next, it is easy to see using the particular form of Q_1 and Q_2 that if $\nu > 0$ and $V(t = 0) \in H^{s-1}$ then we have (as long as v exists) a global solution in $L^\infty(H^{s-1})$ which satisfies $\nabla V \in L^1(0,T;L^\infty)$. On the other hand if $\nu = 0$ and $V(t = 0) \in H^{s-1}$ then we can construct a local (in time) solution in $L^\infty(H^{s-1})$ which satisfies $\nabla V \in L^1(0,T;L^\infty)$ for all $T < T^{**}$. Finally, the energy method is based on the fact that we can write a sort of Gronwall lemma for the following quantity

$$\left\| \sqrt{\rho_\varepsilon}u_\varepsilon - v - \mathcal{L}_2\left(\frac{t}{\varepsilon}\right)V \right\|_{L^2}^2 + \left\| \Phi_\varepsilon - \mathcal{L}_1\left(\frac{t}{\varepsilon}\right)V \right\|_{L^2}^2 .$$

We want to point out that the method of proof is the same for the whole space case and is simpler since we do not have to study all the resonances (the acoustic waves go to the infinity).

5   Study of the Limit $\gamma \rightarrow \infty$

In this section we are going to study the limit γ going to infinity. Depending on the total mass, we will recover at the limit either a mixed model, which behaves as a compressible one if $\rho < 1$ and as an incompressible one if $\rho = 1$ or the classical incompressible Navier-Stokes system. We start with the first case and introduce the limit system, namely the following one

$$\frac{\partial\rho}{\partial t} + \mathrm{div}\,(\rho u) = 0 \text{ in } (0,T) \times \Omega,\ 0 \leq \rho \leq 1 \text{ in } (0,T) \times \Omega , \tag{61}$$

$$\frac{\partial \rho u}{\partial t} + \text{div}\,(\rho u \otimes u) - \mu \Delta u - \xi \nabla \text{div}\, u + \nabla \pi = 0 \text{ in } (0,T) \times \Omega\,, \qquad (62)$$

$$\text{div}\, u = 0 \text{ a.e. on } \left\{\rho = 1\right\}\,, \qquad (63)$$

$$\pi = 0 \text{ a.e. on } \left\{\rho < 1\right\}\,, \; \pi \geq 0 \text{ a.e. on } \left\{\rho = 1\right\}. \qquad (64)$$

Throughout this section, Ω is taken to be the torus, the whole space case or a bounded domain with Dirichlet boundary conditions. Indeed, the proofs given in [26] can also apply to the case of Dirichlet boundary conditions, by using the bounds given in [24] and [10].

Let γ_n be a sequence of nonnegative real numbers that goes to infinity. Let (ρ_n, u_n) be a sequence of solutions of the isentropic compressible Navier-Stokes equations

$$\begin{cases} \dfrac{\partial \rho}{\partial t} + \text{div}\,(\rho u) = 0\,, \qquad \rho \geq 0, \\[4mm] \dfrac{\partial \rho u}{\partial t} + \text{div}\,(\rho u \otimes u) - \mu_n \Delta u - \xi_n \nabla \text{div}\, u + \nabla \rho^{\gamma_n} = 0\,. \end{cases} \qquad (65)$$

where $\mu_n > 0$ and $\mu_n + \xi_n > 0$, μ_n and ξ_n tend respectively to μ and ξ as n goes to the infinity, with $\mu > 0$ and $\mu + \xi > 0$ (in the sequel, we assume for simplicity that $\mu_n = \mu$ and $\xi_n = \xi$). We recall that global weak solutions of the above system are known to exist, if we assume in addition that $\gamma_n > \frac{N}{2}$ if $N \geq 4$, $\gamma_n \geq \frac{3}{2}$ if $N = 2$ and $\gamma_n \geq \frac{9}{5}$ if $N = 3$. These assumptions are true for n large enough. The sequence (ρ_n, u_n) satisfies in addition the following initial conditions and the following bounds,

$$\rho_n u_n\Big|_{t=0} = m_n^0\,, \; \rho_n\Big|_{t=0} = \rho_n^0, \qquad (66)$$

where $0 \leq \rho_n^0$ a.e. , ρ_n^0 is bounded in $L^1(\Omega)$ and $\rho_n^0 \in L^{\gamma_n}$ with $\int (\rho_n^0)^{\gamma_n} \leq C\gamma_n$ for some fixed C, $m_n^0 \in L^{2\gamma_n/(\gamma_n+1)}(\Omega)$, and $\rho_n^0 |u_n^0|^2$ is bounded in L^1, denoting by $u_n^0 = \frac{m_n^0}{\rho_n^0}$ on $\{\rho_n^0 > 0\}$, $u_n^0 = 0$ on $\{\rho_n^0 = 0\}$. In the periodic case or in the Dirichlet boundary condition case, we also assume that $\fint \rho_n^0 = M_n$, for some M_n such that $0 < M_n \leq M < 1$ and $M_n \to M$. Furthermore, we assume that $\rho_n^0 u_n^0$ converges weakly in L^2 to some m^0 and that ρ_n^0 converges weakly in L^1 to some ρ^0. Our last requirement concerns the following energy bounds we impose on the sequence of solutions we consider,

$$E_n(t) + \int_0^t D_n(s)\, ds \leq E_n^0 \quad \text{a.e. } t, \quad \frac{dE_n}{dt} + D_n \leq 0 \quad \text{in } \mathcal{D}'(0,\infty) \qquad (67)$$

where

$$E_n(t) = \int \frac{1}{2}\rho_n|u_n|^2(t) + \frac{a}{\gamma_n - 1}(\rho_n)^{\gamma_n}(t) ,$$

$$D_n(t) = \int \mu|Du_n|^2(t) + \xi(\operatorname{div} u_n)^2(t) ,$$

and

$$E_n^0 = \int \frac{1}{2}\rho_n^0|u_n^0|^2 + \frac{a}{\gamma_n - 1}(\rho_n^0)^{\gamma_n} .$$

We wish to mention an additional estimate which is available but that, however, we shall not use in this proof. Indeed, the proof made in P.-L. Lions [23] (Chapter 7, Section 7.1) yields the following bound for all $T \in (0, \infty)$

$$\int_0^T dt \int \rho_n^{\gamma_n + \theta_n} \leq C\gamma_n \quad \text{where } \theta_n = \frac{2}{N}\gamma_n - 1 . \tag{68}$$

Unfortunately, this estimate is not uniform in n. Instead, we shall use another estimate which can be derived as (68) was in [23] and which is uniform in n, namely an L^1 bound for $(\rho_n)^{\gamma_n}$.

Without loss of generality, extracting subsequences if necessary, we can assume that (ρ_n, u_n) converges weakly to (ρ, u). More precisely we can assume that $\rho_n \rightharpoonup \rho$ weakly in $L^p((0,T) \times \Omega)$ for any $1 \leq p \leq \infty$ and that $\rho \in L^\infty(0,T; L^p)$ (in fact we will show that ρ actually satisfies $0 \leq \rho \leq 1$), $u_n \rightharpoonup u$ weakly in $L^2(0,T; H^1_{\text{loc}})$.

Theorem 14. *Under the above conditions, we have $0 \leq \rho \leq 1$ and*

$$(\rho_n - 1)_+ \to 0 \quad \text{in } L^\infty(0,T; L^p(\Omega)) \text{ for any } 1 \leq p < +\infty .$$

Moreover, $(\rho_n)^{\gamma_n}$ is bounded in $L^1((0,T) \times \Omega)$ (for n such that $\gamma_n \geq N$). Then extracting subsequences again, there exists $\pi \in \mathcal{M}((0,T) \times \Omega)$ such that

$$(\rho_n)^{\gamma_n} \underset{n}{\rightharpoonup} \pi . \tag{69}$$

If in addition ρ_n^0 converges to ρ^0 in $L^1(\Omega)$, then (ρ, u, π) is a solution of (61)–(64) and the following strong convergences hold:

$$\rho_n \to \rho \qquad \text{in } C(0,T; L^p(\Omega)) \text{ for any } 1 \leq p < +\infty ,$$

$$\rho_n u_n \to \rho u \qquad \text{in } L^p(0,T; L^q(\Omega)) \text{ for any } 1 \leq p < +\infty, \ 1 \leq q < 2 ,$$

$$\rho_n u_n \otimes u_n \to \rho u \otimes u \quad \text{in } L^p(0,T; L^1(\Omega)) \text{ for any } 1 \leq p < +\infty .$$

Before proving the theorem, we have to define precisely the notion of weak solutions for the limit system: (ρ, u, π) is called a weak solution of the limit system (61)–(64) if

$$\rho \in L^\infty(0,T; L^\infty(\Omega) \cap L^1(\Omega)) \cap C(0,T; L^p(\Omega)) \text{ for any } 1 \leq p < \infty \tag{70}$$

$$\nabla u \in L^2(0, T, L^2(\Omega)) \text{ and } u \in L^2(0, T; H^1(B)), \tag{71}$$

where $B = \Omega$ if $\Omega = \mathbb{T}^N$ or if Ω is a bounded domain (with Dirichlet boundary conditions) and B is any ball in $\mathbb{R}^N$ if $\Omega = \mathbb{R}^N$, in this second case we also impose that $u \in L^2(0, T, L^{2N/N-2}(\mathbb{R}^N))$, if in addition $N \geq 3$.

Moreover,

$$\rho|u|^2 \in L^\infty(0, \infty; L^1) \quad \text{and} \quad \rho u \in L^\infty(0, \infty; L^2) . \tag{72}$$

Next, equations (61), (62) must be satisfied in the distributional sense. This can be written using a weak formulation (which also incorporate the initial conditions in some weak sense), namely we require that the following identities hold for all $\phi \in C^\infty([0, \infty) \times \Omega)$ and for all $\Phi \in C^\infty([0, \infty) \times \Omega)^N$ compactly supported in $[0, \infty) \times \Omega$ (i.e. vanishing identically for t large enough)

$$-\int_0^\infty dt \int_\Omega \rho \partial_t \phi - \int_\Omega \rho^0 \phi(0) - \int_0^\infty dt \int_\Omega \rho u.\nabla \phi = 0, \tag{73}$$

$$-\int_0^\infty dt \int_\Omega \rho u.\partial_t \Phi - \int_\Omega m^0.\Phi(0) - \int_0^\infty dt \int_\Omega \rho(u.\nabla \Phi).u \tag{74}$$
$$+ \int_0^\infty dt \left\{ \int_\Omega \mu Du.D\Phi + \xi \operatorname{div} u \operatorname{div} \Phi \right\} - \pi \operatorname{div} \Phi = 0 .$$

On the other hand, the equation (64) should be understood in the following way $\rho \pi = \pi \geq 0$. Of course, we have to define the sense of the product $\rho \pi$ since, we only require that $\pi \in \mathcal{M}$. Indeed, the product can be defined by using that

$$\begin{cases} \rho \in C([0, T]; L^p) \cap C^1([0, T]; H^{-1}), \\ \pi \in W^{-1,\infty}(H^1) + L^1(L^{N/(N-2)}) \cap L^\alpha(L^\beta) + L^2(L^2) \end{cases} \tag{75}$$

where $1 < \alpha, \beta < \infty$ and $\frac{1}{\beta} = \frac{1}{\alpha}\frac{N-2}{N} + (1 - \frac{1}{\alpha})$.

Finally, equation (63) is just a consequence of (61), however we incorporate it in the limit system to emphasis the fact that it is a mixed system which behaves like a compressible one if $\rho < 1$ and as an incompressible one if $\rho = 1$. Indeed, from (61), we can deduce that

$$\frac{\partial \rho^k}{\partial t} + \operatorname{div}(\rho^k u) = (1 - k)\rho^k \operatorname{div} u . \tag{76}$$

Since $0 \leq \rho^k \leq 1$, we see that $\partial_t \rho^k$ is bounded in $W^{-1,\infty}((0, T) \times \Omega)$, in addition $div(\rho^k u)$ is bounded in $L^\infty(0, T; H^{-1}_{loc}(\Omega))$, for $|\rho^k u| \leq |\rho u| \in L^\infty(L^2_{loc})$. This yields that $k\rho^k \operatorname{div} u$ is a bounded distribution (in H^{-1}_{loc} for instance). Letting k go to the infinity, we get

$$\rho^k \operatorname{div} u \underset{k}{\rightharpoonup} 0 \quad \text{in } \mathcal{D}'. \tag{77}$$

Besides, we have

$$\rho^k \to 1_{\{\rho=1\}} \quad \text{a.e},$$

so we get

$$\rho^k \operatorname{div} u \to 1_{\{\rho=1\}} \operatorname{div} u \quad \text{a.e},$$

since $|\rho^k \operatorname{div} u| \le |\operatorname{div} u| \in L^2_{\text{loc}}((0,T) \times \Omega)$. Hence, we get that $\operatorname{div} u = 0$ a.e. on $\{\rho = 1\}$.

Proof (of Theorem 14). The proof is divided into three steps.

Step 1: From the energy conservation and the mass conservation, we deduce that for any $1 < p < \infty$, we have for n such that $\gamma_n > p$,

$$\|\rho_n\|_{L^\infty(0,T;L^p)} \le \|\rho_n\|_{L^\infty(0,T;L^1)}^{\theta_n} \|\rho_n\|_{L^\infty(0,T;L^{\gamma_n})}^{1-\theta_n}$$
$$\le M_n^{\theta_n} (C\gamma_n)^{(1-\theta_n)/\gamma_n}$$

where we have used Hölder's inequality and where θ_n is given for any n by $\frac{1}{p} = \theta_n + \frac{1-\theta_n}{\gamma_n}$. Then, letting n go to infinity, we deduce that $\theta_n \to \frac{1}{p}$ and that

$$\|\rho\|_{L^\infty(0,T;L^p)} \le \liminf_{n\to\infty} \|\rho_n\|_{L^\infty(0,T;L^p)} \le M^{\frac{1}{p}} .$$

Hence, letting p go to infinity, we obtain

$$\|\rho\|_{L^\infty(0,T;L^\infty)} \le \liminf_{p\to\infty} \|\rho\|_{L^\infty(0,T;L^p)} \le 1 .$$

We next introduce $\phi_n = (\rho_n - 1)_+$. We are going to show that ϕ_n goes to 0 uniformly in t in all L^p spaces. In fact, from the energy conservation that we have for any t, we deduce

$$\int_\Omega (1 + \phi_n)^{\gamma_n} 1_{\{\phi_n > 0\}} \le \int_\Omega (\rho_n)^{\gamma_n} \le C\gamma_n .$$

Next, for any $p > 1$, there exists a constant a (for instance we can take $a_p = \frac{1}{2p!}$ if p is an integer) such that for any k large enough the following inequality holds,

$$(1 + x)^k \ge 1 + a_p k^p x^p,$$

for any nonnegative x. Hence, we have for n large enough

$$\int_\Omega \phi_n^p \le \frac{C a_p}{(\gamma_n)^{p-1}},$$

which yields the convergence of $(\rho_n - 1)_+$ to 0.

Step 2: Now, we turn to the proof of the L^1 bound on $(\rho_n)^{\gamma_n}$. We begin by treating the whole space case and then explain the necessary modifications we have to make in the periodic case and in the case of Dirichlet boundary condition. Applying as in [23], the operator $(-\Delta)^{-1} \operatorname{div}$ to (65), we obtain

$$(\rho_n)^{\gamma_n} = \partial_t (-\Delta)^{-1} \operatorname{div}(\rho_n u_n) - R_i R_j(\rho(u_n)_i (u_n)_j) + (\mu + \xi) \operatorname{div}(u_n) . \quad (78)$$

Then multiplying (78) by ρ_n, we deduce (we omit the indices n at the right hand side for the sake of clarity),

$$\begin{cases} (\rho_n)^{\gamma_n+1} = \partial_t[\rho(-\Delta)^{-1}\mathrm{div}\,(\rho u)] + \mathrm{div}\,[\rho u(-\Delta)^{-1}\mathrm{div}\,(\rho u)] \\ \qquad + \rho u_i R_i R_j(\rho u_j) - \rho R_i R_j(\rho u_i u_j) + (\mu+\xi)\rho\,\mathrm{div}\,(u)\ . \end{cases} \quad (79)$$

This and essentially the multiplication of ρ_n by $\partial_t(-\Delta)^{-1}\mathrm{div}\,(\rho_n u_n)$ should be justified (see [26] and the computations above). Then, integrating (79) over $(0,T)\times\mathbb{R}^N$, we see that $(\rho_n)^{\gamma_n+1}$ is bounded in $L^1((0,T)\times\mathbb{R}^N)$, uniformly in n. First, we notice that we do not have an L^∞ bound on ρ_n. However, since $\int(\rho_n)^{\gamma_n}(t)\le C\gamma_n$ for a.e. t, we see that there exists a constant C_1 (for instance $C_1=\exp(\frac{C}{e})$), such that for any n, we have $\|\rho_n\|_{L^\infty(0,T;L^{\gamma_n})}\le C_1$, since

$$\sup_{\gamma>0}(C\gamma)^{1/\gamma}=\exp\left(\frac{C}{e}\right)\ .$$

Next, we remark that the norm of $\rho_n u_n$ in $L^\infty(0,T;L^{2\gamma_n/(\gamma_n+1)}\cap L^1)$ is independent of n (we use here the fact that the norm of $\sqrt{\rho_n}$ in $L^\infty(L^{2\gamma_n}\cap L^2)$ is bounded independently of n and that $\sqrt{\rho_n}u_n$ is bounded in $L^\infty(L^2)$). Hence, we may write

$$\left|\int_0^T\int_{\mathbb{R}^N}\partial_t\left[\rho(-\Delta)^{-1}\mathrm{div}\,(\rho u)\right]\right|\le 2\|\rho\|_{L^\infty(L^{q'})}\|\rho u\|_{L^\infty(L^r)},$$

with $(\frac{1}{r}-\frac{1}{N})+\frac{1}{q'}=1$, $1\le r\le\frac{2\gamma_n}{\gamma_n+1}$ and $1\le q'\le\gamma_n$ (such a choice is possible if $\gamma_n\ge N$ for instance).

For the second term, we only need to show that $U=\rho_n u_n(-\Delta)^{-1}\mathrm{div}(\rho_n u_n)$ belongs to $L^\infty(L^1)$ for all n, which can be deduced from the following bound

$$\|U\|_{L^\infty(L^1)}\le\|\rho_n u_n\|_{L^\infty(L^q)}^2,$$

with $q=\frac{2N}{N+1}\le\frac{2\gamma_n}{\gamma_n+1}$, since $\gamma_n\ge N$.

For the third and the fourth term, we must distinguish two cases, namely $N\ge 3$ and $N=2$. In the first case, we have $\rho|u|^2\in L^1(L^s)$, with $\frac{1}{s}=\frac{N-2}{N}+\frac{1}{\gamma_n}<1$. Therefore $R_i R_j(\rho u_i u_j)\in L^1(L^s)$ and hence $\rho R_i R_j(\rho u_i u_j)\in L^1(L^1)$, since $\frac{N-2}{N}+2\frac{1}{\gamma_n}\le 1$. For the third term we recall the fact that $\rho_n u_n, R_i R_j(\rho(u_n)_i(u_n)_j)\in L^2(L^r\cap L^1)$, where $\frac{1}{r}=\frac{N-2}{2N}+\frac{1}{\gamma_n}\ge 2$. The case $N=2$ is treated using the bound on u_n in $L^2(0,T;BMO)$ and on $\rho_n u_n$ in $L^2(L^{3/2})$ (if $\gamma_n\ge 3$), which yield

$$\|[u,R_i R_j](\rho u_j)\|_{L^1(0,T;L^{3/2})}\le C\|u\|_{L^2(0,T;BMO)}\|\rho u\|_{L^2(0,T;L^{3/2})}$$

and, multiplying by ρ, we deduce

$$\|\rho[u,R_i R_j](\rho u_j)\|_{L^1(0,T;L^1)}\le C\|\rho\|_{L^\infty(0,T;L^3)}\|u\|_{L^2(0,T;BMO)}\times$$
$$\times\|\rho u\|_{L^2(0,T;L^{3/2})}\ .$$

Finally for the fifth term, we have the following straightforward computation

$$\left| \int_0^T \int_{\mathbb{R}^N} \rho_n \operatorname{div} u_n \right| \leq \|\rho_n\|_{L^2(0,T;L^2)} \|\operatorname{div} u_n\|_{L^2(0,T;L^2)} . \tag{80}$$

where we use the fact that ρ_n is bounded in $L^2(0,T;L^2)$ since $\gamma_n \geq N \geq 2$.

Using this bound on $(\rho_n)^{\gamma_n+1}$ and the fact that $\rho \in L^\infty(0,T;L^1)$, we deduce the desired bound

$$\int_0^T \int_{\mathbb{R}^N} (\rho_n)^{\gamma_n} \leq \int_0^T \int_{\mathbb{R}^N} (\rho_n)^{\gamma_n+1} + \rho_n . \tag{81}$$

Hence extracting subsequences again, there exists $\pi \in \mathcal{M}((0,T) \times \Omega)$ such that

$$(\rho_n)^{\gamma_n} \underset{n}{\rightharpoonup} \pi . \tag{82}$$

Next, we explain the changes in the above argument we must perform in the periodic case. Equation (78) must be replaced by

$$(\rho_n)^{\gamma_n} - \int (\rho_n)^{\gamma_n} = \partial_t(-\Delta)^{-1}\operatorname{div}(\rho_n u_n) - R_i R_j (\rho(u_n)_i (u_n)_j) \tag{83}$$
$$+ (\mu + \xi)\operatorname{div}(u_n) .$$

Before integrating over $\mathbb{T}^N$, we multiply (as in the case of $\mathbb{R}^N$) by ρ^n. Notice however that we do so for different reasons than in the whole space case. In fact, since $\mathbb{T}^N$ is a bounded domain $\operatorname{div} u \in L^1((0,T) \times \Omega)$ and we can integrate (83) without any problem, but this integration gives no estimates on $(\rho_n)^{\gamma_n}$. Hence, we get

$$\int_0^T \int_{\mathbb{R}^N} (\rho_n)^{\gamma_n+1} - M \int_0^T \int_{\mathbb{R}^N} (\rho_n)^{\gamma_n} \leq C .$$

Then using as above (81) and the fact that $M < 1$, we get the desired bound.

Now, we turn to the case of Dirichlet boundary conditions. We introduce (see [24] and [10]) the following operator S defined by $Sg = p$ where (u,p) solves the following Stokes problem

$$-\Delta u + \nabla p = g \quad in \quad \Omega, \quad \int_\Omega p = 0, \tag{84}$$

$$\operatorname{div} u = 0, \quad u = 0 \quad on \quad \partial\Omega. \tag{85}$$

It is easy to see that the operator S maps $W^{-1,r}$ into L^r and L^r into $W^{1,r}$. Applying the above operator to (65), we get

$$(\rho_n)^{\gamma_n} - \int (\rho_n)^{\gamma_n} = -\partial_t S(\rho_n u_n) - S(\operatorname{div} \rho_n(u_n) \otimes (u_n)) \tag{86}$$
$$+ \xi \operatorname{div}(u_n) - \mu S(-\Delta u_n) .$$

Multiplying as above by ρ_n, we get after simple computations

$$\int_0^T \int_{\mathbb{R}^N} (\rho_n)^{\gamma_n+1} - M \int_0^T \int_{\mathbb{R}^N} (\rho_n)^{\gamma_n} \leq C$$

and we conclude as in the periodic case.

Step 3: Finally, we show that (ρ, u, π) is actually a solution of the initial system and that the strong convergences hold. First, we observe that $\frac{\partial \rho_n}{\partial t}$ is bounded in $L^\infty(0, T; W^{-1,1})$ and that u_n is bounded in $L^2(0, T; H^1)$. Then, using the compactness Lemma 6, we get that $\rho_n u_n$ converges weakly to ρu. On the other hand, using the bound on $(\rho_n)^{\gamma_n}$ in L^1, we deduce that $\frac{\partial \rho_n u_n}{\partial t}$ is bounded in $L^1(0, T; W^{-1,1})$ and hence $\rho_n u_n \otimes u_n \rightharpoonup \rho u \otimes u$. The only point that should be proved is the relation $\overline{\rho\pi} = \pi$. Next, we denote by $s = \rho \log(\rho)$ and $\bar{s} = \overline{\rho \log(\rho)}$. Hence, we get

$$\frac{\partial}{\partial t}(\bar{s} - s) + \operatorname{div}[(\bar{s} - s)u] = -\overline{\rho \operatorname{div} u} + \rho \operatorname{div} u \ . \tag{87}$$

Then applying the operator $(-\Delta)^{-1}\operatorname{div}$ to the momentum equation, multiplying by ρ_n and passing to the limit (extracting subsequences if necessary), we obtain (in the whole space case)

$$\begin{cases} \overline{(\rho_n)^{\gamma_n+1}} - (\mu + \xi)\overline{\rho \operatorname{div} u} = \partial_t \overline{[\rho(-\Delta)^{-1} \operatorname{div}(\rho u)]} \\ \qquad + \operatorname{div} \overline{[\rho u(-\Delta)^{-1}\operatorname{div}(\rho u)]} \\ \qquad + \overline{\rho u_i R_i R_j(\rho u_j)} - \overline{\rho R_i R_j(\rho u_i u_j)} \end{cases} \tag{88}$$

Changing the order of the multiplication by ρ and the passage to the weak limit, we obtain

$$\begin{cases} \rho\pi - (\mu + \xi)\rho \operatorname{div} u = \partial_t[\rho(-\Delta)^{-1}\operatorname{div}(\rho u)] \\ \qquad + \operatorname{div}[\rho u(-\Delta)^{-1}\operatorname{div}(\rho u)] \\ \qquad + \rho u_i R_i R_j(\rho u_j) - \rho R_i R_j(\rho u_i u_j) \ . \end{cases} \tag{89}$$

It is easy to see ([26]) that the second hand sides of (88) and (89) are equal. Hence, we get

$$-\overline{\rho \operatorname{div} u} + \rho \operatorname{div} u = \frac{1}{\mu + \xi}\left[\rho\pi - \overline{(\rho_n)^{\gamma_n+1}}\right] \ . \tag{90}$$

Let us only explain the changes in the above argument in the periodic case or the case of Dirichlet boundary conditions. In the first case, the left hand sides of (88) and (88) should be respectively replaced by

$$\overline{(\rho_n)^{\gamma_n+1}} - \overline{(\rho_n)} \fint (\rho_n)^{\gamma_n} - (\mu + \xi)\overline{\rho \operatorname{div} u} \tag{91}$$

and

$$\rho\pi - \rho \fint \pi - (\mu + \xi)\rho \operatorname{div} u \ . \tag{92}$$

Hence, using that $f(\rho_n)^{\gamma_n}$ is independent of x and that ρ_n is "compact" in t, we deduce that $\overline{(\rho_n) f(\rho_n)^{\gamma_n}} = \rho f \pi$ and (90) holds. In the case of Dirichlet boundary condition, one can deduce (90) just by localizing the above argument as in [23].

Reporting (90) in (87), we get

$$\frac{\partial}{\partial t}(\bar{s} - s) + \mathrm{div}\left[(\bar{s} - s)u\right] = \frac{1}{\mu + \xi}\left[\rho\pi - \overline{(\rho_n)^{\gamma_n+1}}\right] . \tag{93}$$

Next, we notice that $\rho\pi = \overline{\rho(\rho_n)^{\gamma_n}} \leq \overline{(\rho_n)^{\gamma_n+1}}$. Indeed we have

$$\overline{(\rho_n)^{\gamma_n+1}} - \overline{\rho_n(\rho_n)^{\gamma_n}} = \overline{((\rho_n)^{\gamma_n})(\rho_n - \rho)}$$

$$= \overline{\left((\rho_n)^{\gamma_n} - (\rho)^{\gamma_n}\right)\left(\rho_n - \rho\right)}$$

$$\geq 0$$

where we have used that

$$(\rho)^{\gamma_n} \to 1_{\{\rho=1\}}$$

almost everywhere and in $L^p((0,T) \times \Omega)$ for $1 \leq p < \infty$ which yields the following weak convergence

$$(\rho)^{\gamma_n}\left(\rho_n - \rho\right) \to 0$$

Next, integrating (93) in x, we get

$$\frac{\partial}{\partial t}\int_\Omega (\bar{s} - s)\ dx \leq 0 .$$

Then, since $\bar{s} - s|_{t=0} = 0$ and $s \leq \bar{s}$, we see that $s = \bar{s}$. Therefore, we obtain that

$$\rho\pi = \overline{(\rho_n)^{\gamma_n+1}}$$

Next, we see that for any $\varepsilon > 0$ there exists n_0 such that for $n \geq n_0$ and $x \geq 0$, we have

$$x^{\gamma_n+1} \geq x^{\gamma_n} - \varepsilon .$$

Applying this inequality to ρ_n and passing to the weak limit, we get

$$\overline{(\rho_n)^{\gamma_n+1}} \geq \pi - \varepsilon .$$

Then, letting ε go to 0, we get

$$\rho\pi \geq \pi .$$

Next, using that $0 \leq \rho \leq 1$, we obtain

$$\rho\pi \leq \pi .$$

However, since the product $\rho\pi$ is not defined almost everywhere, we must explain the above inequality. We denote by $\omega_k(t,x) = k^{N+1}\omega(kt, kx)$ a smoothing sequence in both variables t and x, where $\omega \in C^\infty(\mathbb{R}^{N+1})$, $\omega \geq 0$, $\int_{\mathbb{R}^{N+1}}\omega = 1$, $\mathrm{Supp}(\omega) \in B_1(\mathbb{R}^{N+1})$. Then, we denote by $\rho_k = \rho * \omega_k$ (resp $\pi_k = \pi * \omega_k$) a sequence of nonnegative smooth functions converging to ρ (resp π)

$$\begin{cases} \rho_k \to \rho & \text{in } C([0,T];L^p) \cap C^1([0,T];H^{-1}), \\ \pi_k \to \pi & \text{in } W^{-1,2}(H^1) + L^1(L^q), \end{cases} \tag{94}$$

for some $q > 1$ and p such that $\frac{1}{p} + \frac{1}{q} = 1$. Hence writing $(\rho-1)\pi$ as

$$(\rho-1)\pi = (\rho_k - 1)\pi_k + (\rho - \rho_k)\pi_k + (\rho-1)(\pi - \pi_k)$$

we conclude by letting k go to the infinity.

Finally, we deduce that

$$\rho\pi = \pi \ .$$

Therefore, (ρ, u, π) is a solution of (61)–(64). The strong convergences stated in the theorem are then deduced easily as in the proof of the compactness theorem.

Remark 15. In the initial system, we can replace the pressure term by $a\rho^{\gamma_n} + p(\rho)$ where $p(\rho)$ is a nondecreasing positive continuous function. The analysis is then exactly the same and we recover at the limit the same system with the following equation

$$\rho(\pi - p(\rho)) = \pi - p(\rho) \geq 0 \ . \tag{95}$$

instead of $\rho\pi = \pi \geq 0$, which means that $\pi = p(\rho)$ on $\{\rho < 1\}$ and $\pi \geq p(1)$ on $\{\rho = 1\}$.

5.1   The Case $M > 1$

In this subsection, we extend the result of [26] to the case of "general" initial data. We are concerned here with the periodic case or the case of Dirichlet boundary conditions. Let (ρ_n, u_n) be a sequence of solutions of (65) satisfying the above requirement but where $\int \rho_n^0 = M > 1$, $\int(\rho_n^0)^{\gamma_n} \leq M^{\gamma_n} + C\gamma_n$ for some fixed C.

Theorem 16. *Under the above assumptions, ρ_n converges to M in space $C([0,T];L^p(\Omega))$ for $1 \leq p < +\infty$, $\sqrt{\rho_n}\, u_n$ converges weakly to $\sqrt{M}u$ in $L^\infty(0,T;L^2(\Omega))$ and Du_n converges weakly to Du in $L^2(0,T;L^2(\Omega))$ for all $T \in (0,\infty)$ where u is a solution of the incompressible Navier-Stokes system*

$$\frac{\partial u}{\partial t} + \mathrm{div}\,(u \otimes u) - \frac{\mu}{M}\Delta u + \nabla p = 0,$$

$$\mathrm{div}\, u = 0, \quad u|_{t=0} = P(m^0) \ .$$

The proof of this result uses the same method introduced in subsection 4.4. Indeed, the major difficulty here is the passage to the limit in the advective term due to the presence of oscillations in time. Besides it turns out that there is no group which describes these oscillations. Let us begin by proving some uniform bounds. Using the conservation of mass, we see that Jensen's inequality yields for almost all t

$$\int_\Omega (\rho_n)^{\gamma_n}(t) \geq |\Omega| \left[\int_\Omega \rho_n\right]^{\gamma_n}(t) = |\Omega|M^{\gamma_n} .$$

Hence, we get the following uniform bounds

$$\int \frac{1}{2}\rho_n|u_n|^2(t) + \frac{a}{\gamma_n - 1}[(\rho_n)^{\gamma_n} - M^{\gamma_n}](t)$$
$$+ \int_0^t \int \mu|Du_n|^2 + \xi(\operatorname{div} u_n)^2 \leq \int \frac{1}{2}\rho_n^0|u_n^0|^2 + C . \tag{96}$$

Next, we can extract (as above) subsequences such that ρ_n converges weakly to some ρ in all the L^p, u_n converges weakly to some u in $L^2(0,T;H^1)$. Using that for all p $(1 < p < +\infty)$ and for n large enough $(\gamma_n > p)$, we have

$$\|\rho_n\|_{L^\infty(0,T;L^p)} \leq \|\rho_n\|_{L^\infty(0,T;L^1)}^{\theta_n}\|\rho_n\|_{L^\infty(0,T;L^{\gamma_n})}^{1-\theta_n}$$
$$\leq M_n^{\theta_n}(C\gamma_n + |\Omega|M^{\gamma_n})^{(1-\theta_n)/\gamma_n}$$

where θ_n is given for any n by $\frac{1}{p} = \theta_n + \frac{1-\theta_n}{\gamma_n}$, which yields uniform bounds for ρ_n in $L^\infty(0,T;L^p)$. Then, letting n go to infinity, we deduce that

$$\|\rho\|_{L^\infty(0,T;L^p)} \leq \liminf_{n\to\infty} \|\rho_n\|_{L^\infty(0,T;L^p)} \leq M .$$

Then, since $\fint \rho(t) = M$ for all t, we get, using the Jensen inequality (or letting p go to infinity) that $\rho \equiv M$. We also deduce the convergence of ρ_n to M in $L^p(0,T;L^p)$. Next, to get the convergence in $L^\infty(0,T;L^p)$, we write that

$$\rho_n^{\gamma_n} \geq M^{\gamma_n}\left[1 + \gamma_n\frac{(\rho_n - M)_+}{M} + \frac{\gamma_n(\gamma_n - 1)}{2}\left(\frac{(\rho_n - M)_+}{M}\right)^2\right] . \tag{97}$$

and deduce easily from the bounds we have on $\rho_n^{\gamma_n}$ that $(\rho_n - M)_+$ converges to 0 in $L^\infty(0,T;L^2)$ and then in $L^\infty(0,T;L^1)$. Using that $\fint \rho_n = M$, we get that $(\rho - M)_-$ goes to 0 in $L^\infty(0,T;L^1)$. Finally, to conclude we remark that we can use the expansion of $\rho_n^{\gamma_n}$ up to the order p as in (97) and deduce that $(\rho_n - M)_+$ tends to 0 in $L^\infty(0,T;L^p)$. Then, we see that

$$\int |\rho_n - M|^p \leq \int (\rho_n - M)_+^p + M^{p-1}(M - \rho_n)_+ . \tag{98}$$

Letting n go to infinity, we deduce that ρ_n converges to M in $L^\infty(0,T;L^p)$. Next we deduce that $\rho_n u_n$ converges weakly to Mu. It remains to show that

u solves the incompressible Navier-Stokes system. To this end, it is sufficient to pass to the limit in the advective term $\operatorname{div}(\rho_n u_n \otimes u_n)$ and recover at the limit $M \operatorname{div}(u \otimes u) + \nabla q$. As in the case where the Mach number goes to 0, we have just to pass to the limit in $\operatorname{div}(Q(\rho_n u_n) \otimes Q(u_n))$. Let us denote by $\phi_n = M^{\gamma_n}(\rho_n - M)$, hence using (97) and that $\int(\phi_n)_+ = \int(\phi_n)_-$, we deduce that $\|\phi_n\|_{L^1} \leq C$. Moreover, (65) can rewritten as

$$\begin{cases} \dfrac{\partial \phi_n}{\partial t} + M^{\gamma_n} \operatorname{div}(\rho_n u_n) = 0 , & \rho \geq 0, \\[2mm] \dfrac{\partial Q(\rho_n u_n)}{\partial t} + F_n + \nabla \rho_n^{\gamma_n} = 0 . \end{cases} \tag{99}$$

where $F_n = Q \operatorname{div}(\rho_n u_n \otimes u_n) - \mu_n \Delta Q u_n - \xi_n \nabla \operatorname{div} u_n$. Using the bound of $D u_n$ in L^2, we see that we can regularize the above system in x and that it is sufficient to pass to the limit in $\operatorname{div}(\nabla \psi_n^\delta \otimes \nabla \psi_n^\delta)$, where $Q(\rho_n u_n) = \psi_n$ and $\psi_n^\delta = \psi_n * K_\delta$. Let us also denote by $\phi_n^\delta = \phi_n * K_\delta$. Hence, we get

$$\begin{aligned} \operatorname{div}\left(\nabla \psi_n^\delta \otimes \nabla \psi_n^\delta\right) &= \frac{1}{2} \nabla |\nabla \psi_n^\delta|^2 + \Delta \psi_n^\delta \nabla \psi_n^\delta \\ &= \frac{1}{2} \nabla\left(|\nabla \psi_n^\delta|^2\right) - \frac{\partial}{\partial t}\left(\frac{\phi_n^\delta}{M^{\gamma_n}} \nabla \psi_n^\delta\right) \\ &\quad + \frac{\phi_n^\delta}{M^{\gamma_n}} F_n^\delta - \frac{\phi_n^\delta}{M^{\gamma_n}} \nabla\left((\rho_n)^{\gamma_n} * K_\delta - M^{\gamma_n}\right) \end{aligned}$$

and we conclude easily.

References

1. A. Babin, A. Mahalov, and B. Nicolaenko: Global splitting, integrability and regularity of 3D Euler and Navier-Stokes equations for uniformly rotating fluids. *European J. Mech. B Fluids*, 15(3): 291–300, 1996.
2. A. Babin, A. Mahalov, and B. Nicolaenko: Regularity and integrability of 3D Euler and Navier-Stokes equations for rotating fluids. *Asymptot. Anal.*, 15(2): 103–150, 1997.
3. C. Bardos: Existence et unicité de l'equation l'Euler en dimension deux. *J. Math. Pures Appl.*, 40: 769–790, 1972.
4. A. J. Bourgeois and J. T. Beale: Validity of the quasigeostrophic model for large-scale flow in the atmosphere and ocean. *SIAM J. Math. Anal.*, 25(4): 1023–1068, 1994.
5. J.-Y. Chemin: Fluides parfaits incompressibles. *Astérisque*, (230): 177, 1995.
6. T. Colin and P. Fabrie: Rotating fluid at high Rossby number driven by a surface stress: existence and convergence. *preprint*, 1996.
7. B. Desjardins and E. Grenier: Low Mach number limit of viscous compressible flows in the whole space. *R. Soc. Lond. Proc. Ser. A Math. Phys. Eng. Sci.*, 455(1986): 2271–2279, 1999.
8. B. Desjardins, E. Grenier, P.-L. Lions, and N. Masmoudi: Incompressible limit for solutions of the isentropic Navier-Stokes equations with Dirichlet boundary conditions. *J. Math. Pures Appl. (9)*, 78(5): 461–471, 1999.

9. P. F. Embid and A. J. Majda: Averaging over fast gravity waves for geophysical flows with arbitrary potential vorticity. *Comm. Partial Differential Equations*, 21(3-4): 619–658, 1996.

10. E. Feireisl and H. Petzeltová: On integrability up to the boundary of the weak solutions of the Navier-Stokes equations of compressible flow. *preprint*, 1999.

11. I. Gallagher: Applications of Schochet's methods to parabolic equations. *J. Math. Pures Appl. (9)*, 77(10): 989–1054, 1998.

12. N. Garofalo and F. Segàla: Another step toward the solution of the Pompeiu problem in the plane. *Comm. Partial Differential Equations*, 18(3-4): 491–503, 1993.

13. H. Greenspan: *The theory of rotating fluids,*. Cambridge monographs on mechanics and applied mathematics, 1969.

14. E. Grenier: Oscillatory perturbations of the Navier-Stokes equations. *J. Math. Pures Appl. (9)*, 76(6): 477–498, 1997.

15. E. Grenier and N. Masmoudi: Ekman layers of rotating fluids, the case of well prepared initial data. *Comm. Partial Differential Equations*, 22(5-6): 953–975, 1997.

16. T. Kato: Nonstationary flows of viscous and ideal fluids in R^3. *J. Functional Analysis*, 9: 296–305, 1972.

17. S. Klainerman and A. Majda: Singular limits of quasilinear hyperbolic systems with large parameters and the incompressible limit of compressible fluids. *Comm. Pure Appl. Math.*, 34(4): 481–524, 1981.

18. S. Klainerman and A. Majda: Compressible and incompressible fluids. *Comm. Pure Appl. Math.*, 35(5): 629–651, 1982.

19. J. Leray: Etude de diverses équations intégrales nonlinéaires et de quelques problèmes que pose l'hydrodynamique. *J. Math. Pures Appl.*, 12: 1–82, 1933.

20. J. Leray: Essai sur les mouvements plans d'un liquide visqueux emplissant l'espace. *Acta. Math.*, 63: 193–248, 1934.

21. J. Leray: Essai sur les mouvements plans d'un liquide visqueux qui limitent des parois. *J. Math. Pures Appl.*, 13: 331–418, 1934.

22. P.-L. Lions: *Mathematical topics in fluid mechanics. Vol. 1*. The Clarendon Press Oxford University Press, New York, 1996. Incompressible models, Oxford Science Publications.

23. P.-L. Lions: *Mathematical topics in fluid mechanics. Vol. 2*. The Clarendon Press Oxford University Press, New York, 1998. Compressible models, Oxford Science Publications.

24. P.-L. Lions: Bornes sur la densité pour les équations de Navier-Stokes compressibles isentropiques avec conditions aux limites de Dirichlet. *C. R. Acad. Sci. Paris Sér. I Math.*, 328(8): 659–662, 1999.

25. P.-L. Lions and N. Masmoudi: Incompressible limit for a viscous compressible fluid. *J. Math. Pures Appl. (9)*, 77(6): 585–627, 1998.

26. P.-L. Lions and N. Masmoudi: On a free boundary barotropic model. *Ann. Inst. H. Poincaré Anal. Non Linéaire*, 16(3): 373–410, 1999.

27. P.-L. Lions and N. Masmoudi: Une approche locale de la limite incompressible. *C. R. Acad. Sci. Paris Sér. I Math.*, 329(5): 387–392, 1999.

28. N. Masmoudi: The Euler limit of the Navier-Stokes equations, and rotating fluids with boundary. *Arch. Rational Mech. Anal.*, 142(4): 375–394, 1998.

29. N. Masmoudi: Ekman layers of rotating fluids: the case of general initial data. *Comm. Pure Appl. Math.*, 53(4): 432–483, 2000.

30. N. Masmoudi: Incompressible, inviscid limit of the compressible navier-Stokes system. *Ann. Inst. H. Poincaré Anal. Non Linéaire*, to appear, 2000.
31. J. Pedlovsky: *Geophysical fluid dynamics,*. Springer, 1979.
32. S. Schochet: Fast singular limits of hyperbolic PDEs. *J. Differential Equations*, 114(2): 476–512, 1994.
33. H. S. G. Swann: The convergence with vanishing viscosity of nonstationary Navier-Stokes flow to ideal flow in R^3. *Trans. Amer. Math. Soc.*, 157: 373–397, 1971.
34. S. Ukai: The incompressible limit and the initial layer of the compressible Euler equation. *J. Math. Kyoto Univ.*, 26(2): 323–331, 1986.

Weighted Spaces with Detached Asymptotics in Application to the Navier-Stokes Equations

Serguëi A. Nazarov

St. Petersburg University, Math. Mech. Department, Theory of Elasticity, Bibliotechnaya pl. 2, 198904 St. Petersburg, Russia
email: `serna@snark.ipme.ru`

Abstract. Function spaces with weighted norms and detached asymptotics naturally appear in the treatment of boundary value problems when linear and nonlinear terms have got same asymptotic behavior either at a singularity point of the boundary, or at infinity. The characteristic feature of these spaces is that their norms are composed from both, norms of angular parts in the detached terms and norms of asymptotic remainders. The developed approach is described for the Navier-Stokes problems in domains with conical (angular) outlets to infinity while the 3-D exterior and 2-D aperture problems imply representative examples. With a view towards compressible and non-Newtonian fluids, the described technique is applied to the transport equation as well.

1 Introduction

We consider the Stokes and Navier-Stokes stationary problems in a domain $\Omega \subset \mathbb{R}^n$, $n = 2, 3$, which has a smooth $(n-1)$–dimensional boundary $\partial\Omega$ and a conical (angular at $n = 2$) outlet to infinity. More precisely, the outlet is formed by either the complete cone $\mathbb{R}^n \setminus \mathcal{O} = \{x : r := |x| > 0\}$, or the half-space $\mathbb{R}^n_+ = \{x : x_n > 0\}$. In the first case Ω becomes the exterior domain

$$\Omega_e = \mathbb{R}^n \setminus G \qquad (1)$$

where G is a compact set with the smooth boundary $\partial G = \partial\Omega$. In the second case Ω is a perturbation Ω_+ of $\mathbb{R}^n_+$ inside the ball $\mathbb{B}_R = \{x : r < R\}$ with the radius $R > 0$; in other words, Ω_+ has a smooth boundary and satisfies the relation

$$\Omega_+ \setminus \mathbb{B}_R = \mathbb{R}^n_+ \setminus \mathbb{B}_R . \qquad (2)$$

We emphasize that the latter type of domains is closely related to the aperture domain

$$\Omega_a = \{x \in \mathbb{R}^n : x_n \neq 0\} \cup \Gamma_0 \qquad (3)$$

where $\Gamma_0 = \{x : x' := (x_1, \ldots, x_{n-1}) \in \Gamma, x_n = 0\}$ and Γ is a subdomain of $\mathbb{R}^{n-1}$ with a smooth boundary $\partial\Gamma$ and a compact closure $\overline{\Gamma}$. In contrast to Ω_+, the domain (3) possesses two outlets $\mathbb{R}^{n-1}_\pm$ to infinity and its boundary $\partial\Omega_a$ is piecewise smooth since it contains the edge $(\partial\Gamma)_0$ (two angular points if $n = 2$). Although $\mathbb{R}^n_+$ and $\mathbb{R}^n_-$, of course, may be treated in the same way and a modification of our approach is evident for domains with those boundary irregularities, in the sequel we deal with Ω_+ only for simplicity.

While rewriting the Stokes problem

$$\begin{aligned}
-\nu\Delta_x v(x) + \nabla_x p(x) &= f'(x), && x \in \Omega, \\
-\nabla_x \cdot v(x) &= f_{n+1}(x), && x \in \Omega, \\
v(x) &= g(x), && x \in \partial\Omega,
\end{aligned} \tag{4}$$

in the condensed one-line form

$$S(\nabla_x)u = f \quad \text{in} \quad \Omega, \qquad u' = g \quad \text{on} \quad \partial\Omega, \tag{5}$$

we get the analogous form for the Navier-Stokes problem

$$S(\nabla_x)u + (N(\nabla_x)u', 0) = f \quad \text{in} \quad \Omega, \qquad u' = g \quad \text{on} \quad \partial\Omega \tag{6}$$

where $\nabla_x = \mathrm{grad}, \nabla_x \cdot = \mathrm{div}, \Delta_x = \nabla_x \cdot \nabla_x$ is the Laplacian,

$$N(\nabla_x)u' := (u' \cdot \nabla_x)u', \tag{7}$$

$\nu > 0$ is the constant viscosity of the fluid,

$$\begin{aligned}
u &= (u', u_{n+1}) = (u_1, \ldots, u_{n+1}), \\
f &= (f', f_{n+1}) = (f_1, \ldots, f_{n+1}), \\
g &= (g_1, \ldots, g_n),
\end{aligned}$$

$v = u' = (u_1, \ldots, u_n)$ and $p = u_{n+1}$ stand for the velocity vector and the pressure, respectively. In what follows the problem (6) in the domain (1) is shortly denoted by $(6)_e$.

The paper focuses on the derivation of asymptotic expansions of solutions to the problem (6) in domains of the aforementioned types. To this end, we employ *weighted Hölder spaces with detached asymptotics* which, by definition, consist of functions taking suitable asymptotic forms. Thus, our task reduces to the verification of existence and uniqueness of solutions to the Navier-Stokes problems in those spaces. With this aim, we apply the contraction principle which requires the problem data to be small.

At the first sight, the asymptotic results obtained in Sections 4–6 do crucially depend on the dimension n and the type of the domain Ω. However, as shown in Section 2 where we discuss and construct *power solutions* to the model Stokes problems in the cones $\mathbb{R}^n \setminus \mathcal{O}$ and $\mathbb{R}^n_+$, a simple analysis of the degrees of possible power solutions detects both, basic properties of the Stokes problem operator in weighted spaces and asymptotic forms admitted by solutions (cf. also Section 3). In particular, this analysis demonstrates that in 3-D problem $(6)_+$ (in 2-D problem $(6)_e$) the nonlinear term turns out to be *weaker* (*stronger*) than the linear one and, therefore, the desired asymptotic expansions can be concluded by rather simple arguments (this cannot be derived by means of tools we describe here). At the same time, in the 3-D problem $(6)_e$ and the 2-D problem $(6)_+$ those terms become of *the same order* $\mathcal{O}(r^{-3})$ as $r \to \infty$ while detaching certain asymptotic terms helps to overcome serious difficulties appearing in cases when usual Sobolev and Hölder spaces with either original, or weighted norms are employed.

In Section 2 and Section 3 we simply treat the Stokes problem as some formally self-adjoint elliptic boundary value problem and we make use of general results in [6,8,9] (see also the survey [19] and the book [14] which, for particular theorems, we refer to). Those results are also sufficient to study the 3-D problem $(6)_+$ in Section 4 (cf. [3]). Weighted spaces with detached asymptotics become necessary in Sections 5–7 where we investigate the 3-D problem $(6)_e$ (see [20,21]), the 2-D problem $(6)_+$ (see [16,18] and [25]), and the transport equation (see [23]).

The papers cited in relation with Sections 5–7 involve weighted Sobolev spaces with detached asymptotics. Advantages of Hölder norms we use here, become evident while processing nonlinear terms and deriving pointwise decay estimates.

Function spaces with detached asymptotics were treated in many papers for different reasons and purposes. In the book [31], they are regarded as original domains of pseudo-differential operators on piecewise smooth manifolds while the corresponding operator calculus is developed. Generalized Green's formulae for elliptic problems in domains with conical points and edges are related to these spaces as well and, in [12–14], they provide domains of self-adjoint extensions of differential operators in Sobolev spaces with both, usual and weighted norms. Some specific aspects of the spaces with detached asymptotics are outlined in [15,5], namely those that are connected with the classical method of matched asymptotic expansions. We also mention the papers [22] and [17] where such techniques maintain peculiar problems of continuum mechanics.

In all the papers cited in the above paragraph, it was sufficient to detach *linear combinations* of power solutions with *fixed* angular parts so that a basic weighted space is amplified with a *finite-dimensional* complement only. In contrast to those, the complements here are *infinite-dimensional* since angular parts in detached asymptotics influenced by the nonlinear term $N(\nabla_x)v$, must be taken *arbitrary*. In other words, the complements become proper Sobolev and Hölder spaces on the intersection of the conical outlet and the unit sphere. For other geometries, namely, for cylindrical and quasi-cylindrical (periodic) outlets, similar spaces are exploited in [26,32].

2   Power Solutions

An important role in our further investigations is played by *power solutions* of the Stokes problem in a cone

$$\mathbb{K} := \{x \in \mathbb{R}^n : r := |x| > 0, \varphi := |x|^{-1}x \in \omega\}$$

where (r, φ) are spherical (polar at $n = 2$) coordinates and ω is a domain on the unit sphere $\mathbb{S}^{n-1}$. In accordance with (1) and (2) we are interested in the specific cases $\omega = \mathbb{S}^{n-1}$ ($\mathbb{K}$ is the punctured space) and $\omega = \mathbb{S}^{n-1}_+ :=$ $\mathbb{S}^{n-1} \cap \mathbb{R}^n_+$ ($\mathbb{K}$ is the half-space). If $\lambda \in \mathbb{C}$ and $\mathcal{U}$ is sufficiently smooth vector function on $\overline{\omega}$, we determine the power solution U by the formulae

$$U(x) = (V(x), P(x)),$$
$$V(x) = r^\lambda \mathcal{V}(\varphi), \quad P(x) = r^{\lambda-1} \mathcal{P}(\varphi), \tag{8}$$
$$\mathcal{U}(\varphi) = (\mathcal{U}'(\varphi), \mathcal{U}_{n+1}(\varphi)) := (\mathcal{V}(\varphi), \mathcal{P}(\varphi)) \ .$$

The exponent λ is called (generalized) *degree* of the power solution $(8)_2$. We stress that, owing to differential structures of the matrix operator $S(\nabla_x)$, the pressure component P gets a smaller exponent $\lambda - 1$ than the velocity V.

Example 1. The constant velocity $(C', 0)^t$ with $C' \in \mathbb{R}^n$ takes the form $(8)_2$ and its degree is equal to 0. The constant pressure $(0, C_{n+1})^t$ is also of the form $(8)_2$, but has degree 1 (t indicates transposition).

In view of (4) we have

$$S(\nabla_x)U(x) = F(x), \quad x \in \mathbb{K},$$
$$F'(x) = r^{\lambda-2}\mathcal{F}'(\varphi), \quad F_{n+1}(x) = r^{\lambda-1}\mathcal{F}_{n+1}(\varphi) \tag{9}$$

and, hence, separating the variable r reduces the equation $(9)_1$ on the sphere,

$$S(\lambda, \varphi, \nabla_\varphi)\mathcal{U}(\varphi) = \mathcal{F}(\varphi) := (\mathcal{F}'(\varphi), \mathcal{F}_{n+1}(\varphi)), \quad \varphi \in \omega \ . \tag{10}$$

The problem (5) is related to the Green formula

$$(S(\nabla_x)u^1, u^2)_\Omega + (T(x, \nabla_x)u^1, v^2)_{\partial\Omega}$$
$$- (u^1, S(\nabla_x)u^2)_\Omega - (v^1, T(x, \nabla_x)u^2)_{\partial\Omega} = 0 \tag{11}$$

where $u^i = (v^i, p^i) \in C_0^\infty(\overline{\Omega})^{n+1}$, $(\ ,\ )_\Xi$ stands for the inner product in $L^2(\Omega)^m$, $\mathbf{n}$ is the unit vector of the outward normal to $\partial\Omega$, $\partial_\mathbf{n} = \mathbf{n}(x) \cdot \nabla_x$,

$$T(x, \nabla_x)u(x) = \nu\partial_\mathbf{n}v(x) - \mathbf{n}(x)p(x) \ . \tag{12}$$

If $\Omega = \mathbb{K}$, then similarly to (9) the power solution (8) satisfies the relation

$$T(x, \nabla_x)U(x) = r^{\lambda-1}\mathcal{T}(\lambda, \varphi, \nabla_\varphi)\mathcal{U}(\varphi), \quad x \in \partial\mathbb{K} \setminus \mathcal{O} \ . \tag{13}$$

Lemma 2. *Let* $\mathcal{U}^i = (\mathcal{V}^i, \mathcal{P}^i) \in C^\infty(\overline{\omega})^{n+1}$, $i = 1, 2$, *and* $\lambda \in \mathbb{C}$. *Then the Green formula*

$$(S(\lambda, \varphi, \nabla_\varphi)\mathcal{U}^1, \mathcal{U}^2)_\omega + (\mathcal{T}(\lambda, \varphi, \nabla_\varphi)\mathcal{U}^1, \mathcal{V}^2)_{\partial\omega}$$
$$- (\mathcal{U}^1, S(2 - n - \overline{\lambda}, \varphi, \nabla_\varphi)\mathcal{U}^2)_\omega$$
$$- (\mathcal{V}^1, \mathcal{T}(2 - n - \overline{\lambda}, \varphi, \nabla_\varphi)\mathcal{V}^2)_{\partial\omega} = 0 \tag{14}$$

is valid on the subdomain ω *of the sphere* $\mathbb{S}^{n-1}$.

Proof. Into Green's formula (11) on $\Omega = \mathbb{K}$, we insert the vector functions

$$u^1(x) = \chi_N(r)(r^\lambda \mathcal{V}^1(\varphi), r^{\lambda-1}\mathcal{P}^1(\varphi)),$$

$$u^2(x) = \chi_N(r)(r^{2-n-\bar\lambda}\mathcal{V}^2(\varphi), r^{1-n-\bar\lambda}\mathcal{P}^2(\varphi))$$

where $N \in \mathbb{R}_+$, $\chi_N(r) = \chi(\log r - N)\chi(-\log r - N)$ and χ is a cut-off function,

$$\chi \in C^\infty(\mathbb{R}), \quad \chi(t) = 1 \quad \text{as} \quad t \leq 1/2, \quad \chi(t) = 0 \quad \text{as} \quad t \geq 1 .$$

By commuting the differential operators $\mathcal{S}(\ldots)$ and $\mathcal{T}(\ldots)$ with the cut-off functions $\chi(\ldots)$, we obtain the relation

$$Q(u^1, u^2) = \int\limits_{\exp(-N)}^{\exp(N)} \mathcal{Q}(\mathcal{U}^1, \mathcal{U}^2)\frac{dr}{r} + \mathcal{O}(1) =$$

$$= 2N\mathcal{Q}(\mathcal{U}^1, \mathcal{U}^2) + \mathcal{O}(1) \quad \text{as} \quad N \to +\infty \quad (15)$$

where $Q(u^1, u^2)$ and $\mathcal{Q}(\mathcal{U}^1, \mathcal{U}^2)$ denote the left-hand sides of (11) and (14), respectively. To derive the formula (14) from (11), it remains to multiply (15) by $(2N)^{-1}$ and perform the limit passage $N \to +\infty$. $\qquad\square$

In view of the representations (9) and (10), the homogeneous Stokes problem

$$S(\nabla_x)U(x) = 0, \ x \in \mathbb{K}; \quad U'(x) = 0, \ x \in \partial\mathbb{K} \setminus \mathcal{O}, \qquad (16)$$

has a nontrivial power solution (8) if and only if λ is an eigenvalue of the spectral problem

$$\mathcal{S}(\lambda, \varphi, \nabla_\varphi)\mathcal{U}(\varphi) = 0, \ \varphi \in \omega; \quad \mathcal{U}'(\varphi) = 0, \ \varphi \in \partial\omega, \qquad (17)$$

and $\mathcal{U} \neq 0$ is the corresponding eigenvector. By virtue of Lemma 2 the problem (17) itself and the problem (17) under the substitution

$$\lambda \mapsto 2 - n - \bar\lambda \qquad (18)$$

are formally adjoint. This means that the spectrum $\sigma(\omega)$ of the problem (17) is invariant with respect to the transformation (18); in other words, the set $\sigma(\omega) \subset \mathbb{C}$ possesses the symmetry centre $\lambda = 1 - n/2$. Here we avoid to discuss other generic properties of $\sigma(\omega)$ (see [14], §1.2, 6.1) because $\sigma(\mathbb{S}^{n-1})$ and $\sigma(\mathbb{S}_+^{n-1})$ will be found explicitly.

For the exterior domain (1), the model Stokes problem (16) is posed on the punctured space. Thus, it consists only of the system

$$S(\nabla_x)U(x) = 0, \ x \in \mathbb{R}^n \setminus \mathcal{O}, \qquad (19)$$

while the boundary condition on $\partial\omega = \emptyset$ disappears from the spectral problem (17).

Lemma 3. *Any nontrivial power solution* (8) *of the model problem* (19) *is either a polynomial, or a linear combination of derivatives of the columns* $\Phi^1(x), \ldots, \Phi^{n+1}(x)$ *which fulfill the systems of differential equations*

$$S(\nabla_x)\Phi^j(x) = \delta(x)\mathrm{e}^j, \quad x \in \mathbb{R}^n, \tag{20}$$

where $\mathrm{e}^j = (\delta_{j,1}, \ldots, \delta_{j,n+1})$; $\delta(x)$ *and* $\delta_{j,k}$ *mean Dirac's* δ-*function and Kronecker's symbol. In other words,* $\Phi = (\Phi^1, \ldots, \Phi^{n+1})$ *implies the fundamental* $(n+1) \times (n+1)$-*matrix for the operator* $S(\nabla_x)$ *in* $\mathbb{R}^n$.

Proof. Let us start with two simple observations. First, if the problem (19) has a nontrivial power solution U of degree $\lambda \neq 0$, then at least one among the derivatives $\partial U/\partial x_1, \ldots, \partial U/\partial x_n$ implies a nontrivial power solution of degree $\lambda - 1$, i.e.,

$$\lambda \in \sigma(\mathbb{S}^{n-1}) \setminus \{0\} \Rightarrow \lambda - 1 \in \sigma(\mathbb{S}^{n-1}) . \tag{21}$$

Second, the representations $(8)_2$ with $\operatorname{Re}\lambda > 1 - n/2$ provide the inclusion $U = (V, P) \in H^1_{\mathrm{loc}}(\mathbb{R}^n)^n \times L^2_{\mathrm{loc}}(\mathbb{R}^n)$. Thus, the equation (19) can be extended at $x = 0$ and U becomes a smooth vector function, namely, a polynomial due to the representation $(8)_2$. In other words,

$$\lambda \in \sigma(\mathbb{S}^{n-1}), \operatorname{Re}\lambda > 1 - n/2 \Rightarrow \lambda \in \mathbb{N}_0 := \{0, 1, \ldots\} . \tag{22}$$

Recalling Example 1 and the symmetry property

$$\lambda \in \sigma(\omega) \Rightarrow 2 - n - \overline{\lambda} \in \sigma(\omega), \tag{23}$$

we derive from (21) and (22) that

$$\sigma(\mathbb{S}^{n-1}) = \mathbb{Z} := \{0, \pm 1, \ldots\}, \quad n = 2, 3 . \tag{24}$$

As known (see e.g. [7]), $\Phi^1(x), \ldots, \Phi^n(x)$ and $\Phi^{n+1}(x)$ are power solutions of degrees $2 - n$ and $1 - n$, respectively. (see Remark 5 for the case $n = 2$). It is worth to note that power solutions are defined on the punctured space and, hence, the equation (19) follows from (20) as well as the pressure component $\Phi^{n+1}_{n+1}(x) = \nu\delta(x)$ vanishes while the column $\Phi^{n+1}(x)$ is treated as a power solution. Differentiating equation (20) we recognize any derivative of Φ^j as a power solution, too. To finish the proof, it is now sufficient to mention that the eigenvalues λ and $2 - n - \overline{\lambda}$ in (23) have the same multiplicity and, for any polynomial $U(x)$ of generalized degree m, we get the power solution

$$U(\nabla_x) \cdot \Phi(x) = \sum_{j=1}^{n+1} U_j(\nabla_x)\Phi^j(x)$$

of degree $2 - n - m$ which is nontrivial provided $U \neq 0$ satisfies (19) (see [14], §6.4 for details). $\square$

Arguing in a similar way, one can conclude with the following assertion (cf. [3]).

Lemma 4. *Any nontrivial power solution* (8) *of the model problem*

$$S(\nabla_x)U(x) = 0, \quad x \in \mathbb{R}^{n+1}_+,$$
$$U'(x',0) = 0, \quad x' := (x_1,\ldots,x_{n-1}) \in \mathbb{R}^{n-1} \setminus \mathcal{O}, \tag{25}$$

is either a polynomial, or a linear combination of tangential derivatives of the columns $\Psi^1(x),\ldots,\Psi^n(x)$ *which fulfill the boundary value problems*

$$(\nabla_x)\Psi^j(x) = 0, \quad x \in \mathbb{R}^n_+, \quad \Psi^{j'}(x',0) = \delta(x')\,e^{j'}, \quad x' \in \mathbb{R}^{n-1} . \tag{26}$$

In other words, $\Psi = (\Psi^1,\ldots,\Psi^n)$ *implies the Poisson kernel of the Dirichlet problem for the operator* $S(\nabla_x)$ *in the half-space.*

Owing to the Dirichlet conditions $(25)_2$, a power solution U of degree 0 becomes trivial (see Example 1). The model problem (25) admits polynomial solutions of generalized degree 1, namely,

$$(0,0,1)^t, \ (x_2,0,0)^t \quad \text{if} \quad n = 2,$$
$$(0,0,0,1)^t, \ (x_3,0,0,0)^t, \ (0,x_3,0,0)^t \quad \text{if} \quad n = 3 . \tag{27}$$

The first vector in $(27)_1$ and $(27)_2$ is nothing but a constant pressure, the other the Couette flows. We mention also that, as known (see e.g. [10]), the Poisson kernel columns are power solutions of degree $1 - n = 2 - n - 1$ (cf. (23)). Thus, by Lemma 4 we arrive at the equality

$$\sigma(\mathbb{S}^{n-1}_+) = \mathbb{Z} \setminus \Sigma(\mathbb{S}^{n-1}_+) \tag{28}$$

where $\Sigma(\mathbb{S}^{n-1}_+) = (1 - n, 1)$. The degree set (28) splits naturally into the couple of sets

$$\sigma_i(\mathbb{S}^{n-1}_+) = \mathbb{N} := \{1, 2, \ldots\},$$
$$\sigma_e(\mathbb{S}^{n-1}_+) = \{1 - n, -n, \ldots\} . \tag{29}$$

The set $(29)_1$ corresponds to polynomial power solutions which are involved in Taylor expansions at a boundary point $\mathcal{O}$ of smooth solutions $u(x)$ to the interior Stokes problem in Ω_i with no-slip condition $u' = 0$ on $\partial\Omega_i$. On the other hand, power solutions of degrees $\lambda \in \sigma_e(\mathbb{S}^{n-1}_+)$ compose asymptotics at infinity of a decaying solution to the problem $(5)_+$ with compactly supported right-hand sides (see Theorem 13 below). We emphasize that between the sets (29) there is a non-empty gap $\Sigma(\mathbb{S}^{n-1}_+)$.

Let us divide the degree set (24) into the subsets

$$\sigma_i(\mathbb{S}^{n-1}) = \mathbb{N}_0,$$
$$\sigma_e(\mathbb{S}^{n-1}) = \{2 - n, 1 - n, \ldots\} . \tag{30}$$

The first subset is related to polynomial power solutions and Taylor expansions of smooth solutions at the interior point $\mathcal{O}$ of the bounded domain Ω_i. The second set consists of degrees of power solutions occurring in the asymptotic forms of solutions to the problem $(5)_e$ with $f = 0$ (see [7] and Theorem 11 below).

Remark 5. As in the case (29), the sets (30) for $n = 3$ are separated by the interval $\Sigma(\mathbb{S}^2) = (-1,0)$. At the same time, the lower bound of $\sigma_i(\mathbb{S}^1)$ and the upper bound of $\sigma_e(\mathbb{S}^1)$ coincide since solutions to the 2-D problem $(5)_e$ with $f = 0$ do not decay at infinity but stabilize to a constant velocity vector. Hence, the interval $\Sigma(\mathbb{S}^{n-1}) = (2 - n, 0)$ degenerates for $n = 2$ and that requires a modification of the approach employed in Section 3 for linear problems (see [14], §6.4). The 2-D problems $(5)_e$ and $(6)_e$ also possess other distinct features. First, the columns Φ^1 and Φ^2 gain logarithmic factors but nevertheless can be treated as power solutions of degree 0 (indeed, according to [14] the angular part $\mathcal{U} = (\mathcal{V}, \mathcal{P})$ of the power solution $(8)_2$ may polynomially depend on $\log r$ and we avoided to mention this around (8) because henceforth such power solutions do not appear in our paper). Second, the nonlinear term in the 2-D exterior Navier-Stokes problem, as we shall see, happens to be stronger than the linear one and, consequently, our approach based on a linearization does not work for this problem at all.

Similarly to (9), the nonlinear term (7) calculated for the power solution (8), satisfies the relations

$$N(\nabla_x)U'(x) = H'(x) = r^{2\lambda-1}\mathcal{H}(\varphi), \quad x \in \mathbb{K},$$
$$\mathcal{N}(\lambda, \varphi, \nabla_\varphi)\mathcal{U}'(\varphi) = \mathcal{H}'(\varphi), \quad \varphi \in \omega. \tag{31}$$

By comparing formulae $(31)_1$ and (9) where $\lambda \in \mathbb{R}$, we observe that for $\lambda < -1$ $(\lambda > -1)$ the linear term $S(\nabla_x)U(x)$ attains a larger (smaller) exponent and becomes weaker (stronger) as $r \to \infty$ than the convective term $N(\nabla_x)U(x)$. If $\lambda = -1$, both terms are of the same order.

The entries in $\sigma_e(\mathbb{S}^2_+)$ fulfill the inequality $\lambda \leq -2$ and, hence, a fast decay of solutions to 3-D problem $(5)_+$ provides the 3-D problem $(6)_+$ with a weak nonlinearity and reduces the nonlinear asymptotic structures to linear ones. Both $\sigma_e(\mathbb{S}^2)$ and $\sigma_e(\mathbb{S}^1_+)$ include the critical exponent $\lambda = -1$ equalizing the asymptotic orders of $S(\nabla_x)u(x)$ and $N(\nabla_x)u'(x)$ in the 3-D problem $(6)_e$ and the 2-D problem $(6)_+$ which attract the most attention in the paper. On the contrary, the set $\sigma_e(\mathbb{S}^1)$ contains the degree $\lambda = 0$ which exceeds the critical one and makes the nonlinear term in the 2-D problem $(6)_e$ dominate over the linear one.

3   The Stokes Problem in Weighted Hölder Spaces

The change of variables

$$x \mapsto (r, \varphi) \mapsto (t, \varphi) := (\log r, \varphi) \tag{32}$$

transforms the cone $\mathbb{K}$ into the cylinder $\mathbb{Q} = \mathbb{R} \times \omega$. With $l \in \mathbb{N}_0$ and $\alpha \in (0,1)$, we determine the Hölder space $C^{l,\alpha}(\mathbb{Q})$ as a completion of $C_0^\infty(\overline{\mathbb{Q}})$ with respect to the norm

$$\|w; C^{l,\alpha}(\mathbb{Q})\| = \sum_{k=0}^{l} \sup_{\mathbb{Q}} |\nabla_{(t,\varphi)}^k w(t,\varphi)|$$

$$+ \sup_{\cdots} \left\{ [|\varphi^1 - \varphi^2| + |t_1 - t_2|]^{-\alpha} |\nabla_{(t,\varphi)}^l w(t_1,\varphi^1) - \nabla_{(t,\varphi)}^l w(t_2,\varphi^2)| \right\} \quad (33)$$

where $\nabla^k w$ denotes the collection of all order k derivatives of the function w, the dots stand for the set $\{\varphi^1, \varphi^2 \in \omega, t_1, t_2 \in \mathbb{R} : |t_1 - t_2| < 1\}$ and $|\varphi^1 - \varphi^2|$ for the distance between φ^1 and φ^2 on the sphere $\mathbb{S}^{n-1}$. We preserve symbol w for the function written in any coordinate system and perform the inverse change of variables (32). Since $\partial_t := \partial/\partial t = r\partial_r$ and $\nabla_x = r^{-1}(r\partial_r, \nabla_\varphi)$, we obtain that the norm (33) is equivalent to

$$\sum_{k=0}^{l} \sup_{\mathbb{R}_+ \times \omega} |(r\partial_r, \nabla_\varphi)^k w(r,\varphi)|$$

$$+ \sup_{\cdots} \left\{ \left[|\varphi^1 - \varphi^2| + \left| \log \frac{r_1}{r_2} \right| \right]^{-\alpha} |(r\partial_r, \nabla_\varphi)^l w(r_1,\varphi^1) - (r\partial_r, \nabla_\varphi)^l w(r_2,\varphi^2)| \right\}$$

$$\sim \sum_{k=0}^{l} \sup_{\mathbb{K}} \left| |x|^k \nabla_x^k w(x) \right|$$

$$+ \sup_{\cdots} \left\{ |x^1 - x^2|^{-\alpha} \left| |x^1|^{l+\alpha} \nabla_x^l w(x^1) - |x^2|^{l+\alpha} \nabla_x^l w(x^2) \right| \right\} . \quad (34)$$

In (34) the dots mean the following set:

$$\left\{ r_1, r_2 \in \mathbb{R}_+, \varphi^1, \varphi^2 \in \omega : \left| \log \frac{r_1}{r_2} \right| < 1 \right\} = \left\{ x^1, x^2 \in \mathbb{K} : \frac{1}{e} < \frac{|x^1|}{|x^2|} < e \right\} . \quad (35)$$

Remark 6. The Napier number e may be replaced in (34), (35) by any constant $c > 1$. The equivalence of the norms in (34) is provided by the evident inequalities

$$[|\varphi^1 - \varphi^2| + |\log \frac{r_1}{r_2}|]^{-\alpha} |W_1 - W_2| \leq$$

$$\leq \left| \log \frac{r_1}{r_2} \right|^{-\alpha} \left| 1 - \frac{r_2^\alpha}{r_1^\alpha} \right| |W_1| + \left[r_1 |\varphi^1 - \varphi^2| + r_1 \left| \log \frac{r_1}{r_2} \right| \right]^{-\alpha} |r_1^\alpha W_1 - r_2^\alpha W_2|,$$

$$|r_1 \varphi^1 - r_2 \varphi^2| \leq r_1 |\varphi^1 - \varphi^2| + r_1 |1 - r_1^{-1} r_2|,$$

$$\left| \log \frac{r_1}{r_2} \right|^{-\kappa} \left| 1 - \frac{r_2^\kappa}{r_1^\kappa} \right| \leq C, \ \kappa \in (0,1],$$

where $W_i = r_i^l \nabla_x^l w(x^i)$ and $|\log \frac{r_1}{r_2}| < 1$.

In accordance with (33) and (34) the weighted Hölder space $\Lambda_{l+\alpha}^{l+\alpha}(\mathbb{K})$ appears as a completion of $C_0^\infty(\overline{\mathbb{K}} \setminus \mathcal{O})$ with respect to the norm $(34)_2$. Let $\Lambda_\beta^{l,\alpha}(\mathbb{K})$ be composed by functions w such that $r^{l+\alpha-\beta} w \in \Lambda_{l+\alpha}^{l,\alpha}(\mathbb{K})$ while

$$\|w; \Lambda_\beta^{l,\alpha}(\mathbb{K})\| = \sum_{k=0}^l \sup_{\mathbb{K}} \left\{ r^{\beta-l-\alpha+k} |\nabla_x^k w(x)| \right\}$$
$$+ \sup_{\cdots} \left\{ |x^1 - x^2|^{-\alpha} |r_1^\beta \nabla_x^l w(x^1) - r_2^\beta \nabla_x^l w(x^2)| \right\} \quad (36)$$

where $r_i = |x^i|$ and the dots stand for the set (35).

The weight index β in (36) governs the behavior of w both, as x approaches $\mathcal{O}$ and x goes to infinity. To eliminate the influence of weights at the point $\mathcal{O} \subset \overline{\Omega}$, we change $r, r_i, \mathbb{K}$ in (36) for $1+r, 1+r_i, \Omega$ and obtain the space $\Lambda_\beta^{l,\alpha}(\Omega)$ equipped with the norm

$$\|w; \Lambda_\beta^{l,\alpha}(\Omega)\| = \sum_{k=0}^l \sup_\Omega \left\{ (1+r)^{\beta-l-\alpha+k} |\nabla_x^k w(x)| \right\}$$
$$+ \sup_{\cdots} \left\{ |x^1 - x^2|^{-\alpha} |(1+r_1)^\beta \nabla_x^l w(x^1) - (1+r_2)^\beta \nabla_x^l w(x^2)| \right\} \quad (37)$$

where the last supremum is calculated over $x^1, x^2 \in \Omega$ such that $c^{-1}|x^1| < |x^2| < c|x^1|$, $c > 1$. Note that elements of $\Lambda_\beta^{l,\alpha}(\Omega)$ belong to $C^{l,\alpha}(\Omega \cap \mathbb{B}_R)$ with any $R > 0$ and the norms (36) with $\mathbb{K} = \Omega$ and (37) are equivalent in the case $\mathcal{O} \notin \overline{\Omega}$. The trace space $\Lambda_\beta^{l,\alpha}(\partial\Omega)$ is equipped with the norm (37) where Ω and ∇_x are replaced by the boundary $\partial\Omega$ and the surface gradient ∇_s, respectively. For the exterior domain (1), the boundary $\partial\Omega_e = \partial G$ is compact and, therefore, $\Lambda_\beta^{l,\alpha}(\partial G)$ coincides with $C^{l,\alpha}(\partial G)$.

Each derivation reduces the exponent of the weight $1+r$ in (37) by 1. In other words, for $w \in \Lambda_\beta^{l,\alpha}(\Omega)$ and $q \leq l$, we have $\nabla_x^q w \in \Lambda_\beta^{l-q,\alpha}(\Omega)$ and

$$\|\nabla_x^q w; \Lambda_\beta^{l-q,\alpha}(\Omega)\| \leq C \|w; \Lambda_\beta^{l,\alpha}(\Omega)\| .$$

This observation yields that the operator $A_\beta^{l,\alpha}$ associated to the Stokes problem (5),

$$\mathcal{D}_\beta^{l,\alpha} \Lambda(\Omega) \ni u \mapsto A_\beta^{l,\alpha} u := (S(\nabla_x)u, u'|_{\partial\Omega}) \in \mathcal{R}_\beta^{l,\alpha} \Lambda(\Omega),$$
$$\mathcal{D}_\beta^{l,\alpha} \Lambda(\Omega) := \Lambda_\beta^{l+1,\alpha}(\Omega)^n \times \Lambda_\beta^{l,\alpha}(\Omega), \quad (38)$$
$$\mathcal{R}_\beta^{l,\alpha} \Lambda(\Omega) := \Lambda_\beta^{l-1,\alpha}(\Omega)^n \times \Lambda_\beta^{l,\alpha}(\Omega) \times \Lambda_\beta^{l+1,\alpha}(\partial\Omega)^n$$

is continuous for any $l \in \mathbb{N}$, $\alpha \in (0,1)$, and $\beta \in \mathbb{R}$.

The norm (37) is adapted for power solutions (8). To smooth out their possible singularities at the coordinate origin $\mathcal{O}$, we set $U_X(x) = X(x)U(x)$ where $X(x) = 1 - \chi(r)$ with χ taken from the proof of Lemma 2. A direct calculation shows that in the case $\operatorname{Re} \lambda < l+1+\alpha-\beta$ ($\operatorname{Re}\alpha > l+1+\alpha-\beta$) the vector function U_X lives inside (outside) the space $\mathcal{D}_\beta^{l,\alpha}\Lambda(\Omega)$. For $\operatorname{Re}\lambda = l+1+\alpha-\beta$, the norm $\|U; \mathcal{D}_\beta^{l,\alpha}\Lambda(\Omega)\|$ remains formally finite but U_X cannot be approximated by elements in $C_0^\infty(\overline{\Omega})^{n+1}$ and, hence, $U_X \notin \mathcal{D}_\beta^{l,\alpha}\Lambda(\Omega)$. Properties of the mapping $(38)_1$ derived from general results [6,9], reflect the latter fact.

Theorem 7. *(see Theorems 4.1.2, 3.5.4 [14]). The operator $A_\beta^{l;\alpha}$ is Fredholm if and only if the model problem (16) has only trivial power solutions $(8)_2$ such that $\operatorname{Re}\lambda = l+1+\alpha-\beta$. This condition is equivalent to*

$$\{\lambda \in \sigma(\omega) : \operatorname{Re}\lambda = l+1+\alpha-\beta\} = \emptyset \ . \tag{39}$$

A weight index β, at which $A_\beta^{l;\alpha}$ loses the Fredholm property, is called *forbidden*.

Remark 8. If $A_\beta^{l;\alpha}$ is a Fredholm operator, its kernel $\ker A_\beta^{l;\alpha} \subset \mathcal{D}_\beta^{l,\alpha}\Lambda(\Omega)$ and co-kernel $\operatorname{coker} A_\beta^{l;\alpha} \subset \mathcal{R}_\beta^{l,\alpha}\Lambda(\Omega)^*$ are finite-dimensional and, moreover, the range $\operatorname{Im} A_\beta^{l;\alpha} = A_\beta^{l;\alpha}\mathcal{D}_\beta^{l,\alpha}\Lambda(\Omega)$ is closed. By virtue of the evident inclusions $\Lambda_\beta^{s,\alpha}(\Omega) \subset \Lambda_\gamma^{s,\alpha}(\Omega)$ and $\Lambda_\beta^{s,\alpha}(\Omega)^* \subset \Lambda_\sigma^{s,\alpha}(\Omega)^*$ with $\gamma < \beta < \sigma$, we observe that

$$\dim \ker A_\beta^{l,\alpha} \leq \dim \ker A_\gamma^{l,\alpha}, \quad \dim \operatorname{coker} A_\beta^{l,\alpha} \leq \dim \operatorname{coker} A_\sigma^{l,\alpha}, \tag{40}$$

and, therefore, under a proper choice of γ, σ, Theorem 7 yields finite dimensions (40) even in the case of forbidden β. Thus, for such β, $\operatorname{Im} A_\beta^{l,\alpha}$ cannot be closed. To verify this directly, we suppose $\operatorname{Im} A_\beta^{l,\alpha}$ to be a subspace, take the power solution $U \neq 0$ with $\operatorname{Re}\lambda = l+1+\alpha-\beta$, and set

$$U^N(x) = \chi^N(x)U_X(x), \qquad \chi^N(x) = \chi(N^{-1}\log r) \ .$$

Since $S(\nabla_x)U = 0$ and $|\nabla_x \chi^N(x)| \leq cN^{-1}r^{-1}$, we detect that, first,

$$(f_X, g_X) := (S(\nabla_x)XU, XU'|_{\partial\Omega})$$
$$= \lim_{N\to\infty}(S(\nabla_x)U^N, U^{N'}|_{\partial\Omega}) \in \operatorname{Im} A_\beta^{l,\alpha} \tag{41}$$

and, second,

$$\|U^N - u_X; \mathcal{D}_\beta^{l,\alpha}\Lambda(\Omega)\| \geq c > 0,$$
$$\|A_\beta^{l,\alpha}(U^N - u_X); \mathcal{R}_\beta^{l,\alpha}\Lambda(\Omega)\| = \mathcal{O}(N^{-1}) \quad \text{as} \quad N \to \infty \tag{42}$$

where $u_X \in \mathcal{D}_\beta^{l,\alpha}\Lambda(\Omega)$ is a particular solution of the problem (5) with the right-hand side (17). The relations (42) contradict our assumption that $\operatorname{Im} A_\beta^{l,\alpha}$ is closed.

The next assertion, proven in [8], connects attributes of the operators $A_\beta^{l,\alpha}$ and $A_{2(l+\alpha)-\beta+n}^{l,\alpha}$ supplied with so-called *adjoint weight indices*. We stress that the transformation (23) and the condition (39) show that those indices can be forbidden simultaneously and, furthermore,

$$2(l+\alpha) - \gamma + n =: \beta \mapsto 2(l+\alpha) - \beta + n = \gamma \ . \tag{43}$$

Theorem 9. *(See Theorems 4.2.4, 8.3.3 [14]). If β is not forbidden, then*

$$\mathrm{coker} A_\beta^{l,\alpha} = \{(u, Tu|_{\partial\Omega}) : u \in \ker A_{2(l+\alpha)-\beta+n}^{l,\alpha}\} \ . \tag{44}$$

In other words, the problem (5) with a right-hand side $(f,g) \in \mathcal{R}_\beta^{l,\alpha}\Lambda(\Omega)$ admits a solution in $\mathcal{D}_\beta^{l,\alpha}\Lambda(\Omega)$ if and only if there holds the compatibility condition

$$(f,u)_\Omega + (g, Tu)_{\partial\Omega} = 0, \quad \forall u \in \ker A_{2(l+\alpha)-\beta+n}^{l,\alpha} \ . \tag{45}$$

Remark 10. The relations (44) and (45) imply the Fredholm alternative for the operator $A_\beta^{l,\alpha}$. In view of (43), $A_\beta^{l,\alpha}$ and $A_{2(l+\alpha)-\beta+n}^{l,\alpha}$ turn out to be adjoint with respect to Green's formula (11). If the fixed point $\beta^0 = l + \alpha + n/2$ of (43) is not a forbidden index, then among the family $(38)_1$ of Stokes problem operators there appears the formally self-adjoint operator $A_{\beta^0}^{l,\alpha}$ and

$$\mathrm{Ind} A_{\beta^0}^{l,\alpha} := \dim \ker A_{\beta^0}^{l,\alpha} - \dim \mathrm{coker} A_{\beta^0}^{l,\alpha} = 0 \ .$$

The last general result of further use concerns asymptotic representations of solutions at infinity. Its proof can be found in [6,9] (see also Theorem 4.2.1 [14]).

Theorem 11. *Let $u \in \mathcal{D}_\beta^{l,\alpha}\Lambda(\Omega)$ be a solution of the problem (5) with right-hand side $(f,g) \in \mathcal{R}_\sigma^{l,\alpha}\Lambda(\Omega)$ while β and σ are not forbidden indices and $\sigma > \beta$. Then the asymptotic representation*

$$u(x) = X(x) \sum_{k=1}^m b_k U^k(x) + \tilde{u}(x) \tag{46}$$

is valid where $\tilde{u} \in \mathcal{D}_\sigma^{l,\alpha}\Lambda(\Omega), U^1, \ldots, U^m$ is a basis in the linear space $\mathcal{L}(\beta, \sigma)$ of all power solutions to the model problem (16) with $\mathrm{Re}\,\lambda \in (l+1+\alpha-\sigma, l+1+\alpha-\beta)$, and $b_k \in \mathbb{R}$. Moreover, the following inequality holds:

$$\|\tilde{u}; \mathcal{D}_\sigma^{l,\alpha}\Lambda(\Omega)\| + \sum_{k=1}^m |b_k| \le c \left(\|(f,g); \mathcal{R}_\sigma^{l,\alpha}\Lambda(\Omega)\| + \|u; \mathcal{D}_\beta^{l,\alpha}\Lambda(\Omega)\| \right) \ . \tag{47}$$

Remark 12. Since $u \in \ker A_\gamma^{l,\alpha}$ satisfies the problem (5) with $f = 0$, $g = 0$, Theorem 11 yields

$$u \in \mathcal{D}_{\gamma+\varepsilon}^{l,\alpha} \Lambda(\Omega), \quad \|u; \mathcal{D}_{\gamma+\varepsilon}^{l,\alpha} \Lambda(\Omega)\| \leq c \, \|u; \mathcal{D}_\gamma^{l,\alpha} \Lambda(\Omega)\| \tag{48}$$

with any $\varepsilon > 0$ such that the closed interval $[\gamma, \gamma + \varepsilon]$ is free of forbidden indices. If β and γ are connected by (43), then we, owing to (48), observe that the left-hand side of (45) is majorized by the product

$$c \, \|u; \mathcal{D}_{\gamma+\varepsilon}^{l,\alpha} \Lambda(\Omega)\| \times \|(f,g); \mathcal{R}_\beta^{l,\alpha} \Lambda(\Omega)\| \int_\Omega (1+r)^{-\beta-\gamma-\varepsilon+2(l+\alpha)} \, dx \; .$$

The exponent at $(1+r)$ is equal to $-n - \varepsilon$, the last integral converges, and the left-hand side of (45) is a continuous functional on $\mathcal{R}_\beta^{l,\alpha} \Lambda(\Omega) \ni (f,g)$.

Everything is now prepared to examine the solvability of various Stokes problems.

Connecting degrees of power solutions and weight indices, the relationship (39) converts the interval $\Sigma(\omega)$, introduced in Section 2, into the following intervals,

$$\begin{aligned}
\Sigma(\mathbb{S}_+^2) = (-2, 1) &\Rightarrow \quad \Upsilon(\mathbb{S}_+^2) = (l + \alpha, l + \alpha + 3), \\
\Sigma(\mathbb{S}^2) = (-1, 0) &\Rightarrow \quad \Upsilon(\mathbb{S}^2) = (l + \alpha + 1, l + \alpha + 2), \\
\Sigma(\mathbb{S}_+^1) = (-1, 1) &\Rightarrow \quad \Upsilon(\mathbb{S}_+^1) = (l + \alpha, l + \alpha + 2) \; .
\end{aligned} \tag{49}$$

We emphasize that the intervals Σ and Υ in (48) are invariant with respect to transformations of λ and β indicated in (18) and (43), respectively.

Since in all the cases the linear space $\mathcal{L}(\Upsilon)$ is trivial and the sum in (46) vanishes, Theorem 11 leads to that the subspace $\ker A_\beta^{l,\alpha}$ does not depend on $\beta \in \Upsilon$. Referring to the formula (44) and recalling the invariance property of Υ, we conclude the same about the subspace $\operatorname{coker} A_\beta^{l,\alpha}$. Moreover, $\beta^0 \in \Upsilon$ and, by Remark 10, we arrive at the equality

$$\dim \ker A_\beta^{l,\alpha} = \dim \operatorname{coker} A_\beta^{l,\alpha} = \text{const}, \quad \forall \beta \in \Upsilon \; . \tag{50}$$

We take $u \in \ker A_\beta^{l,\alpha}$ and, under the condition $\beta > \beta^0$ ensuring convergence of integrals in the Green formula

$$0 = \int_\Omega v \cdot (-\nu \Delta_x v + \nabla_x p) \, dx - \int_\Omega p \nabla_x \cdot v \, dx = \nu \int_\Omega |\nabla_x v|^2 \, dx,$$

we deduce that $\nabla_x v = 0$. Owing to the equations $(4)_{1,3}$ with $f = 0$, $g = 0$, we now have $v = 0$ and $p = \text{const}$. Since $1 \notin \Lambda_\beta^{l,\alpha}(\Omega)$ provided $\beta > l + \alpha$ (cf. (49)), we see that $p = 0$ and thus $\ker A_\beta^{l,\alpha} = \{0\}$, i.e. $A_\beta^{l,\alpha}$ is an isomorphism.

Theorem 13. *Let $A_\beta^{l,\alpha}$ be the operator (38)$_1$ of the problem (5) in the domain $\Omega_e(n = 3)$ or $\Omega_+(n = 2, 3)$. This operator is an isomorphism if and only if the weight index β belongs to the interval $\Upsilon = (\beta_-, \beta_+)$ indicated in (49). In other words, for $\beta \in (\beta_-, \beta_+)$ the problem (5) with the right-hand side $(f, g) \in \mathcal{R}_\beta^{l,\alpha}\Lambda(\Omega)$ has the unique solution $u \in \mathcal{D}_\beta^{l,\alpha}\Lambda(\Omega)$ and there holds the estimate*

$$\|u; \mathcal{D}_\beta^{l,\alpha}\Lambda(\Omega)\| \le c\,\|(f, g); \mathcal{R}_\beta^{l,\alpha}\Lambda(\Omega)\| \tag{51}$$

where c is a constant, independent of u and f, g. Existence (uniqueness) of solutions is lost in the case $\beta > \beta_+$ ($\beta < \beta_-$) and the estimate (51) is spoiled at $\beta = \beta_\pm$.

Proof. It remains only to confirm the loss of the isomorphism property. We have explained in Remark 8 how to derive relations (42) in case $\beta = \beta_\pm$ which contradict to both, the closedness of Im $A_\beta^{l,\alpha}$ and the estimate (51).

Let now $\gamma < \beta_-$ and let $U^- \ne 0$ denote a power solution with $\lambda_- = l + 1 + \alpha - \beta_-$. Clearly, $\lambda_- < l + 1 + \alpha - \gamma$ and, hence, $U_X^- \in \mathcal{D}_\gamma^{l,\alpha}\Lambda(\Omega) \setminus \mathcal{D}_\beta^{l,\alpha}\Lambda(\Omega)$ with any $\beta \in (\beta_-, \beta_+)$. Since U^- fulfills the model problem (19) or (25), $(f_X^-, g_X^-) = (S(\nabla_x)U_X^-, U_X^{-\prime}|_{\partial\Omega})$ is a smooth vector function with a compact support. Denoting by $u_X^- \in \mathcal{D}_\beta^{l,\alpha}\Lambda(\Omega)$ the unique solution to the problem (5) with the right-hand side $(f_X, g_X) \in \mathcal{R}_\beta^{l,\alpha}\Lambda(\Omega)$, we meet the solution $U_X^- - u_X^- \in \mathcal{D}_\gamma^{l,\alpha}\Lambda(\Omega)$ of the homogeneous problem (5). The latter solution cannot vanish because U_X^- and u_X^- live in different spaces. Thus, dim ker $A_\gamma^{l,\alpha} > 0$.

If $\sigma > \beta_+$, we get $\gamma = 2(l + \alpha) - \sigma + n < \beta_-$ and, by Theorem 9, dim coker $A_\sigma^{l,\alpha} > 0$. □

Remark 14. 1) For the 2-D exterior Stokes problem, the interval $\Upsilon(\mathbb{S}^1)$ is empty and the operator (38)$_1$ is not an isomorphism with any $\beta \in \mathbb{R}$. In [14], §8.4, a modification of weighted norms can be found that provides the Stokes problem operator with the isomorphism property.

2) The boundary $\partial\Omega_a$ of the aperture domain (3) has the irregularity set $(\partial\Gamma)_0$, namely, angular points if $n = 2$ and an edge if $n = 3$. If function spaces are adapted to those irregularities by introducing into the norms (37) additional weights (powers of dist$\{x, (\partial\Gamma)_0\}$), the operator of the problem (5)$_a$ still is an isomorphism (see [18]).

4   The Navier-Stokes Problem in Weighted Hölder Spaces

Owing to (37) and (38)$_2$, $u \in \mathcal{D}_\beta^{l,\alpha}\Lambda(\Omega)$ satisfies the inequalities

$$(1 + r)^{\beta - l - 1 - \alpha + k}|\nabla_x^k u'(x)| \le \|u; \mathcal{D}_\beta^{l,\alpha}\Lambda(\Omega)\|, \quad k = 0, \ldots, l + 1, \tag{52}$$

which make the following assertion evident.

Proposition 15. *The nonlinear mapping*

$$\Lambda_\beta^{l+1,\alpha}(\Omega)^n \ni v \mapsto N(\nabla_x)v \in \Lambda_\beta^{l-1,\alpha}(\Omega)^n, \tag{53}$$

generated by the convective term (7), is continuous if and only if

$$\beta \in \Upsilon_N := [l+\alpha+2,+\infty) . \tag{54}$$

Under the condition (54) the estimate

$$\|N(\nabla_x)v^1 - N(\nabla_x)v^2; \Lambda_\beta^{l-1,\alpha}(\Omega)\| \tag{55}$$

$$\leq c\|v^1 - v^2; \Lambda_\beta^{l+1,\alpha}(\Omega)\| \left(\|v^1; \Lambda_\beta^{l+1,\alpha}(\Omega)\| + \|v^2; \Lambda_\beta^{l+1,\alpha}(\Omega)\|\right)$$

is valid for any $v^i \in \Lambda_\beta^{l+1,\alpha}(\Omega)^n$, $i=1,2$.

Remark 16. The following simple calculations comment Proposition 15:

$$(v^1 \cdot \nabla_x)v^1 - (v^2 \cdot \nabla_x)v^2 = ((v^1 - v^2) \cdot \nabla_x)v^1 + (v^2 \cdot \nabla_x)(v^1 - v^2),$$

$$(1+r)^{\beta-(l-1+\alpha)}(v \cdot \nabla_x)v = R\left\{(1+r)^{\beta-(l+1+\alpha)}v \cdot (1+r)^{\beta-(l+\alpha)}\nabla_x\right\}v,$$

$$R := (1+r)^{2+l+\alpha-\beta} \leq \text{const} \Leftrightarrow \beta \geq l+\alpha+2 . \tag{56}$$

Moreover, they prove that $N(\nabla_x)v \in \Lambda_{\beta+\delta}^{l-1,\alpha}(\Omega)^n$ and

$$\|N(\nabla_x)v; \Lambda_{\beta+\delta}^{l-1,\alpha}(\Omega)\| \leq c\|v; \Lambda_\beta^{l+1,\alpha}(\Omega)\| \tag{57}$$

where $v \in \Lambda_\beta^{l+1,\alpha}(\Omega)^n$ and $\delta = \beta - l - \alpha - 2$.

Among the intervals $\Upsilon(\ldots)$ in (49), only one has non-empty intersection with the interval Υ_N in (54), namely, $\Upsilon(\mathbb{S}_+^2)$ which corresponds to the 3-D problem $(5)_+$. We take $\beta \in [l+\alpha+2, l+\alpha+3)$ and, using Theorem 13, we rewrite the above-mentioned problem as the equation

$$u = Mu := (A_\beta^{l,\alpha})^{-1}(f' - N(\nabla_x)u', f_4, g) . \tag{58}$$

By Proposition 15, the operator M is a contraction on the ball

$$\{u \in \mathcal{D}_\beta^{l,\alpha}\Lambda(\Omega_+) : \|u; \mathcal{D}_\beta^{l,\alpha}\Lambda(\Omega_+)\| \leq \rho_D\} \tag{59}$$

with a small radius $\rho_D > 0$. Thus, Banach's contraction principle leads to the following assertion.

Theorem 17. *Let $l \in \mathbb{N}$, $\alpha \in (0,1)$ and $\beta \in [l+\alpha+2, l+\alpha+3)$. Then there exists $\rho_R > 0$ such that the 3-D problem $(5)_+$ with the right-hand side satisfying the inequality*

$$\|(f,g); \mathcal{R}_\beta^{l,\alpha}\Lambda(\Omega_+)\| \leq \rho_R, \tag{60}$$

has a unique solution in the ball (59) with radius $\rho_D = c_0\rho_R$. Moreover, the following estimate holds:

$$\|u; \mathcal{D}_\beta^{l,\alpha}\Lambda(\Omega_+)\| \leq c_0\|(f,g); \mathcal{R}_\beta^{l,\alpha}\Lambda(\Omega_+)\| . \tag{61}$$

Let assume a faster decay of the right-hand side than prescribed by Theorem 54, i.e.,

$$(f,g) \in \mathcal{R}_\sigma^{l,\alpha} \Lambda(\Omega_+), \ \sigma \in (l+\alpha+3, l+\alpha+4) \ . \tag{62}$$

Since the definition (37) furnishes the estimate

$$\|(f,g); \mathcal{R}_\beta^{l,\alpha} \Lambda(\Omega_+)\| \le \|(f,g); \mathcal{R}_\sigma^{l,\alpha} \Lambda(\Omega_+)\| \tag{63}$$

with any $\beta \le \sigma$, we choose the weight index β as follows:

$$\beta = \frac{1}{2}[\sigma + l + \alpha + 2] \in \left(l + \alpha + \frac{5}{2}, l + \alpha + 3\right) \subset \Upsilon(\mathbb{S}_+^2) \cap \Upsilon_N \ . \tag{64}$$

Recalling Remark 16, we now have

$$\beta + \delta = 2\beta - l - \alpha - 2 = \sigma,$$
$$(f - N(\nabla_x)u', f_4, g) \in \mathcal{R}_\sigma^{l,\alpha} \Lambda(\Omega_+) \ . \tag{65}$$

Thus, we may apply Theorem 11 to the 3-D problem $(5)_+$ with right-hand side $(65)_2$. Owing to Lemma 3 and the restrictions (64), (62) on β, σ, the linear space $\mathcal{L}(\beta, \sigma)$ contains power solutions of degree $\lambda = -2$, i.e. linear combinations of the Poisson kernel columns Ψ^j.

Proposition 18. *Let β, σ and (f,g) be subject to the conditions (64) and (62). Then a solution $u \in \mathcal{D}_\beta^{l,\alpha} \Lambda(\Omega_+)$ of 3-D problem $(6)_+$ takes the asymptotic form*

$$u(x) = X(x) \sum_{j=1}^3 b_j \Psi^j(x) + \tilde{u}(x), \tag{66}$$

where $b_j \in \mathbb{R}$, $\tilde{u} \in \mathcal{D}_\sigma^{l,\alpha} \Lambda(\Omega_+)$ and

$$\|\tilde{u}; \mathcal{D}_\sigma^{l,\alpha} \Lambda(\Omega_+)\| + \sum_{j=1}^3 |b_j| \le c \left(\|(f,g); \mathcal{R}_\sigma^{l,\alpha} \Lambda(\Omega_+)\| + \|u; \mathcal{D}_\beta^{l,\alpha} \Lambda(\Omega_+)\| \right) \ . \tag{67}$$

The estimate (67) is deduced from the estimates (57) and (47). If the inequality (60) is valid, the solution u in the ball (59) fulfills both the estimates (61), (67) and, therefore, by virtue of (63) the majorant in (67) may be changed for $c_1 \|(f,g); \mathcal{R}_\sigma^{l,\alpha} \Lambda(\Omega_+)\|$. This simple observation provides a starting point for considerations in the next section: Under the conditions (62) and (60) a solution of the 3-D problem $(6)_+$ can be found in the space composed from vector functions in the form (66) and supplied with the norm written in the left-hand side of (67). The new space is to be identified with the direct product $\mathcal{D}_\sigma^{l,\alpha} \Lambda(\Omega) \times \mathbb{R}^3$. Since detached asymptotic terms $b_j \Psi^j$ satisfy the model problem (25), the range of constructed Stokes problem operator does still coincide with weighted Hölder space $\mathcal{R}_\sigma^{l,\alpha} \Lambda(\Omega)$. This analysis outlines a simple structure of detached asymptotics and gives rise to a finite-dimensional complement of the domain $\mathcal{D}_\sigma^{l,\alpha} \Lambda(\Omega)$ (cf. [14,31,15] and others). More complicated structures detached shall appear in next sections.

5   The 3-D Exterior Problems in Weighted Spaces with Detached Asymptotics

The approach used to derive Theorem 17 and Proposition 18, does not work for the 3-D problem $(6)_e$ because the interval $\Upsilon(\mathbb{S}^2)$ in $(49)_2$ does not contain a weight index β, at which the mapping (53) is continuous. To ensure the continuity of the convective term (7), we have to diminish the domain of the Stokes problem operator. Theorems 13 and 11 specified by Lemma 2, prove that the 3-D Stokes problem $(5)_e$ with $(f,g) \in \mathcal{R}_\gamma^{l,\alpha}\Lambda(\Omega)$, $\gamma \in (l+\alpha+2, l+\alpha+3)$ has a unique solution admitting the representation

$$u(x) = X(x)\,U(x) + \tilde{u}(x) \tag{68}$$

where $\tilde{u} \in \mathcal{D}_\gamma^{l,\alpha}\Lambda(\Omega)$ and $U = b_1\Phi^1 + b_2\Phi^2 + b_3\Phi^3$ is a power solution of degree -1 (see (20)). We preserve the asymptotic form for a solution for the 3-D Navier-Stokes problem $(6)_e$ but introduce arbitrary angular parts of the detached asymptotic term U_X. In other words, we compose the space $\mathfrak{D}_\gamma^{l,s,\alpha}(\Omega)$ from vector functions $u = (v,p)$,

$$v(x) = X(x)r^{-1}\mathcal{V}(\varphi) + \tilde{v}(x), \qquad p(x) = X(x)r^{-2}\mathcal{P}(v) + \tilde{p}(x), \tag{69}$$

where, denoting the sphere $\mathbb{S}^2$ by ω, we set

$$\mathcal{U} = (\mathcal{V}, \mathcal{P}) \in \mathcal{D}^{s,\alpha}C(\omega) := C^{s+1,\alpha}(\omega)^n \times C^{s,\alpha}(\omega),$$
$$\tilde{u} = (\tilde{v}, \tilde{p}) \in \mathcal{D}_\gamma^{l,\alpha}\Lambda(\Omega), \tag{70}$$
$$\gamma \in (l+\alpha+2, l+\alpha+3), \; l,s \in \mathbb{N}, \; s \geq l, \; \alpha \in (0,1) \;.$$

We equip $\mathfrak{D}_\gamma^{l,s,\alpha}(\Omega)$ with the natural norm

$$\|u; \mathfrak{D}_\gamma^{l,s,\alpha}(\Omega)\| = \|\tilde{u}; \mathcal{D}_\gamma^{l,\alpha}\Lambda(\Omega)\| + \|\mathcal{U}; \mathcal{D}^{s,\alpha}C(\omega)\|$$

while identifying the space with the direct product $\mathcal{D}_\gamma^{l,\alpha}\Lambda(\Omega) \times \mathcal{D}^{s,\alpha}C(\omega)$. Note that $\mathfrak{D}_\gamma^{l,s,\alpha}(\Omega)$ does not depend on the choice of the cut-off function X neither algebraically, nor topologically.

Based on $(70)_3$ and $(49)_2$, a simple calculation yields

$$\mathfrak{D}_\gamma^{l,s,\alpha}(\Omega) \subset \mathcal{D}_\beta^{l,\alpha}\Lambda(\Omega), \quad \forall \beta \in \Upsilon(\mathbb{S}^2) \;. \tag{71}$$

Thus, we can restrict on $\mathfrak{D}_\gamma^{l,s,\alpha}(\Omega)$ the isomorphism $A_\beta^{l,\alpha}$ (Theorem 13) and, recalling the relation (9), we come across the continuous mapping

$$\mathfrak{S}_\gamma^{l,s,\alpha} : \mathfrak{D}_\gamma^{l,s,\alpha}(\Omega) \to \mathfrak{R}_\gamma^{l,s,\alpha}(\Omega) \tag{72}$$

where $\mathfrak{R}_\gamma^{l,s,\alpha}(\Omega)$ consists of couples (f,g) such that

$$f(x) = X(x)F(x) + \tilde{f}(x), \; g(x) = X(x)G(x) + \tilde{g}(x),$$
$$F'(x) = r^{-3}\mathcal{F}'(\varphi), \; F_{n+1}(x) = r^{-2}\mathcal{F}_{n+1}(\varphi), \; G(x) = r^{-1}\mathcal{G}(\varphi),$$
$$(\mathcal{F}, \mathcal{G}) \in \mathcal{R}^{s,\alpha}C(\omega) := C^{s-1,\alpha}(\omega)^n \times C^{s,\alpha}(\omega) \times C^{s+1,\alpha}(\partial\omega)^n,$$
$$(\tilde{f}, \tilde{g}) \in \mathcal{R}_\gamma^{l,\alpha}\Lambda(\Omega) \;. \tag{73}$$

The norm induced by (72), reads as follows:

$$\|(f,g);\mathfrak{R}_\gamma^{l,s,\alpha}(\Omega)\| = \|(\tilde{f},\tilde{g});\mathcal{R}_\gamma^{l,\alpha}\Lambda(\Omega)\| + \|(\mathcal{F},\mathcal{G});\mathcal{R}^{s,\alpha}C(\omega)\| \ .$$

We emphasize that, for 3-D exterior domain Ω_e, we have to put $n = 3$, $\omega = \mathbb{S}^2$, $\partial\omega = \Omega$ and $G = 0$, $\mathcal{G} = 0$, $\tilde{g} = g$ in (73). However, the definitions (69), (70) and (73) will be used in the next sections also for domains of other types.

We confirm the above introduction of weighted Hölder space with detached asymptotics by considering the convective term (7).

Proposition 19. *Under the condition* $(70)_3$, *the nonlinear mapping*

$$\mathfrak{D}_\gamma^{l,s,\alpha}(\Omega) \ni (v,0) \mapsto (N(\nabla_x)v,0,0) \in \mathfrak{R}_\gamma^{l,s,\alpha}(\Omega) \tag{74}$$

is continuous. The inequality

$$\|(N(\nabla_x)v^1 - N(\nabla_x)v^2,0,0);\mathfrak{R}_\gamma^{l,s,\alpha}(\Omega)\|$$
$$\leq c\,\|(v^1 - v^2,0);\mathfrak{D}_\gamma^{l,s,\alpha}(\Omega)\|\left(\|(v^1,0);\mathfrak{D}_\gamma^{l,s,\alpha}(\Omega)\| + \|(v^2,0);\mathfrak{D}_\gamma^{l,s,\alpha}(\Omega)\|\right)$$

is valid for any $(v^i,0) \in \mathfrak{D}_\gamma^{l,s,\alpha}(\Omega)$, $i = 1,2$.

Proof. In view of $(56)_1$ it is sufficient to treat the expression

$$\begin{aligned}
(v^1(x) \cdot \nabla_x)v^2(x) = {}&(\tilde{v}^1(x) \cdot \nabla_x)\tilde{v}^2(x) + X(x)r^{-1}(\mathcal{V}^1(\varphi) \cdot \nabla_x)\tilde{v}^2(x) \\
&+ (\tilde{v}^1(x) \cdot \nabla_x)X(x)r^{-1}\mathcal{V}^2(\varphi) \\
&+ X(x)r^{-1}(\mathcal{V}^1(\varphi) \cdot \nabla_x)X(x)r^{-1}\mathcal{V}^2(\varphi) \ .
\end{aligned} \tag{75}$$

Similarly to the estimate (55) with $\beta = \gamma \in \Upsilon_N$, we obtain

$$\|(\tilde{v}^1 \cdot \nabla_x)\tilde{v}^2; \Lambda_\gamma^{l-1,\alpha}(\Omega)\| \leq c\|\tilde{v}^1; \Lambda_\gamma^{l+1,\alpha}(\Omega)\| \times \|\tilde{v}^2; \Lambda_\gamma^{l+1,\alpha}(\Omega)\| \ .$$

By appealing to the embeddings $\Lambda_\gamma^{l+1,\alpha}(\Omega) \subset \Lambda_{\gamma-1}^{l,\alpha}(\Omega) \subset \Lambda_{\gamma-2}^{l-1,\alpha}(\Omega)$ provided by the definition (37), we deduce that

$$\begin{aligned}
\|Xr^{-1}(\mathcal{V}^1 \cdot \nabla_x)\tilde{v}^2; \Lambda_\gamma^{l-1,\alpha}(\Omega)\| &\leq c\|\mathcal{V}^1; C^{l-1,\alpha}(\omega)\| \times \|\tilde{v}^2; \Lambda_{\gamma-1}^{l,\alpha}(\Omega)\| \\
&\leq c\|\mathcal{V}^1; C^{s+1,\alpha}(\omega)\| \times \|\tilde{v}^2; \Lambda_\gamma^{l+1,\alpha}(\Omega)\|, \\
\|(\tilde{v}^1 \cdot \nabla_x)Xr^{-1}\mathcal{V}^2; \Lambda_\gamma^{l-1,\alpha}(\Omega)\| &\leq c\|\tilde{v}^1; \Lambda_{\gamma-2}^{l-1,\alpha}(\Omega)\| \times \|\mathcal{V}^2; C^{l,\alpha}(\omega)\| \\
&\leq c\|\tilde{v}^1; \Lambda_\gamma^{l+1,\alpha}(\Omega)\| \times \|\mathcal{V}^2; C^{s+1,\alpha}(\omega)\| \ .
\end{aligned}$$

Moreover, we have

$$\begin{aligned}
Xr^{-1}(\mathcal{V}^1 \cdot \nabla_x)Xr^{-1}\mathcal{V}^2 = {}&Xr^{-1}(\mathcal{V}^1 \cdot \nabla_x)r^{-1}\mathcal{V}^2 - \\
&- Xr^{-1}(\mathcal{V}^1 \cdot \nabla_x)(1-X)r^{-1}\mathcal{V}^2 =: Xr^{-3}\mathcal{K} + K
\end{aligned}$$

and, since $\operatorname{supp} K$ is compact and $l \leq s$,

$$\|K; C^{s-1,\alpha}(\omega)\| + \|K; \Lambda_\gamma^{l-1,\alpha}(\Omega)\|$$
$$\leq c\|\mathcal{V}^1; C^{s-1,\alpha}(\omega)\| \times \|\mathcal{V}^2; C^{l,\alpha}(\omega)\|$$
$$\leq c\|\mathcal{V}^1; C^{s+1,\alpha}(\omega)\| \times \|\mathcal{V}^2; C^{s+1,\alpha}(\omega)\| .$$

All majorants in the derived inequalities do not exceed the expression

$$C\|(v^1, 0); \mathfrak{D}_\gamma^{l,s,\alpha}(\Omega)\| \times \|(v^2, 0); \mathfrak{D}_\gamma^{l,s,\alpha}(\Omega)\| \tag{76}$$

and, therefore, detaching the term $Xr^{-3}\mathcal{K}$, we see that the norm

$$\|((v^1 \cdot \nabla_x)v^2, 0, 0); \mathfrak{R}_\gamma^{l,s,\alpha}(\Omega)\|$$

does not exceed (76), too. $\qquad\qquad\square$

Now we investigate properties of the Stokes problem operator $\mathfrak{S}_\gamma^{l,s,\alpha}$.

Theorem 20. *Under the condition* $(70)_3$ *the operator* (72) *implies a Fredholm monomorphism with a co-kernel of dimension 3. The 3-D problem* $(5)_e$ *with right-hand side* $(f, g) \in \mathfrak{R}_\gamma^{l,s,\alpha}(\Omega_e)$ *has a solution in* $\mathfrak{D}_\gamma^{l,s,\alpha}(\Omega_e)$ *if and only if the compatibility conditions*

$$\int_{\mathbb{S}^2} \mathcal{F}_j(\varphi)\, ds_\varphi = 0, \quad j = 1, 2, 3, \tag{77}$$

are valid where $\mathcal{F}_j$ *stand for components of the angular part* $\mathcal{F}'$ *in* $(73)_2$. *This solution is unique and satisfies the estimate*

$$\|u; \mathfrak{D}_\gamma^{l,s,\alpha}(\Omega)\| \leq c\|(f, g); \mathfrak{R}_\gamma^{l,s,\alpha}(\Omega)\| . \tag{78}$$

Proof. We proceed with solving the system on the sphere

$$\mathcal{S}(-1, \varphi, \nabla_\varphi)\mathcal{U}(\varphi) = \mathcal{F}(\varphi), \quad \varphi \in \mathbb{S}^2 . \tag{79}$$

According to Green's formula (14) on $\omega = \mathbb{S}^2$, the compatibility condition for this system is nothing but the orthogonality

$$(\mathcal{F}, \mathcal{W})_{\mathbb{S}^2} = 0 \tag{80}$$

of the right-hand side to any solution of the system

$$\mathcal{S}(0, \varphi, \nabla_\varphi)\mathcal{W}(\varphi) = 0, \quad \varphi \in \mathbb{S}^2 . \tag{81}$$

Recalling Lemma 2 and Remark 1, we observe that a solution of (81) is the angular part of a polynomial of generalized degree 0, i.e. a constant velocity vector. Hence, the condition (80) turns into (77). Moreover, a solution of the system (79) is unique up to the summand

$$b_1^0 \Phi^1(\varphi) + b_2^0 \Phi^2(\varphi) + b_3^0 \Phi^3(\varphi) \tag{82}$$

where $b_j^0 \in \mathbb{R}$ and $\Phi^j(\varphi)$ implies the trace on $\mathbb{S}^2$ of the fundamental matrix column Φ^j, i.e. a power solution of degree -1. We fix somehow the linear inverse operator $\mathcal{R}^0$ for the epimorphism

$$\mathcal{S}(-1, \varphi, \nabla_\varphi) : \mathcal{D}^{s,\alpha} C(\mathbb{S}^2) \to \mathcal{R}^{s,\alpha} C(\mathbb{S}^2)_\perp :=$$
$$:= \{\mathcal{F} \in \mathcal{R}^{s,\alpha} C(\mathbb{S}^2) : \mathcal{F} \text{ satisfies } (77)\} .$$

The particular solution $\mathcal{U}^0 = \mathcal{R}^0 \mathcal{F}$ of the system (79) enjoys the estimate

$$\|\mathcal{U}^0; \mathcal{D}^{s,\alpha} C(\mathbb{S}^2)\| \le c \|\mathcal{F}; \mathcal{R}^{s,\alpha} C(\mathbb{S}^2)\| . \tag{83}$$

We now reduce the problem $(5)_e$ to the following one:

$$S(\nabla_x) u^1(x) = f^1(x) := f(x) - S(\nabla_x) U_X^0(x), \quad x \in \Omega_e,$$
$$u^{1\prime}(x) = g^1(x) := g(x) - U_X^{0\prime}(x), \quad x \in \partial\Omega_e, \tag{84}$$

where $U_X^0 = X U^0$ and $U^0(x) = (r^{-1} \mathcal{U}^{0\prime}(\varphi), r^{-2} \mathcal{U}_4^0(\varphi))$. Since

$$f^1 = f - X S(\nabla_x) U^0 - [S(\nabla_x), X] U^0 = \tilde{f} - [S(\nabla_x), X] U^0$$

and the commutator $[S(\nabla_x), X] U^0$ has a compact support, the new right-hand side fulfills the inequality

$$\|(f^1, g^1); \mathcal{R}_\gamma^{l,\alpha} \Lambda(\Omega_e)\| \le c \left(\|(\tilde{f}, g); \mathcal{R}_\gamma^{l,\alpha} \Lambda(\Omega_e)\| + \|\mathcal{U}^0; \mathcal{D}^{l,\alpha} C(\mathbb{S}^2)\| \right) . \tag{85}$$

Applying Theorems 13 (see $(49)_2$) and 11 to the problem (84) we obtain the unique solution $u^1 \in \mathcal{D}_\beta^{l,\alpha} \Lambda(\Omega_e)$ which takes the form

$$u^1(x) = \sum_{j=1}^3 b_j \Phi^j(x) + \tilde{u}^1(x) . \tag{86}$$

Besides, the estimates (47) and (51) furnish that

$$\|\tilde{u}^1; \mathcal{D}_\gamma^{l,\alpha} \Lambda(\Omega_e)\| + \sum_{j=1}^3 |b_j^1| \le c \|(f^1, g^1); \mathcal{R}_\gamma^{l,\alpha} \Lambda(\Omega_e)\| . \tag{87}$$

Finally we determine

$$\tilde{u} := \tilde{u}^1, \quad U := U^0 + b_1^1 \Phi^1 + b_2^1 \Phi^2 + b_3^1 \Phi^3 \tag{88}$$

and obtain the solution (69) of the whole problem $(5)_e$ in the space $\mathfrak{D}_\gamma^{l,s,\alpha}(\Omega_e)$. This solution is unique by virtue of the embedding (71) and Theorem 13 (note that, in particular, a perturbation of $\mathcal{U}^0$ by the combination (82) replaces also b_j^1 by $b_j^1 - b_j^0$ in (86) while the vector functions (88) remain the same). Collecting the estimates (83), (85), (87), we arrive at the inequality (78). To finish the proof, we mention that due to Lemma 3.5.11 [14] violation of the conditions (77) brings a logarithmic factor into the power solution

$$U^0(x) = \log r \sum_{j=1}^{3} c_j \Phi^j(x) + U^{\#}(x) \tag{89}$$

which was used to cancel the detached term $F(x)$ in $(73)_1$. This logarithm leads the sum $XU + \tilde{u}^1$ out of the space $\mathfrak{D}_\gamma^{l,s,\alpha} \Lambda(\Omega_e)$ and makes the compatibility conditions (77) necessary. $\qquad\square$

We are in position to examine the 3-D problem $(6)_e$.

Theorem 21. *Let* (f,g) *belong to* $\mathfrak{R}_\gamma^{l,s,\alpha}(\Omega_e)$ *with the indices taken from* $(70)_3$ *and let the conditions* (77) *and*

$$\mathcal{F}_4 = 0 \tag{90}$$

hold true. Then there exists $\rho_\mathcal{R} > 0$ *such that in the case*

$$\|(f,g); \mathfrak{R}_\gamma^{l,s,\alpha}(\Omega)\| \leq \rho_\mathcal{R} \tag{91}$$

the 3-D problem $(6)_e$ *has the unique solution* u *in the ball*

$$\left\{ u \in \mathfrak{D}_\gamma^{l,s,\alpha}(\Omega) : \|u; \mathfrak{D}_\gamma^{l,s,\alpha}(\Omega)\| \leq \rho_\mathcal{D} \right\} \tag{92}$$

with the radius $\rho_\mathcal{D} = c_0 \rho_\mathcal{R}$. *The solution enjoys the estimate* (78) *with the constant* $c = c_0$.

Proof. With the aim to employ the contraction principle we, owing to Theorem 20 and Proposition 19, need to ensure the mapping

$$\mathfrak{D}_\gamma^{l,s,\alpha}(\Omega)_\nabla \ni (v,0) \mapsto (N(\nabla_x)v, 0, 0) \in \mathfrak{R}_\gamma^{l,s,\alpha}(\Omega)_\perp \tag{93}$$

where in accordance with (90) and (77)

$$\mathfrak{D}_\gamma^{l,s,\alpha}(\Omega)_\nabla := \left\{ u \in \mathfrak{D}_\gamma^{l,s,\alpha}(\Omega) : \nabla_x \cdot U'(x) = 0 \right\}, \tag{94}$$

$$\mathfrak{R}_\gamma^{l,s,\alpha}(\Omega)_\perp := \left\{ (f,g) \in \mathfrak{R}_\gamma^{l,s,\alpha}(\Omega) : (\mathcal{F}_j, 1)_{s^2} = 0, j = 1, 2, 3, \mathcal{F}_4 = 0 \right\} .$$

By referring to (69), (8) and (31),

$$N(\nabla_x)U'(x) = r^{-3} \mathcal{N}(-1, \varphi, \nabla_\varphi) \mathcal{U}'(\varphi) .$$

Integrating by parts in the annulus $\Xi = \{x : b > |x| > a > 0\}$ yields

$$\log \frac{b}{a} \int_{\mathbb{S}^2} \mathcal{N}(-1, \varphi, \nabla_\varphi) U'(\varphi) \, ds_\varphi = \int_{\mathbb{S}^2} \int_a^b r^{-3} \mathcal{N}(-1, \varphi, \nabla_\varphi) \mathcal{U}'(\varphi) r^2 \, dr \, ds_\varphi$$

$$= \int_\Xi (U'(x) \cdot \nabla_x) U'(x) \, dx = - \int_\Xi U'(x)(\nabla_x \cdot U'(x)) \, dx$$

$$+ \int_{\mathbb{S}_b^2} \left(U'(x) \cdot \frac{x}{|x|} \right) U'(x) \, ds_x - \int_{\mathbb{S}_a^2} \left(U'(x) \cdot \frac{x}{|x|} \right) U'(x) \, ds_x \quad (95)$$

where $|x|^{-1}x$ is the unit normal vector to the sphere $\mathbb{S}_R^2 = \{x : |x| = R\}$. The last integral over Ξ vanishes according to the equality $\nabla_x \cdot U' = 0$ imposed in $(94)_1$. Since the integrand

$$x \mapsto \left(U'(x) \cdot \frac{x}{|x|} \right) U'(x) = r^{-2}(\mathcal{U}'(\varphi) \cdot \varphi) \mathcal{U}'(\varphi)$$

is a homogeneous function of order -2, the integrals over the spheres $\mathbb{S}_b^2$ and $\mathbb{S}_a^2$ are mutually canceled in (95). Thus,

$$\int_{\mathbb{S}^2} \mathcal{N}(-1, \varphi, \nabla_\varphi) \mathcal{U}'(\varphi) \, ds_\varphi = 0$$

and, indeed, $(N(\nabla_x)u', 0, 0)$ belongs to the subspace $(94)_2$. It remains to reformulate the problem $(6)_e$ as the equation

$$u = \mathfrak{M}u := (\mathfrak{S}_\gamma^{l,s,\alpha})^{-1}(f' - N(\nabla_x)u', f_4, g) \in \mathfrak{D}_\gamma^{l,s,\alpha}(\Omega)_\nabla$$

and to mention that $\mathfrak{M}$ is a contraction on the ball (92) with a small radius ρ_D. □

Remark 22. 1) Let us assume that

$$f'(x) = \nabla_x \mathfrak{f}(x), \quad f_4 = 0, \quad g \in C^{l+1,\alpha}(\partial\Omega)^3, \quad (96)$$
$$\mathfrak{f}(x) = X(x)r^{-2}\mathfrak{F}(\varphi) + \tilde{\mathfrak{f}}(x), \quad \mathfrak{F} \in C^{s,\alpha}(\mathbb{S}^2), \quad \tilde{\mathfrak{f}} \in \Lambda_\gamma^{l,\alpha}(\Omega) \ .$$

Comparing (96) with (73), we see that $(f, g) \in \mathfrak{R}_\gamma^{l,s,\alpha}(\Omega)$. Moreover, similarly to (95), we have

$$\log \frac{b}{a} \int_{\mathbb{S}^2} \mathcal{F}'(\varphi) \, ds_\varphi = \int_\Xi r^{-3} \mathcal{F}'(\varphi) \, dx = \int_\Xi \nabla_x (r^{-2} \mathfrak{F}(\varphi)) \, dx$$

$$= b^{-2} \int_{\mathbb{S}_b^2} \varphi \mathfrak{F}(\varphi) \, ds_\varphi - a^{-2} \int_{\mathbb{S}_a^2} \varphi \mathfrak{F}(\varphi) \, ds_\varphi = 0 \ .$$

This proves that a right-hand side (f, g) of the form (96) is in $\mathfrak{R}^{l,s,\alpha}_\gamma(\Omega)_\perp$. In other words, the gradient form of f' ensures the validity of the compatibility conditions (77) and Theorem 5.3 delivers the small solenoidal solution $u \in \mathfrak{D}^{l,s,\alpha}_\gamma(\Omega)$ of the 3-D problem $(6)_e$.

2) The inequality (78) for the solution $u = (v, p)$ can be simplified and reduced to pointwise estimates of the velocity, the pressure, and their derivatives. To this end, we take into account that, by virtue of (37) and (70),

$$(1 + r)^{\gamma - l - 1 - \alpha + k} |\nabla^k_x \tilde{v}(x)| \le \|\tilde{v}; \Lambda^{l+1,\alpha}_\gamma(\Omega)\|, \quad \gamma - l - 1 - \alpha > 1, \qquad (97)$$

$$|\nabla^k_x r^{-1} \mathcal{V}(\varphi)| \le c_k r^{-1-k} \|\mathcal{V}; C^{l+1,\alpha}(\mathbb{S}^2)\|, \quad k = 0, \ldots, l+1 .$$

We now eliminate detaching of asymptotics in (69) and combine (97) to obtain the pointwise estimates

$$|\nabla^k_x v(x)| \le C_{f,g}(1 + r)^{-1-k}, \ k = 0, \ldots, l+1 . \qquad (98)$$

The analogous estimates for the pressure read as follows:

$$|\nabla^j_x p(x)| \le C_{f,g}(1 + r)^{-2-j}, \ j = 0, \ldots, l .$$

Here $C_{f,g} = c \|(f, g); \mathfrak{R}^{l,s,\alpha}_\gamma(\Omega)\|$. Based on the inequalities (98), the same arguments as in [4] prove that, under the assumption (91) with a sufficiently small $\rho_\mathcal{R}$, any solution u of the 3-D exterior Navier-Stokes problem $(6)_e$ with

$$\nabla_x v \in L_2(\Omega), \ (1 + r)^{-1} v, \ p \in L_2(\Omega) \qquad (99)$$

belongs to the ball (92).

We point out that the estimates (98) at $k = 0, 1$ were derived in [29,2] for a small solution of the 3-D problem $(6)_e$ under certain conditions on the matrix $\mathfrak{f}$ in the right-hand side $(96)_1$.

6   The Stokes and Navier-Stokes Problems in the 2-D Domain Ω_+

To determine the function spaces $\mathfrak{D}^{l,s,\alpha}_\gamma(\Omega_+)$ and $\mathfrak{R}^{l,s,\alpha}_\gamma(\Omega_+)$ for the domain Ω_+ indicated in (2), we recall the formulae (69), $(70)_{1,2}$ and (73) where $\omega = \mathbb{S}^1_+ := (0, \pi) \ni \varphi$. Arguing in the same way as in the proof of Theorem 20, we deduce the following assertion (see [18] for details).

Theorem 23. *Under the restrictions* $(70)_3$, *the operator* (72) *becomes a Fredholm monomorphism with a co-kernel of dimension 2. The 2-D problem* $(5)_+$ *with the right-hand side* $(f, g) \in \mathfrak{R}^{l,s,\alpha}_\gamma(\Omega_+)$ *has a solution in* $\mathfrak{D}^{l,s,\alpha}_\gamma(\Omega_+)$ *if and only if the compatibility conditions*

$$\int_0^\pi \mathcal{F}_3(\varphi) \, d\varphi + \mathcal{G}_2(0) + \mathcal{G}_2(\pi) = 0,$$

$$\qquad (100)$$

$$\int_0^\pi \sin \varphi \, \mathcal{F}_1(\varphi) \, d\varphi + \nu(\mathcal{G}_1(0) + \mathcal{G}_1(\pi)) = 0$$

are valid where $\mathcal{F}_p(\varphi), \mathcal{F}_3(\varphi)$ and $\mathcal{G}_p(\varphi)$ stand for components of angular parts in the representations (73) of f and g. This solution u is unique and satisfies the estimate (78).

Remark 24. To confirm the conditions (100), one may appeal to Green's formula (14) where $\lambda = -1, 2 - n - \overline{\lambda} = 1$ and ω means the arc $\mathbb{S}^1_+$. Indeed, if $\mathcal{U}^1$ is replaced by the solution $\mathcal{U}$ of the problem

$$\mathcal{S}(-1, \varphi, \partial_\varphi)\,\mathcal{U}(\varphi) = \mathcal{F}(\varphi), \quad \varphi \in (0, \pi); \quad \mathcal{U}'(0) = \mathcal{G}(0), \quad \mathcal{U}'(\pi) = \mathcal{G}(\pi) \quad (101)$$

and $\mathcal{U}^2$ by the angular parts $(0, 0, 1)^t$ and $(\sin\varphi, 0, 0)^t$ of the polynomials $(27)_1$, the equality (14) turns into $(100)_1$ and $(100)_2$, respectively.

As in the previous section, we regard the Navier-Stokes problem as the Stokes problem with right-hand side $(f' - N(\nabla_x)u', f_3, g)$. The condition $(100)_1$ does not touch the nonlinear term at all. The second condition $(6.1)_2$ leads to the equality

$$\int\limits_0^\pi \sin\varphi\,\mathcal{N}_1(-1, \varphi, \partial_\varphi)\,\mathcal{U}'(\varphi)\,d\varphi = 0 \quad (102)$$

but there is no occasion for its validity (one may find this out by trying to repeat the proof of Theorem 21). In [16] it was noticed that (102) is provided by the assumption on the data to be symmetric with respect to the x_2-axis

$$\Omega^s_+ = \left\{ x = (x_1, x_2) : (-x_1, x_2) \in \Omega^s_+ \right\},$$
$$f^s_1(x) = -f^s_1(-x_1, x_2), \ f^s_2(x) = f^s_2(-x_1, x_2), \ f^s(x) = f^s_3(-x_1, x_2), \quad (103)$$
$$g^s_1(x) = -g^s_1(-x_1, x_2), \ g^s_2(x) = g^s_2(-x_1, x_2) \ .$$

The superscript s means "symmetric". The solution $u^s = (v^s, p^s)$ keeps the symmetry property $(103)_2$,

$$v^s_1(x) = -v^s_1(-x_1, x_2), \ v^s_2(x) = v^s_2(-x_1, x_2), \ p^s(x) = p^s(-x_1, x_2) \quad (104)$$

and, as a result, the function

$$x \mapsto N_1(\nabla_x)v(x) := (v(x) \cdot \nabla_x)v_1(x) = v_1(x)\frac{\partial v_1}{\partial x_1}(x) + v_2(x)\frac{\partial v_1}{\partial x_2}(x)$$

becomes odd in x_1. The latter, indeed, yields (102) because $x \mapsto \sin\varphi$ is even in x_1!

This simple observation together with Theorem 23, Proposition 19, and the contraction principle leads us to the following assertion (see [16,18]).

Proposition 25. *Let (f^s, g^s) belong to $\mathfrak{R}^{l,s,\alpha}_\gamma(\Omega^s_+)$ with the indices taken from $(70)_3$. Let also the conditions $(100)_1$ and (103) be fulfilled. Then there exists $\rho_R > 0$ such that in the case (91) the 2-D problem $(6)^s_+$ has the unique symmetric (i.e. (104)) solution u^s in the ball (92) with the radius $\rho_D = c_0\rho_R$. The solution satisfies the estimate (78) with the constant $c = c_0$.*

With the help of another approach, similar asymptotic representations were derived in the paper [4] which, in principle, contains also the following uniqueness result: If the radius ρ_R in (91) is sufficiently small, any solution u of the Navier-Stokes problem $(6)_+^s$ satisfying the inclusions (99), coincides with the solution given by Proposition 25.

Assuming a faster decay of the right-hands sides,

$$\mathcal{F}^s = 0, \ \mathcal{G}^s = 0, \ (f^s, g^s) \in \mathcal{R}_\gamma^{l,\alpha}(\Omega_+^s), \tag{105}$$

the angular part $\mathcal{U}^s = (\mathcal{V}^s, \mathcal{P}^s)$ in the representation (69) can be found explicitly. To this end, it is convenient to use polar components of the velocity vector v,

$$v_r(r, \varphi) = v_1(x) \cos \varphi + v_2(x) \sin \varphi,$$
$$v_\varphi(r, \varphi) = -v_1(x) \sin \varphi + v_2(x) \cos \varphi \ .$$

Note that $x_1 = r \cos \varphi$, $x_2 = r \sin \varphi$. In polar coordinates, the continuity equation $\nabla_x \cdot V(x) = 0$ takes the form

$$\partial_r V_r + r^{-1} V_r + r^{-1} \partial_\varphi V_\varphi = 0 \ .$$

Hence, taking into account that in view of (104) and (101) $\mathcal{V}^s(0) = \mathcal{V}^s(\pi) = 0$, we deduce the equalities $\partial_\varphi \mathcal{V}_\varphi^s = 0$ and

$$\mathcal{V}_\varphi^s(\varphi) = 0, \ \varphi \in (0, \pi) \ . \tag{106}$$

Substituting $r^{-1} \mathcal{V}_r^s(\varphi)$, $r^{-1} \mathcal{V}_\varphi^s(\varphi) = 0$ and $r^{-2} \mathcal{P}^s(\varphi)$ into the Navier-Stokes equations written in polar coordinates, direct calculations (see [4]) show that

$$\mathcal{P}^s(\varphi) = 2\nu \mathcal{V}_r^s(\varphi) + 2^{-1} \nu C^s, \ C^s \in \mathbb{R} \tag{107}$$

and $\mathcal{U}_r^s$ satisfies the Jeffrey-Hamel equation

$$-\partial_\varphi^2 \mathcal{V}_r^s(\varphi) - 4\mathcal{V}_r^s(\varphi) = \nu^{-1} \mathcal{V}_r^s(\varphi)^2 + C^s, \ \varphi \in (0, \pi),$$
$$\mathcal{V}_r^s(0) = \mathcal{V}_r^s(\pi) = 0 \ . \tag{108}$$

The constant C^s is connected with the total flux to infinity, namely, C^s must be chosen such that

$$\mu := \int_\Omega f_3^s(x) \, ds + \int_{\partial\Omega} \mathbf{n}(x) \cdot g^s(x) \, ds_x = -\int_0^\pi \mathcal{V}_r^s(\varphi) \, d\varphi \ . \tag{109}$$

To derive the relationship (109), we integrate the equation $(4)_2$ by parts in $\Omega \cap \mathbb{B}_R$, use the boundary condition $(4)_3$ on $\partial\Omega \cap \mathbb{B}_R$ and the representation $(69)_1$ on $\partial\mathbb{B}_R \cap \Omega$, and pass to limit as $R \to +\infty$. All integrals in (109) converge by virtue of (105) and $(70)_3$. Since μ is small, elementary arguments lead us to the unique small solution $V_r^s \in C^\infty[0, \pi]$ of the Jeffrey-Hamel problem (108), (109). Moreover, V_r^s takes the asymptotic form

$$V_r^s(\varphi) = 2\pi^{-1}\mu(\cos\varphi)^2 + \tilde{V}_r^s(\varphi), \; |\partial_\varphi^k \tilde{V}_r^s(\varphi)| \leq c_k \mu^2, \; k \in \mathbb{N}_0 \; . \tag{110}$$

Let us now make use of an idea stated in [25]. If the Navier-Stokes equations are linearized at the solution $v^s(x)$ of the 2-D problem $(6)_+^s$ in Ω_+^s, there appears the perturbed problem (5) with the operator $\hat{S}(x, \nabla_x)$ and the perturbed spectral problem (17) on the arc $\mathbb{S}_+^1$,

$$\hat{S}(\lambda, \varphi, \partial_\varphi)\mathcal{U}(\varphi) = 0, \; \varphi \in (0, \pi); \; \mathcal{U}'(0) = \mathcal{U}'(\pi) = 0 \; . \tag{111}$$

It *may happen* that Theorem 23 reformulated for the problem in the domain Ω_+ with the operator $\hat{S}(x, \nabla_x)$, gives rise to less compatibility conditions and *does not maintain* the condition $(100)_2$ which rejects the approach employed. We shall see that all those possibilities realize for a small *positive* μ. Note that $\mu > 0$ means that the flow drives flux *from infinity*.

The differential matrix operator $\hat{S}(x, \nabla_x)$ originates from the system of linear differential equations

$$-\nu\Delta_x v(x) + \nabla_x p(x)$$
$$+ (v^s(x) \cdot \nabla_x)v(x) + (v(x) \cdot \nabla_x)v^s(x) = f'(x), \; x \in \Omega_+,$$
$$-\nabla_x \cdot v(x) = f_3(x), \; x \in \Omega_+, \tag{112}$$

which occurs to be not self-adjoint. If v^s is changed for its main asymptotic term $V^s(x)$ found from (106)–(109), $\hat{S}(x, \nabla_x)$ turns into its main part $\hat{S}^0(x, \nabla_x)$ at infinity. The crucial point is that, owing to the relation $V^s(tx) = t^{-1}V^s(x)$ with $t > 0$, the perturbation terms $(V^s \cdot \nabla_x)v$ and $(v \cdot \nabla_x)V^s$ are again homogeneous as required in (9). That is why we still have the formulae of type (9), (10),

$$\hat{S}^0(x, \nabla_x)U(x) = F(x), \; x \in \mathbb{R}_+^2,$$
$$F'(x) = r^{\lambda-2}\mathcal{F}'(\varphi), \; F_3(x) = r^{\lambda-1}\mathcal{F}_3(x),$$
$$\hat{S}(\lambda, \varphi, \partial_\varphi)\mathcal{U}(\varphi) = \mathcal{F}(\varphi)$$

which lead to the spectral problem (111). Based on the representation (100), one can construct the asymptotics of eigenvalues of the problem (111). Necessary calculations in the framework of the classical perturbation theory are demonstrated in [25] and we only formulate the result.

Lemma 26. *There exists $\mu_0 > 0$ such that, if $|\mu| < \mu_0$, the strip $\{\lambda \in \mathbb{C} : \mathrm{Re}\,\lambda \in [-3/2, 0]\}$ contains only two eigenvalues $\lambda_1 = -1$ and*

$$\hat{\lambda}(\mu) = -1 - (2\pi\nu)^{-1}\mu + \mathcal{O}(\mu^2) \tag{113}$$

of the spectral problem (111). If $\mu \neq 0$, then the problem

$$\widehat{\mathcal{S}}(-1, \varphi, \partial_\varphi) \mathcal{U}(\varphi) = \mathcal{F}(\varphi), \quad \varphi \in (0, \pi); \quad \mathcal{U}'(0) = \mathcal{G}(0), \quad \mathcal{U}'(\pi) = \mathcal{G}(\pi) \quad (114)$$

meets the only compatibility condition $(100)_1$ while the solution $\mathcal{U} = (\mathcal{V}, \mathcal{P}) \in \mathcal{D}^{s,\alpha}(\mathbb{S}^1_+)$ subject to the orthogonality condition

$$\int_0^\pi \mathcal{P}(\varphi) \cos 2\varphi \, d\varphi = 0 \qquad (115)$$

becomes unique and admits the estimate

$$\| \mathcal{U}; \mathcal{D}^{s,\alpha} C(\mathbb{S}^1_+) \| \leq c |\mu|^{-1} \| (\mathcal{F}, \mathcal{G}); \mathcal{R}^{s,\alpha} C(\mathbb{S}^1_+) \| . \qquad (116)$$

Remark 27. The majorant in (116) has got the factor $|\mu|^{-1} = \mathcal{O}(|\lambda_1 - \widehat{\lambda}(\mu)|^{-1})$ due to the eigenvalue (113) in the vicinity of λ_1. Treating the eigenvector $\mathcal{U}^1 = (\mathcal{V}^1, \mathcal{P}^1)$ of the problem (111) at $\lambda = \lambda_1$, one obtains the asymptotic form $\mathcal{P}^1(\varphi) = \cos 2\varphi + \mathcal{O}(\mu)$, consistent with the choice of the orthogonality condition (115).

In what follows we assume $\mu \in (0, \mu_0)$, keep the indices l, s and α in $(70)_3$, and specify the weight index $\gamma = \sigma$ by

$$\sigma(\mu) \in (l + \alpha + 2 + (6\pi\nu)^{-1}\mu, l + \alpha + 2 + (3\pi\nu)^{-1}\mu) . \qquad (117)$$

Note that Lemma 26 recognizes $\gamma_- = l + \alpha + 2$ and $\gamma_+ = l + \alpha + 1 - \widehat{\lambda}(\mu)$ as forbidden indices but the distances between $\gamma_\pm$ and any $\sigma(\mu)$ in (117) are larger than $(6\pi\nu)^{-1}\mu + \mathcal{O}(\mu^2)$.

The second of the following assertions holds true in the case $\mu > 0$ only.

Theorem 28. *1) Let $\beta \in (l + \alpha, l + \alpha + 2) = \Upsilon(\mathbb{S}^1_+)$ and $(f, g) \in \mathcal{R}^{l,\alpha}_\beta \Lambda(\Omega^s_+)$. Then there exist $\mu(\beta) > 0$ such that, if $\mu \in (0, \mu(\beta))$, the problem (112), $(4)_3$ has a unique solution $u \in \mathcal{D}^{l,\alpha}_\beta \Lambda(\Omega^s_+)$ and the estimate (51) is valid with the constant $c = c(\beta, l, \alpha, \Omega^s_+)$ independent of u, f, g and μ.*

2) Let (117) be fulfilled and let $(f, g) \in \mathfrak{R}^{l,s,\alpha}_{\sigma(\mu)}(\Omega^s_+)$ satisfy the condition $(100)_1$. Then there exists $\mu_0 > 0$ such that, if $\mu \in (0, \mu_0)$, the problem (112), $(4)_3$ has the unique solution $u \in \mathcal{D}^{l,s,\alpha}_{\sigma(\mu)}(\Omega^s_+)$. There holds the estimate

$$\| \tilde{u}; \mathcal{D}^{l,\alpha}_{\sigma(\mu)} \Lambda(\Omega^s_+) \| + \| \mathcal{U}; \mathcal{D}^{s,\alpha} C(\mathbb{S}^1_+) \| \leq$$

$$\leq c\mu^{-1} \left(\| (\tilde{f}, \tilde{g}); \mathcal{R}^{l,\alpha}_{\sigma(\mu)} \Lambda(\Omega^s_+) \| + \mu^{-1} \| (\mathcal{F}, \mathcal{G}); \mathcal{R}^{s,\alpha} C(\mathbb{S}^1_+) \| \right) \quad (118)$$

where $c = c(l, s, \alpha, \Omega^s_+)$ does not depend on either $\mu \in (0, \mu_0)$, or the entries $\tilde{u}, \mathcal{U}$ and $(\tilde{f}, \tilde{g})$, $(\mathcal{F}, \mathcal{G})$ of the representations (69) and $(73)_{1,2}$.

Remark 29. 1) In view of (110) the norm of the mapping

$$\widehat{S} - S : \mathcal{D}_\beta^{l,\alpha} \Lambda(\Omega_+^s) \to \Lambda_\beta^{l-1,\alpha}(\Omega_+^s)^2 \times \Lambda_\beta^{l,\alpha}(\Omega_+^s)$$

is infinitesimal as $\mu \to 0$. Hence, the first assertion follows immediately from Theorem 13. The second assertion can be proven in the same way as Theorem 20. For a particular solution $\mathcal{U}^0$ of the problem (114), the estimate (83) must be replaced by the estimate (84) introducing the factor μ^{-1} at $\|(\mathcal{F}, \mathcal{G}) \dots \|$ in (86). The other factor μ^{-1} results from a modification of Theorem 11 applied to the problem (112), $(4)_3$ because the constant c in (47) is inversely proportional to the distance $\mathcal{O}(\mu)$ between $l + 1 + \alpha - \gamma$ and the nearest eigenvalue (see e.g. [14], §1.2). The latter also explains our choice of the interval (117).

2) If $\mu < 0$, then $\widehat{\lambda}(\mu) > -1$ and, by the theorem on asymptotics, the solution $u^1 \in \mathcal{D}_\beta^{l,\alpha} \Lambda(\Omega_+^s)$ of the problem (112), $(4)_3$ gains a supplementary power solution $\widehat{cU}(x)$ of degree $\widehat{\lambda}(\mu)$ which abolishes the necessary representation (69).

Let us consider the 2-D problem $(6)_+$ in the domain $\widehat{\Omega}_+$ which is not symmetric but implies a small perturbation of a symmetric domain. In other words, $\Omega_+^s = \kappa(\widehat{\Omega}_+)$ with the diffeomorphism κ which turns into identity outside the circle $\mathbb{B}_R$. The corresponding change of variables

$$\widehat{\Omega}_+ \ni \widehat{x} \mapsto x = \kappa(\widehat{x}) \in \Omega_+^s$$

converts the Navier-Stokes problem in $\widehat{\Omega}_+$ into the problem

$$\begin{aligned} S(\nabla_x)u + (N(\nabla_x)u', 0) + K(x, \nabla_x)u &= f &&\text{in } \Omega_+^s, \\ u' &= g &&\text{on } \partial\Omega_+^s \end{aligned} \tag{119}$$

where K is compactly supported nonlinear differential operator. Proximity of the boundaries $\partial\widehat{\Omega}_+$ and $\partial\Omega_+^s$ will be expressed by a smallness assumption on the constant δ_K in the inequality

$$\|(Ku, 0); \mathcal{R}_\gamma^{l,\alpha} \Lambda(\Omega_+^s)\| \leq \delta_K \|u; \mathcal{D}_\gamma^{l,\alpha} \Lambda(\Omega_+^s)\| \left(1 + \|u; \mathcal{D}_\gamma^{l,\alpha} \Lambda(\Omega_+^s)\|\right)$$

where γ means now a fixed weight index in the interval $(l + \alpha + 2, l + \alpha + 3)$.

Representing the right-hand side in (119) as the sum

$$(f, g) = (f^s, g^s) + (\widehat{f}, \widehat{g}) \in \mathcal{R}_\gamma^{l,\alpha} \Lambda(\Omega_+^s),$$

we assume that the first couple (f^s, g^s) is symmetric and has the small positive flux μ (see $(103)_{2,3}$ and (109)) while the second one can be arbitrary but small,

$$\|(\widehat{f}, \widehat{g}); \mathcal{R}_\gamma^{l,\alpha} \Lambda(\Omega_+^s)\| = \widehat{\delta} .$$

Owing to Theorem 27 (2) and Proposition 19, the contraction principle delivers the following assertion which is formulated for the problem (119) but really serves the Navier-Stokes problem $(6)_+$ in the 2-D asymmetric domain $\widehat{\Omega}_+$.

Theorem 30. *Let the above assumptions holds true. Then there exist positive* $\rho_\mathcal{R}, \rho_\mathcal{D}$ *and* μ_0, δ_0 *such that in the case*

$$\|(f^s, g^s); \mathcal{R}^{l,\alpha}_\gamma \Lambda(\Omega^s_+)\| \le \rho_\mathcal{R},$$

$$\widehat{\delta} \le \mu^3 \delta_0, \quad \delta_K \le \mu \delta_0 \min\{1, \mu^2 \rho_\mathcal{R}^{-1}\}, \quad \mu \in (0, \mu_0),$$

the problem (119) *has a unique solution in the ball*

$$\left\{ u \in \mathfrak{D}^{l,s,\alpha}_{\sigma(\mu)}(\Omega^s_+) : \|u - u^s; \mathfrak{D}^{l,s,\alpha}_{\sigma(\mu)}(\Omega^s_+)\| \le \mu^2 \rho_\mathcal{D} \right\}$$

where $u^s \in \mathfrak{D}^{l,s,\alpha}_\gamma(\Omega^s_+)$ *is the symmetric solution of the problem* (6)$^s_+$ *indicated in Proposition 25.*

Theorem 30 provides an asymmetric solution with the asymptotics of Jeffrey-Hamel type only as a small perturbation of the symmetric solution u^s which sucks the flux $\mu > 0$ *from infinity*. If u^s drives the flux *to infinity*, the detailed asymptotic procedure does not work (see Remark 29 (2)) and, moreover, as shown in [24], a perturbation of a *completely asymmetric* solution may avoid the Jeffrey-Hamel asymptotic forms. Asymptotic structures in the case $\mu < 0$ are not known yet and, therefore, the 2-D asymmetric aperture problem is still open, too. As for problem (6)$_a$ in the symmetric domain (3), a direct modification of Proposition 25 ensures existence of a small symmetric solution (see [16,4,18] and Remark 14 (2)).

Asymptotic structures of solutions have not been developed also for a 2-D domain Ω_θ possessing an angular outlet to infinity $\mathbb{K} = \{x : r > 0, \varphi \in (0, \theta)\}$ with the opening $\theta \in (\pi, 2\pi]$. For $\theta \in (0, \pi)$, an analog of Theorem 23 about the linear problem (5)$_\theta$ loses obstructive compatibility condition (100)$_2$ that makes the problem (6)$_\theta$ with $\theta \in (0, \pi)$ rather simple (see [4,18]).

7 The Transport Equation

Miscellaneous reduction schemes intended to find solutions of problems related to compressible and non-Newtonian fluids (cf. [28,11,30] and others), include as an element the transport equation

$$z(x) + (w(x) \cdot \nabla_x) z(x) = h(x), \quad x \in \Omega, \tag{120}$$

with a small vector field $w = (w_1, \dots, w_n)$ such that

$$\mathbf{n}(x) \cdot w(x) = 0, \quad x \in \partial\Omega . \tag{121}$$

We assume Ω to have a smooth boundary $\partial\Omega$ and the conical outlet to infinity $\mathbb{K} = \{k : r > 0, \varphi \in \omega\}$. Furthermore, in order to employ the results [1,27] on solvability of (120) in usual Sobolev and Hölder spaces, we suppose that there exists an extension operator acting from $C^{l,\alpha}(\Omega)$ onto $C^{l,\alpha}(\mathbb{R}^n)$ continuously.

Let $\mathfrak{T}_\tau^{l,s,\alpha}(\Omega)$ denote the space composed from functions of the form

$$z(x) = X(x)r^{-q}\mathcal{Z}(\varphi) + \tilde{z}(x),$$
$$\mathcal{Z} \in C^{s,\alpha}(\omega), \ \tilde{z} \in \Lambda_\tau^{l,\alpha}(\Omega), \tag{122}$$
$$q \in \mathbb{R}_+, \ l,s \in \mathbb{N}_0, \ s \geq l+1, \ \alpha \in (0,1), \ \tau \in (l+\alpha+q, l+\alpha+q+1) \ .$$

We equip the space with the natural norm

$$\|z; \mathfrak{T}_\tau^{l,s,\alpha}(\Omega)\| = \|\tilde{z}; \Lambda_\tau^{l,\alpha}(\Omega)\| + \|\mathcal{Z}; C^{s,\alpha}(\omega)\| \ .$$

Let $h \in \mathfrak{T}_\tau^{l,s,\alpha}(\Omega)$ with the attributes $\mathcal{H}, \tilde{h}$ and

$$w \in \Lambda_\beta^{l+1,\alpha}(\Omega)^n, \ \beta \geq l+1+\alpha \ . \tag{123}$$

Setting $\mathcal{Z} = \mathcal{H}$ reduces the equation (120) to the following one:

$$\tilde{z}(x) + (w(x) \cdot \nabla_x)\tilde{z}(x) \tag{124}$$
$$= \mathfrak{H}(x) := \tilde{h}(x) - (w(x) \cdot \nabla_x)X(x)r^{-q}\mathcal{Z}(\varphi), \quad x \in \Omega \ .$$

As observed in [23], we now may take advantage of the difference between the smoothness indices s and l for angular parts and asymptotic remainders (see $(70)_3$ and $(122)_3$). By using $s \geq l+1$, we deduce that, similarly to the proof of Proposition 19,

$$(w \cdot \nabla_x)Xr^{-q}\mathcal{Z} \in \Lambda_{(\beta-1)+(q+1)}^{l,\alpha}(\Omega) \subset \Lambda_\tau^{l,\alpha}(\Omega)$$

(see $(122)_3$). Hence, we have

$$\|\mathcal{Z}; C^{l,\alpha}(\omega)\| + \|\mathfrak{H}; \Lambda_\tau^{l,\alpha}(\Omega)\| \leq c\|h; \mathfrak{T}_\tau^{l,s,\alpha}(\Omega)\|$$

and it becomes sufficient to solve (124) in the space $\Lambda_\tau^{l,\alpha}(\Omega)$.

Theorem 31. *Let the norm* $\|w; \Lambda_\beta^{l+1,\alpha}(\Omega)\|$ *be sufficiently small, the indices taken from* $(122)_3$ *and* (123). *Then the equation* (120) *with right-hand side* $h \in \mathfrak{T}_\tau^{l,s,\alpha}(\Omega)$ *has a unique solution* $z \in \mathfrak{T}_\tau^{l,s,\alpha}(\Omega)$ *and there holds the estimate*

$$\|z; \mathfrak{T}_\tau^{l,s,\alpha}(\Omega)\| \leq c\|h; \mathfrak{T}_\tau^{l,s,\alpha}(\Omega)\| \ .$$

Proof. It remains to verify the assertion with $\mathfrak{T}_\tau^{l,s,\alpha}(\Omega)$ replaced by $\Lambda_\tau^{l,\alpha}(\Omega)$. We introduce the new unknown $\mathbf{z} = \mathcal{R}z$ and rewrite the equation (120) in the form

$$\mathbf{z}(x) + (\mathbf{w}(x) \cdot \nabla_x)\mathbf{z}(x) = \mathbf{h}(x), \ x \in \Omega, \tag{125}$$

where $\mathcal{R}(x) = (1 + |x|^2)^{(\tau-l-\alpha)/2}$,

$$\mathbf{h} = \rho^{-1}\mathcal{R}h \in \Lambda_{\tau-(\tau-l-\alpha)}^{l,\alpha}(\Omega) = \Lambda_{l+\alpha}^{l,\alpha}(\Omega) \subset C^{l,\alpha}(\Omega),$$
$$\mathbf{w} = \rho^{-1}w \in \Lambda_\beta^{l+1,\alpha}(\Omega)^n \subset C^{l+1,\alpha}(\Omega)^n,$$
$$\rho(x) = 1 - \mathcal{R}(x)^{-1}(w(x) \cdot \nabla_x)\mathcal{R}(x) \geq c_0 > 0 \ .$$

Referring to [27], we get the unique solution $\mathbf{z} \in C^{l,\alpha}(\Omega)$ of the equation (125) and the inequality

$$\|\mathbf{z}; C^{l,\alpha}(\Omega)\| \leq c \|\mathbf{h}; C^{l,\alpha}(\Omega)\| \leq C \|h; \Lambda_\tau^{l,\alpha}(\Omega)\| . \qquad (126)$$

This yields the correct estimate of the function z itself,

$$|\mathbf{z}(x)| = (1 + r^2)^{(\tau-l-\alpha)/2}|z(x)| \leq c\|h; \Lambda_\tau^{l,\alpha}(\Omega)\|,$$

however, estimates of its derivatives, provided by (126), do contain weaker weights than we need and it is necessary to verify the stronger estimate

$$\|\mathbf{z}; \Lambda_{l+\alpha}^{l,\alpha}(\Omega)\| \leq c\|\mathbf{h}; \Lambda_{l+\alpha}^{l,\alpha}(\Omega)\| . \qquad (127)$$

To this end, we observe that, multiplying by proper cut-off functions, the equation (125) is restricted to the set $\mathbb{K}_R = \{x \in \mathbb{K} : r > R\} = \Omega \setminus \mathbb{B}_R$,

$$\chi_z \mathbf{z} + (\chi_w \mathbf{w} \cdot \nabla_x)\chi_z \mathbf{z} = \chi_z \mathbf{h} + \mathbf{z} \mathbf{w} \cdot \nabla_x \chi_z \quad \text{in} \quad \mathbb{K}_R . \qquad (128)$$

We stress that here $\chi_w \chi_z = \chi_w$ and $\chi_w(R) = 0$, $\chi_w(r) = 1$ as $r > 2R$. Applying now the Euler change of variables (32), we convert (128) into the following equation on the semicylinder $\mathbb{Q}_R = \{(t, \varphi) : t > e^R, \varphi \in w\}$:

$$\mathbf{z}_\chi(t, \varphi) + e^{-t}(\mathbf{w}_\chi(t, \varphi) \cdot (\partial_t, \nabla_\varphi))\mathbf{z}_\chi(t, \varphi) = \mathbf{H}_\chi(t, \varphi), \ (t, \varphi) \in \mathbb{Q}_R, \quad (129)$$

where $\mathbf{z}_\chi(t, \varphi) = \chi_z(e^t)\mathbf{z}(e^t \varphi)$ and so on. Bearing in mind the equation (129), we repeat word by word the proofs in [27] with the only change $\nabla_x \mapsto (\partial_t, \nabla_\varphi)$. Finally, we derive that

$$\|\mathbf{z}_\chi; C^{l,\alpha}(\mathbb{Q}_R)\| \leq c\|\mathbf{H}_\chi; C^{l,\alpha}(\mathbb{Q}_R)\| . \qquad (130)$$

The calculations (33), (34) turn (130) into the estimate

$$\|\chi_z \mathbf{z}; \Lambda_{l+\alpha}^{l,\alpha}(\mathbb{K}_R)\| \leq c\|\chi_w \mathbf{H}; \Lambda_{l+\alpha}^{l,\alpha}(\mathbb{K}_R)\|$$
$$\leq C \left(\|\chi_z \mathbf{h}; \Lambda_{l+\alpha}^{l,\alpha}(\mathbb{K}_R)\| + \|\mathbf{z}; C^{l,\alpha}(\Omega \cap \mathbb{B}_{2R})\| \right)$$

which, together with (126), leads us to the inequality (127). Putting $z = \mathcal{R}^{-1}\mathbf{z}$, we arrive at a solution of the equation (120) and the estimate

$$\|z; \Lambda_\tau^{l,\alpha}(\Omega)\| \leq c\|h; \Lambda_\tau^{l,\alpha}(\Omega)\| .$$

Since uniqueness of z is equivalent with uniqueness of $\mathbf{z}$, this completes the proof. $\qquad \square$

References

1. H.Beirao da Veiga: On a stationary transport equation, Ann. Univ. Ferrara, Nuova Ser., Sez. VII, 32 (1986), 79–91.

2. W.Borchers, T.Miyakawa: On stability of exterior stationary Navier-Stokes flows, Acta Mathematica, 174 (1995), 311–382.

3. W.Borchers, K.Pileckas: Existence, uniqueness and asymptotics of steady jets, Arch. Rat. Mech. Analysis, 120 (1992), 1–49.

4. G.P.Galdi, M.Padula, V.A.Solonnikov: Existence, uniqueness and asymptotic behavior of solutions of steady-state Navier-Stokes equations in a plane aperture domain, Indiana Univ. Math. J., 45: 4 (1996), 961–995.

5. I.V.Kamotskii, S.A.Nazarov: Spectral problems in singularly perturbed domains and self-adjoint extensions of differential operators, Trudy Sankt-Peterburg. Mat. Obshch., 6 (1998), 151–212 (Russian).

6. V.A.Kondrat'ev: Boundary value problems for elliptic equations in domains with conical or angular points, Trudy Moskov. Mat. Obshch., 16 (1967), 209–292; English transl. in Trans. Moscow Math. Soc., 16 (1967), 227–313.

7. O.A.Ladyzhenskaya: The mathematical theory of viscous incompressible flow, Gordon and Breach, New-York, London, Paris, 1969.

8. V.G.Maz'ya, B.A.Plamenevskii: On the coefficients in the asymptotics of solutions of elliptic boundary value problems in domains with conical points, Math. Nachr., 76 (1977), 29–60; English transl. in Amer. Math. Soc. Transl., 123 (2) (1984), 57–88.

9. V.G.Maz'ya, B.A.Plamenevskii: Estimates in L_p and Hölder classes and Miranda-Agmon maximum principle for solutions of elliptic boundary value problems in domains with singular points on the boundary, Math. Nachr., 81 (1978), 25–82; English transl. in Amer. Math. Soc. Transl., 123 (2) (1984), 1–56.

10. V.G.Maz'ya, B.A.Plamenevskii, L.I.Stupyalis: The three dimensional problem of steady-state motion of a fluid with a free surface, Differentsial'nye Uravneniya i Primenen. Trudy Sem. Protsessy Upravleniya. I Sektsiya, 23 (1979), 1–155; English transl. in Amer. Math. Soc. Transl. 123 (2) (1984), 171–268.

11. I.Mogilevskii, V.A.Solonnikov: Problem on stationary flow of second grade fluid in Hölder spaces of functions, Zapiski nauchn. seminar POMI, 243 (1997), 154-165 (Russian).

12. S.A.Nazarov: Self-adjoint extensions of the operator of the Dirichlet problem in weighted function spaces, Mat. sbornik, 137 (1988), 224–241; English transl. in Math. USSR Sbornik. 65 (1990), 229–247.

13. S.A.Nazarov, B.A.Plamenevskii: Selfadjoint elliptic problems with radiation conditions on the edges of the boundary, Algebra i Analiz, 4:3 (1992), 196–225; English transl. in St.Petersburg Math. J. 4 (1993), 569–594.

14. S.A.Nazarov, B.A.Plamenevskii: Elliptic Problems in domains with Piecewise Smooth Boundaries, Walter de Gruyter, Berlin, 1994.

15. S.A.Nazarov: Asymptotic conditions at a point, selfadjoint extensions of operators and the method of matched asymptotic expansions, Trudy Sankt-Peterburg. Mat. Obshch., 5 (1996), 112–183; English transl. in Amer. Math. Soc. Transl., 193 (2) (1999), 77–126.

16. S.A.Nazarov: On the two-dimensional aperture problem for Navier-Stokes equations, C.R. Acad. Sci. Paris. Sér. 1, 323 (1996), 699–703.

17. S.A.Nazarov: The operator of a boundary value problem with Chaplygin-Zhukovskii-Kutta type conditions on an edge of the boundary has the Fredhom property, Funkt. Analiz i Ego Prilozheniya, 31:3 (1997), 44–56; English transl. in Functional Analysis and Its Applications, 31:3 (1997), 183–192.

18. S.A.Nazarov: The Navier-Stokes problem in a two-dimensional domain with angular outlets to infinity, Zapiski nauchn. seminar. POMI, 257 (1999), 207–227 (Russian).

19. S.A.Nazarov: The polynomial property of self-adjoint elliptic problems and an algebraic description of their attributes, Uspekhi Mat. Nauk, 54:5 (1999) 77–142 (Russian).

20. S.A.Nazarov, K.Pileckas: Asymptotics of solutions to the Navier-Stokes equations in the exterior of a bounded body, Doklady RAN, 367:4 (1999), 461–463; English tranl. in Doklady Math, 60:1 (1999), 133-135.

21. S.A.Nazarov, K.Pileckas: On steady Stokes and Navier-Stokes problems with zero velocity at infinity in a three-dimensional exterior domain, Kyoto University Math. J. (to appear).

22. S.A.Nazarov, O.R.Polyakova: Rupture criteria, asymptotic conditions at crack tips, and selfadjoint extensions of the Lamé operator, Trudy Moskov. Mat. Obshch., 57 (1996), 16–75; English transl. in Trans. Moscow Math. Soc., 57 (1996), 13–66.

23. S.A.Nazarov, A.Sequeira, J.H.Videman: Asymptotic behavior at infinity of three dimensional steady viscoelastic flows, Pacific Journal (submitted).

24. S.A.Nazarov, A.Sequeira, J.H.Videman: Steady flows of Jeffrey-Hamel type from the half-plane into an infinite channel. 1. Linearization on an asymmetric solution (in preparation).

25. S.A.Nazarov, A.Sequeira, J.H.Videman: Steady flows of Jeffrey-Hamel type from the half-plane into an infinite channel. 2. Linearization on a symmetric solution (in preparation).

26. S.A.Nazarov, M.Specovius-Neugebauer, G.Thäter: Quiet flows for Stokes and Navier-Stokes problems in domains with cylindrical outlets to infinity, Kyshu J. Math, 53:2 (1999), 369–394.

27. A.Novotny: About steady transport equation. II — Shauder estimates in domains with smooth boundaries, Portugaliae Mathematica, 54:3 (1997), 317–333.

28. A.Novotny, M.Padula: L_p-approach to steady flows of viscous compressible fluids in exterior domains, Arch. Rat. Mech. Analysis, 126 (1994), 243–297.

29. A.Novotny, M.Padula: Note on decay of solutions of steady Navier-Stokes equations in 3-D exterior domains, Differential Integral Equations, 8 (1995), 1833–1844.

30. K.Pileckas, A.Sequeira, J.H.Videman: A note on steady flows of non-Newtonian fluids in channels and pipes, in: L.Magalhães, L.Sanchez, C.Rocha (eds), EQUADIFF-95, World Scientific (1998), 458–467.

31. B.-W.Schulze: Boundary Value Problems and Singular Pseudo-differential Operators, John Wiley & Sons, Chichester, New-York, 1999.

32. G.Thäter: Quiet flows for the steady Navier-Stokes problem in domains with quasicylindrical outlets to infinity, in: H.Amann, G.P.Galdi et all. (eds.) Navier-Stokes Equations and Related Nonlinear problems, TEV/VSP. Vilnius/Utrecht (1998), 412–438.

On the Mathematical Theory of Fluid Dynamic Limits to Conservation Laws

Athanasios E. Tzavaras*

Department of Mathematics, University of Wisconsin, Madison, WI 53706, USA
e-mail: `tzavaras@math.wisc.edu`

In memory of John A. Nohel

Abstract. These lectures discuss topics in the theory of hyperbolic systems of conservation laws focusing on the mathematical theory of fluid-dynamic limits. First, we discuss the emergence of the compressible Euler equations for an ideal gas in the fluid-dynamic limit of the Boltzmann equation or of the BGK model. Then we survey the current state of the mathematical theory of fluid-dynamic limits for BGK systems and for discrete velocity models of relaxation type. This is done for the case that the limit is a scalar conservation law or a system of two equations.

1   Introduction

These lecture notes deal with the subject of fluid dynamic limits from kinetic equations to conservation laws. The subject is motivated by the formal derivation of the compressible Euler equations for a mono-atomic gas as the zero mean-free-path limit of the Boltzmann equation. While the rigorous justification of the fluid-dynamic limit for the Boltzmann equation is a challenging open problem, it has prompted recent work on the derivation of hyperbolic systems of conservation laws from kinetic models, in simpler situations where the limits are scalar equations or systems of two conservation laws. The objective of the present article is to describe some of these recent results.

We begin in Section 2 with a discussion of the structure of the fluid dynamic limit for the Boltzmann equation. On the one hand the formalism of the fluid limit for Boltzmann circumscribes the framework that the mathematical analysis is challenged to elucidate. On the other hand this formalism introduces some fundamental notions like the collision invariants and associated evolutions of moments, the H-theorem, and the closure of the conservation laws in the fluid dynamic limit. These notions have been instrumental in the development of an analytical theory where such a theory is currently present.

---

* Research partially supported by the TMR project HCL #ERBFMRXCT960033 and the National Science Foundation.

In Section 3 we discuss certain kinetic models that are equipped with one conservation law. In certain cases those models generate L^1-contractions and the Kruzhkov theory can be adapted to develop a rigorous derivation of the fluid limit to the entropy solution of a scalar multi-dimensional conservation law. This theory is currently well understood and applies to a variety of kinetic or discrete kinetic (of relaxation type) models. We also take the opportunity to discuss the zero mean free path limit of a special kinetic model that motivates the so-called kinetic formulation of scalar conservation laws. The kinetic formulation, proposed by Lions, Perthame and Tadmor, provides a notion of solution for hyperbolic equations (and systems) that is equivalent to the notion of entropy solution and connects the theory of conservation laws with the theory of transport equations.

In the final Section 4 we discuss some examples of kinetic models that converge to systems of two conservation laws in one space dimension. We emphasize the fact that some kinetic models are equipped with stronger dissipative structures than the H-theorem, what allows the use of compensated compactness to effect the derivation of the fluid limit.

These lectures were presented at the VIth School on Mathematical Theory in Fluid Mechanics, at Paseky of the Czech Republic. I would like to thank the organizers J. Málek, J. Nečas and M. Rokyta for providing a very pleasant and stimulating atmosphere.

2   The Boltzmann Equation and its Fluid-Dynamical Limit

In the kinetic theory of gases the statistical description of a gas is given by its distribution function f. For a monoatomic gas f is a function of the positions $x \in \mathbb{R}^d$ and the momenta $\xi \in \mathbb{R}^d$ in the phase space. The product $f(t, x, \xi) \, dx \, d\xi$ describes the mean number of molecules in an element of the phase space centered at (x, ξ) of range dx and $d\xi$.

The Boltzmann equation describes the evolution of the distribution function f for a dilute gas and has the form

$$\partial_t f + \xi \cdot \nabla_x f = Q(f, f) \ . \tag{1}$$

The collision operator $Q(f, f)$ describes the gain and loss of the distribution function due to collisions and is given by

$$Q(f, f) = \int_{\mathbb{R}^d} \int_{S^{d-1}} b(\xi - \xi_*, \omega) \times \tag{2}$$

$$\times \left[f(t, x, \xi') f(t, x, \xi_*') - f(t, x, \xi) f(t, x, \xi_*) \right] d\xi_* \, d\omega \ .$$

One arrives at the form of Q as follows: The incoming velocities ξ and ξ_* and outgoing velocities ξ' and ξ_*' entering a collision satisfy microscopic balances of mass, momentum and energy

$$\xi + \xi_* = \xi' + \xi'_*$$
$$|\xi|^2 + |\xi_*|^2 = |\xi'|^2 + |\xi'_*|^2 \ . \tag{3}$$

These microscopic balances can be expressed with regard to the center of mass of the collision, noticing that (3) are equivalent to

$$\xi + \xi_* = \xi' + \xi'_* , \quad |\xi - \xi_*| = |\xi' - \xi'_*| \ .$$

A pair of incoming velocities ξ, ξ_* defines a sphere of center $\frac{\xi + \xi_*}{2}$ and radius $\frac{|\xi - \xi_*|}{2}$. Each outgoing pair of velocities ξ', ξ'_* compatible with (3) lies on this sphere and can be obtained by reflection with respect to a plane passing through the collision center. The planes are parametrized by their normal $\omega \in S^{d-1}$. The outgoing velocities are determined through the formulas

$$\xi' = \xi + \omega \cdot (\xi_* - \xi)\omega$$
$$\xi'_* = \xi_* - \omega \cdot (\xi_* - \xi)\omega \ .$$

The process defines a map $T_\omega : \mathbb{R}^d \times \mathbb{R}^d \to \mathbb{R}^d \times \mathbb{R}^d$ taking the incoming into the outgoing velocity pairs

$$(\xi, \xi_*) \mapsto (\xi', \xi'_*) = T_\omega(\xi, \xi_*) \ .$$

T_ω has the properties: For $\omega \in S^{d-1}$,

    (i) $T_\omega \circ T_\omega = Id$ (microreversibility)
    (ii) $|\det T_\omega| = 1$
    (iii) $T_\omega(\xi_*, \xi) = (\xi'_*, \xi')$ (symmetry).

The term $Q(f, f)$ describes the gains and losses in the distribution function due to collisions. Gains occur from collisions at x of the type $(\xi', \xi'_*) \mapsto (\xi, \xi_*)$, and losses from collisions at x of the type $(\xi, \xi_*) \mapsto (\xi', \xi'_*)$. In the form of (2), it is factored the assumption that only binary collisions are admitted and the hypothesis of detailed balancing, stating that the number of collisions $(\xi, \xi_*) \mapsto (\xi', \xi'_*)$ is equal, in equilibrium, to the number of collisions $(\xi', \xi'_*) \mapsto (\xi, \xi_*)$ (see [21]). The factor b is called Boltzmann collision kernel and models the microscopic physics of the collisional process. For mechanical reasons it has to satisfy $b \geq 0$, b is locally integrable and the symmetries

$$b(\xi - \xi_*, \omega) = b(|\xi - \xi_*|, |(\xi - \xi_*) \cdot \omega|) \ . \tag{5}$$

As an example, for collisions associated to a hard sphere potential $b(g, \omega) = |g \cdot \omega|$.

We outline without proof certain properties of the collision operator that are indicative of the structure of the equations. A function $\varphi = \varphi(\xi)$ is called a collision invariant if, along the collisions described by T_ω, it satisfies

$$\varphi(\xi) + \varphi(\xi_*) = \varphi(\xi') + \varphi(\xi'_*) \ . \tag{6}$$

This is written in the shorthand notation $\varphi + \varphi_* = \varphi' + \varphi'_*$.

Proposition 1. *The collision invariants are given by*

$$\varphi(\xi) = a + b \cdot \xi + c|\xi|^2, \qquad a, c \in \mathbb{R}, \ b \in \mathbb{R}^d .$$

Proposition 2. *The collision operator satisfies:*

(i)

$$\int_{\mathbb{R}^d} Q(f, f)(\xi)\varphi(\xi) \, d\xi$$

$$= \frac{1}{4} \int_{\mathbb{R}^d \times \mathbb{R}^d \times S^{d-1}} (f'f'_* - ff_*)(\varphi + \varphi_* - \varphi' - \varphi'_*) b \, d\omega \, d\xi_* \, d\xi$$

$$= \frac{1}{2} \int_{\mathbb{R}^d \times \mathbb{R}^d \times S^{d-1}} ff_*(\varphi' + \varphi'_* - \varphi - \varphi_*) b \, d\omega \, d\xi_* \, d\xi$$

(ii)

$$\int_{\mathbb{R}^d} Q(f, f)(\xi) \ln f(\xi) \, d\xi \leq 0$$

(iii) *If $b > 0$ then the equilibria f_{eq} of the collision operator, $Q(f_{eq}, f_{eq}) = 0$, are*

$$f_{eq}(\xi) = \frac{\rho}{(2\pi\theta)^{\frac{d}{2}}} \exp\left\{ -\frac{|\xi - u|^2}{2\theta} \right\} \tag{7}$$

for some $\rho, \theta > 0$ and $u \in \mathbb{R}^d$.

Macroscopic Balance Laws – H-Theorem

Let ρ, u and E stand for the macroscopic density, velocity and energy respectively, defined through the moments of f

$$\rho(t, x) = \int f(t, x, \xi) \, d\xi$$

$$\rho u(t, x) = \int \xi f(t, x, \xi) \, d\xi \tag{8}$$

$$E(t, x) = \int \frac{1}{2} |\xi|^2 f(t, x, \xi) \, d\xi .$$

Since 1, ξ and $|\xi|^2$ are collision invariants, ρ, u and E evolve according to the moment equations

$$\frac{\partial \rho}{\partial t} + \operatorname{div}(\rho u) = 0$$

$$\frac{\partial}{\partial t} \rho u + \operatorname{div}(\rho u \otimes u + P) = 0 \tag{9}$$

$$\frac{\partial E}{\partial t} + \operatorname{div}(Eu + P \cdot u + Q) = 0$$

where $(u \otimes u)_{ij} = u_i u_j$ and the pressure (tensor) P, heat flux Q, total energy E and internal energy e are determined through the formulas

$$P_{ij} = \int (\xi_i - u_i)(\xi_j - u_j) f \, d\xi$$

$$Q_i = \frac{1}{2} \int (\xi_i - u_i)|\xi - u|^2 f \, d\xi \qquad (10)$$

$$E = \int \frac{1}{2}|\xi - u|^2 f \, d\xi + \frac{1}{2}\rho|u|^2 =: e + \frac{1}{2}\rho|u|^2 \; .$$

The equations (9) describe the evolution of the moments ρ, u and E and are not a closed system of equations. Because of the close connection with the balance equations of continuum physics they are called macroscopic balance laws.

The Boltzmann equation is equipped with an imbalance law, considered to capture the entropy dissipation in a rarefied gas. It is obtained by multiplying (1) by $(1 + \ln f)$ and using Proposition 2. The resulting identity yields the celebrated Boltzmann H-Theorem:

$$\partial_t \int_{\mathbf{R}^d} f \ln f \, d\xi + \text{div}_x \int_{\mathbf{R}^d} \xi f \ln f \, d\xi = \int_{\mathbf{R}^d} Q(f, f) \ln f \, d\xi =: S$$

$$= -\frac{1}{4} \iiint_{\mathbf{R}^d \times \mathbf{R}^d \times S^{d-1}} (f' f'_* - f f_*)(\ln f' f'_* - \ln f f_*) \, b \, d\omega \, d\xi_* \, d\xi \leq 0. \quad (11)$$

The term S is called entropy dissipation (rate) and is negative. Moreover $S = 0$ iff $f' f'_* = f f_*$ or equivalently if f is a Maxwellian.

Maxwellians

The solutions of $Q(f_{eq}, f_{eq}) = 0$ are called Maxwellians. They are determined by solving $f' f'_* = f f_*$ and thus $\ln f_{eq}$ is a collision invariant

$$\ln f_{eq} = a + b \cdot \xi + c|\xi|^2$$

and f_{eq} is expressed in the form (7). The constants ρ, u and θ entering in (7) are determined from the moments of $\mathcal{M}$ through the relations:

$$\int \mathcal{M} = \rho$$

$$\int (\xi - u)\mathcal{M} = 0 \quad \text{implies} \quad \int \xi \mathcal{M} = \rho u \qquad (12)$$

$$\int \frac{1}{2}|\xi - u|^2 \mathcal{M} = \frac{d}{2}\rho\theta \quad \text{implies} \quad E = \int \frac{1}{2}|\xi|^2 \mathcal{M} = \frac{1}{2}\rho|u|^2 + \frac{d}{2}\rho\theta \; .$$

Maxwellians are completely determined by their moments and are denoted $\mathcal{M} = \mathcal{M}(\rho, u, \theta, \xi)$. The pressure and heat flux can also be computed along

Maxwellians by

$$P_{ij} = \int (\xi_i - u_i)(\xi_j - u_j)\mathcal{M}_{(\rho,u,\theta)} = \rho\theta\delta_{ij}$$

$$Q_i = \frac{1}{2}\int (\xi_i - u_i)|\xi - u|^2 \mathcal{M}_{(\rho,u,\theta)} = 0 \ . \tag{13}$$

Another characterization of Maxwellians is that they arise as minima of the H-functional for the constrained minimization problem:

$$\min \int f \ln f \, d\xi \qquad \text{over all } f \geq 0 \text{ such that}$$

$$\int f = \rho, \quad \int \xi f = \rho u, \quad \int \frac{|\xi|^2}{2} f = E \ . \tag{14}$$

The minimum of this minimization problem is attained at $\mathcal{M}(\rho, u, \theta, \xi)$.

It is instructive to outline a proof of this statement. The minimization problem leads to computing the critical points of the functional

$$J(f) = \int f \ln f \, d\xi + a\left(\rho - \int f\right) + b \cdot \left(m - \int \xi f\right) + c\left(E - \int \frac{|\xi|^2}{2} f\right) \tag{15}$$

where a, b, c are Lagrange multipliers. The critical points satisfy the equation

$$0 = J'(f)v = \int_{\mathbf{R}^d} \left(\ln f - (a-1) - b \cdot \xi - c\frac{|\xi|^2}{2}\right)v$$

for all test functions v, which implies

$$\ln f = (a-1) + b \cdot \xi + c\frac{|\xi|^2}{2}$$

or that f is a Maxwellian $\mathcal{M}(\rho, u, \theta)$ with ρ, u, θ computed from the moments of f.

Another proof is obtained via the inequality: $z \ln z - z \ln y + y - z \geq 0$ for $y, z > 0$. For $f \geq 0$ with moments as in (14) and for $\mathcal{M}$ a Maxwellian with the same moments, we have

$$\int f \ln f \geq \int f \ln \mathcal{M} + \int f - \mathcal{M}$$

$$\geq \int (f - \mathcal{M}) \ln \mathcal{M} + \int \mathcal{M} \ln \mathcal{M} = \int \mathcal{M} \ln \mathcal{M} \ .$$

The Euler Limit

Next, we describe the formal fluid limit of the Boltzmann equation. We present the Euler limit which is the limit as $\varepsilon \to 0$ of the scaled Boltzmann equation

$$\partial_t f^\varepsilon + \xi \cdot \nabla_x f^\varepsilon = \frac{1}{\varepsilon} Q(f^\varepsilon, f^\varepsilon) \ . \tag{16}$$

In this scaling ε stands for the Knudsen number the ratio of "the mean free path" (a measure of the average distance between successive collisions) over a macroscopic length scale.

The moments ρ^ε, u^ε, E^ε, P^ε and Q^ε satisfy the macroscopic balance laws (9). In addition the H-theorem reads

$$\partial_t \int_{\mathbf{R}^d} f^\varepsilon \ln f^\varepsilon \, d\xi + \operatorname{div}_x \int_{\mathbf{R}^d} \xi f^\varepsilon \ln f^\varepsilon \, d\xi$$
$$+ \frac{1}{4\varepsilon} \iiint_{\mathbf{R}^d \times \mathbf{R}^d \times S^{d-1}} (f^{\varepsilon\prime} f^{\varepsilon\prime}_* - f^\varepsilon f^\varepsilon_*) \Big(\ln \frac{f^{\varepsilon\prime} f^{\varepsilon\prime}_*}{f^\varepsilon f^\varepsilon_*} \Big) b \, d\omega \, d\xi_* \, d\xi = 0 \ . \quad (17)$$

The last term is positive and vanishes only along Maxwellians. It is then conceivable that in the fluid limit $\varepsilon \to 0$ the kinetic function approaches a Maxwellian $\mathcal{M}(\rho, u, \theta, \xi)$.

Using the relations (12)–(13), we see that the formal fluid limit becomes

$$\partial_t \rho + \operatorname{div}_x(\rho u) = 0$$
$$\partial_t(\rho u) + \operatorname{div}_x(\rho u \otimes u + \rho \theta \delta_{ij}) = 0 \quad (18)$$
$$\partial_t \Big(\frac{1}{2}\rho|u|^2 + \frac{d}{2}\rho\theta \Big) + \operatorname{div}_x \Big((\frac{1}{2}\rho|u|^2 + \frac{d}{2}\rho\theta)u + \rho\theta u \Big) = 0$$

which are the compressible Euler equations for an ideal monoatomic gas. In the fluid limit the H-theorem (17) gives the macroscopic entropy inequality

$$\partial_t \Big(\rho \ln \frac{\rho}{\theta^{\frac{d}{2}}} \Big) + \operatorname{div}_x \Big(\rho u \ln \frac{\rho}{\theta^{\frac{d}{2}}} \Big) \leq 0 \ . \quad (19)$$

The justification of the fluid-limit for weak solutions is a challenging open problem, due to the presence of shocks and the poor understanding of the theory of weak solutions for the compressible Euler equations. In the forthcoming sections we discuss mathematical results for fluid-limits in simpler situations.

BGK Models

A class of collision models sharing some of the properties of the Boltzmann collision operator are the so-called BGK models (after Bhatganar, Gross and Krook). In the BGK model the collision operator is replaced by relaxation to a Maxwellian

$$\partial_t f + \xi \cdot \nabla_x f = -\frac{1}{\varepsilon} \Big(f - \mathcal{M}_{(\rho,u,\theta)} \Big) \quad (20)$$

where $\mathcal{M}_{(\rho,u,\theta)}(\xi)$ is a local Maxwellian with moments

$$\rho = \int f, \quad \rho u = \int \xi f, \quad \rho|u|^2 + d\rho\theta = \int |\xi|^2 f \ .$$

The BGK-collision operator $Q_{BGK}(f) = \mathcal{M}_{(\rho,u,\theta)} - f$ has the properties

$$\int Q_{BGK}(f) \begin{pmatrix} 1 \\ \xi \\ |\xi|^2 \end{pmatrix} = \int (\mathcal{M}_{(\rho,u,\theta)} - f) \begin{pmatrix} 1 \\ \xi \\ |\xi|^2 \end{pmatrix} = 0$$

$$\int Q_{BGK}(f) \ln f = \int (\mathcal{M}_{(\rho,u,\theta)} - f)(\ln f - \ln \mathcal{M}_{(\rho,u,\theta)}) \leq 0$$

with equality iff $f = \mathcal{M}_{(\rho,u,\theta)}$. As a result, (20) is equipped with the same macroscopic balance laws as the Boltzmann equation and also with an analog of the H-theorem:

$$\partial_t \int_{\mathbf{R}^d} f \ln f + \operatorname{div}_x \int_{\mathbf{R}^d} \xi f \ln f + \frac{1}{\varepsilon} \int_{\mathbf{R}^d} (f - \mathcal{M}_{(\rho,u,\theta)}) \left(\ln \frac{f}{\mathcal{M}_{(\rho,u,\theta)}} \right) = 0 .$$

The fluid limit for this model is again the compressible Euler equations (18)–(19) for an ideal mono-atomic gas.

Bibliographic remarks. The books of Lifshitz and Pitaevskii [21] and Cercignani, Illner and Pulvirenti [6] contain detailed accounts on the derivation of the Boltzmann equation. The surveys of Perthame [32] and Golse [14] contain proofs of several of the listed properties of the Boltzmann equation and detailed discussions of the formalism of the fluid limits.

3   Kinetic Models with One Conservation Law

In this section we consider a kinetic model that is equipped with one conservation law and develop the mathematical theory of its fluid dynamic limit. The presented results follow ideas developed in [30,18,27,4] for a series of kinetic and discrete-kinetic (of relaxation type) models. The equation reads

$$\partial_t f + a(\xi) \cdot \nabla_x f = \frac{1}{\varepsilon} C(f(t,x,\cdot),\xi)$$

$$f(0,x,\xi) = f_0(x,\xi)$$

(21)

where $x \in \mathbf{R}^d$ and $C(f,\xi)$ is a functional on $f(t,x,\cdot)$ (depending on ξ) that encodes the detailed properties of a collision process. The variable ξ may be continuous ($\xi \in \mathbf{R}$) or it may take discrete values; in the latter case (21) becomes a discrete velocity kinetic model. Both cases are treated simultaneously and we retain a common notation. A theory is developed for the fluid limit based on structural assumptions of C without recourse to its detailed properties.

It is assumed that C satisfies: $C(0(\cdot),\xi) = 0$, and

$$\int_{\mathbf{R}} C(f,\xi) \, d\xi = 0$$

(hyp1)

so that (21) is equipped with a macroscopic balance law

$$\partial_t u + \operatorname{div}_x \int a(\xi) f \, d\xi = 0$$

(22)

for the "mass" u defined by

$$u := \int f \, d\xi \; . \tag{23}$$

Second, the equilibria of (21) are parametrized in terms of exactly one scalar parameter w, which may be associated to the mass u. More precisely, the solutions of $C(f, \xi) = 0$ are

$$f_{eq} = \mathcal{M}(w, \xi) \qquad \text{where} \qquad \int_\xi \mathcal{M}(w, \xi) \, d\xi = u = b(w) \tag{hyp2}$$

where b is a strictly increasing function, so that the map $w \to u$ is invertible.

3.1   Kinetic Models that Generate L^1-Contractions

It is assumed that the collision operator satisfies

$$\int_{\mathbf{R}} \left(C(f(\cdot), \xi) - C(\bar{f}(\cdot), \xi) \right) \operatorname{sgn} (f - \bar{f})(\xi) \, d\xi \leq 0 \; , \tag{hyp3}$$

for all $f(\cdot)$, $\bar{f}(\cdot)$. This hypothesis guarantees that the space-homogeneous variant of (hyp3) is a contraction in L_ξ^1.

The property is also preserved in the space-nonhomogeneous case and, as a consequence, the kinetic model (21) is endowed with a class of "kinetic entropies".

Theorem 3. *Under hypotheses* (hyp1)–(hyp3):

(i) *The kinetic model is a contraction in* $L^1(\mathbb{R}_x^d \times \mathbb{R}_\xi)$.
(ii) *For all* $\kappa \in \mathbb{R}$, *we have*

$$\partial_t \int_\xi |f - \mathcal{M}(\kappa, \xi)| + \operatorname{div}_x \int_\xi a(\xi)|f - \mathcal{M}(\kappa, \xi)| \leq 0 \quad \text{in } \mathcal{D}'. \tag{24}$$

(iii) *If for some* a, b *it is* $\mathcal{M}(a, \xi) \leq \mathcal{M}(b, \xi)$ *for all* ξ, *then the domain* $\prod_\xi [\mathcal{M}(a, \xi), \mathcal{M}(b, \xi)]$ *is positively invariant.*

Proof. Let f and $\bar{f}$ be two solutions. By subtracting the corresponding equations, multiplying by $\operatorname{sgn} (f - \bar{f})$, and using (hyp3), we obtain

$$\partial_t \int |f - \bar{f}| \, d\xi + \operatorname{div} \int a(\xi)|f - \bar{f}| = \frac{1}{\varepsilon} \int \left(C(f(\cdot), \xi) - C(\bar{f}(\cdot), \xi) \right) \operatorname{sgn} (f - \bar{f}) \leq 0. \tag{25}$$

This shows that any two solutions f and $\bar{f}$ satisfy the L^1-contraction property:

$$\int_x \int_\xi |f - \bar{f}|(t, x, \xi) \, dx \, d\xi \quad \text{is decreasing in } t.$$

Since $\int_x \int_\xi (f - \bar{f}) \, dx \, d\xi$ is a conserved quantity, we have

$$\int_x \int_\xi (f - \bar{f})^+ (t, x, \xi) \, dx \, d\xi \quad \text{is decreasing in } t,$$

and as a result

$$\text{if } f_0 \leq \bar{f}_0 \text{ then } f \leq \bar{f} .$$

A special class of solutions of (21) are the global Maxwellians $\mathcal{M}(\kappa, \xi)$. These may be used as comparison functions. For instance

$$\text{if } f_0(x, \xi) \leq \mathcal{M}(a, \xi), \text{ for some } a \in \mathbf{R}, \text{ then } \quad f(t, x, \xi) \leq \mathcal{M}(a, \xi) .$$

From this property part (iii) follows. Finally, if $\bar{f} = \mathcal{M}(\kappa, \xi)$ in (25) then

$$\int_\xi (\partial_t + a(\xi) \cdot \nabla_x) |f - \mathcal{M}(\kappa, \xi)| \, d\xi = \frac{1}{\varepsilon} \int C(f(\cdot), \xi) \text{sgn} \, (f - \mathcal{M}(\kappa, \xi)) \leq 0$$

which shows (24).   □

We present next two specific models that satisfy hypotheses (hyp1)–(hyp3).

I. A Discrete Velocity Model. Consider the system

$$\partial_t f_0 + a_0 \cdot \nabla_x f_0 = \frac{1}{\varepsilon} \sum_{i=1}^{d} (f_i - h_i(f_0))$$

$$\partial_t f_i + a_i \cdot \nabla_x f_i = -\frac{1}{\varepsilon} (f_i - h_i(f_0)) \qquad i = 1, \ldots, d. \tag{26}$$

for the evolution of $f = (f_0, f_1, \ldots, f_d)$ where $a_0, a_1, \ldots, a_d \in \mathbf{R}^d$. This discrete velocity model of relaxation type is developed in [18] as a relaxation approximation for the scalar multi-dimensional conservation law.

We assume that

$$h_i(w) \text{ are strictly increasing, } i = 1, \ldots, d, \tag{A}$$

and let $u = f_0 + \sum_k f_k$. The Maxwellian functions are

$$f_{eq} = \mathcal{M}(w, j)_{j=0,1,\ldots,d} = (w, h_1(w), \ldots, h_d(w)),$$
$$\text{where } u = w + \sum_i h_i(w) =: b(w) .$$

Clearly (hyp1) and (hyp2) are satisfied. To see Hypothesis (hyp3), note that

$$I = \sum_{j=0}^{d} \Big(C(f,j) - C(\bar{f},j) \Big) \operatorname{sgn}(f_j - \bar{f}_j)$$

$$= \Big[\sum_{i=1}^{d} (f_i - \bar{f}_i - (h_i(f_0) - h_i(\bar{f}_0))) \Big] \operatorname{sgn}(f_0 - \bar{f}_0)$$

$$- \sum_{i=1}^{d} (f_i - \bar{f}_i - (h_i(f_0) - h_i(\bar{f}_0))) \operatorname{sgn}(f_i - \bar{f}_i)$$

$$= \sum_{i=1}^{d} \Big(f_i - \bar{f}_i - (h_i(f_0) - h_i(\bar{f}_0)) \Big) \big(\operatorname{sgn}(f_0 - \bar{f}_0) - \operatorname{sgn}(f_i - \bar{f}_i) \big) \leq 0$$

where the last inequality follows from (A).

Under (A) the model (26) is also equipped with a globally defined entropy function

$$\partial_t \Big(\frac{1}{2} f_0^2 + \sum_{i=1}^{d} \Psi_i(f_i) \Big) + \operatorname{div} \Big(a_0 \frac{1}{2} f_0^2 + \sum_{i=1}^{n} a_i \Psi_i(f_i) \Big) + \frac{1}{\varepsilon} \sum_{i=1}^{d} \phi_i(f_0, f_i) = 0 \quad (27)$$

where

$$\Psi_i(f_i) = \int_0^{f_i} h_i^{-1}(\tau)\, d\tau ,$$

is positive and strictly convex, while

$$\phi_i(f_0, f_i) = (f_0 - h_i^{-1}(f_i))(h_i(f_0) - f_i)$$

satisfies $\phi_i \geq 0$ and $\phi_i = 0$ if and only if f is a Maxwellian: $f_j = \mathcal{M}(w,j)$ for some w. The identity provides control of the distance of solutions from equilibria: If $\frac{dh_i}{dw} \geq c$ then $\phi_i \geq c(h_i(f_0) - f_i)^2$ and (27) leads to

$$\int_0^\infty \int_{\mathbf{R}^d} \sum_{i=1}^{d} (h_i(f_0) - f_i)^2\, dx\, dt \leq O(\varepsilon) .$$

II. A BGK Model. Consider next the kinetic model of BGK type

$$\partial_t f + a(\xi) \cdot \nabla_x f = -\frac{1}{\varepsilon}(f - \mathcal{M}(u,\xi))$$

$$\text{with } u = \int_\xi f \tag{28}$$

where $x \in \mathbf{R}^d$ and $\xi \in \mathbf{R}$. The model is introduced in [30] for the special choice of Maxwellian function $\mathcal{M}(u,\xi) = \mathbb{1}(u,\xi)$. The general case is developed in [27,4].

It is here assumed that $\mathcal{M}(u, \xi)$ is smooth and satisfies

$$\mathcal{M}(\cdot, \xi) \text{ is strictly increasing}, \qquad u = \int \mathcal{M}(u, \xi) . \qquad \text{(B)}$$

Then (hyp1) and (hyp2) are clearly fulfilled. The monotonicity of $\mathcal{M}$ states $\mathcal{M}(u, \xi) > \mathcal{M}(\bar{u}, \xi)$ iff $u > \bar{u}$, and, hence,

$$\int_\xi |\mathcal{M}(u, \xi) - \mathcal{M}(\bar{u}, \xi)| = \text{sgn}\, (u - \bar{u}) \left(\int_\xi \mathcal{M}(u, \xi) - \mathcal{M}(\bar{u}, \xi) \right)$$

$$= |u - \bar{u}| = \left| \int_\xi f - \bar{f} \right| \leq \int_\xi |f - \bar{f}| .$$

In turn that implies (hyp3):

$$I = \int_\xi \left[C(f(\cdot), \xi) - C(\bar{f}(\cdot), \xi) \right] \text{sgn}\, (f - \bar{f})$$

$$= - \int_\xi |f - \bar{f}| + \int_\xi (\mathcal{M}(u, \xi) - \mathcal{M}(\bar{u}, \xi))\text{sgn}\, (f - \bar{f}) \leq 0 .$$

The model also possesses an analog of the H-theorem. If we multiple the equation (28) by $(\mathcal{M}^{-1}(f, \xi) - u)$, integrate over $\xi \in \mathbb{R}$ and denote by

$$\mu(f, \xi) = \int_0^f \mathcal{M}^{-1}(g, \xi)\, dg$$

then $\mu(\cdot, \xi)$ is convex and we have

$$\partial_t \int \mu(f) + div_x \int a(\xi)\mu(f) - u \left(\partial_t u + div_x \int a(\xi) f \right)$$

$$+ \frac{1}{\varepsilon} \int (\mathcal{M}^{-1}(f, \xi) - u)(f - \mathcal{M}(u, \xi)) = 0 . \qquad (29)$$

The third term vanishes due to the conservation law, the last term is positive due to the monotonicity assumption. If we further assume that $\partial_u \mathcal{M} \geq c$ then the last equation yields the bound

$$\int_0^T \int_x \int_\xi c|f - \mathcal{M}(u, \xi)|^2 \leq \int_0^T \int_x \int_\xi (\mathcal{M}^{-1}(f, \xi) - u)(f - \mathcal{M}(u, \xi)) \leq O(\varepsilon) \qquad (30)$$

stating that the Maxwellians are enforced in the fluid limit $\varepsilon \to 0$.

We next consider a family of solutions f^ε of (21) and study their limiting behavior $\varepsilon \to 0$. Let $u^\varepsilon = \int f^\varepsilon$ and set

$$w^\varepsilon = b^{-1} \left(\int f^\varepsilon \right)$$

From (22) we obtain

$$\partial_t b(w^\varepsilon) + \operatorname{div}_x \int_\xi a(\xi) \mathcal{M}(w^\varepsilon, \xi) = \operatorname{div}_x \left(\int_\xi a(\xi) \big(\mathcal{M}(w^\varepsilon, \xi) - f^\varepsilon \big) \right) .$$

We will see in the next section that the L^1-contraction property and the conservation laws allow to conclude that $\{u^\varepsilon\}$ is precompact in $L^1_{loc,x,t}$ and (along a subsequence) $u^\varepsilon \to u$ and, since b is strictly increasing, $w^\varepsilon \to w = b^{-1}(u)$ a.e.

To conclude a hypothesis is needed dictating that Maxwellian distributions are enforced in the limit $\varepsilon \to 0$:

$$\int_\xi a(\xi)(f^\varepsilon - \mathcal{M}(w^\varepsilon, \xi)) \to 0 \quad \text{in } \mathcal{D}', \quad \text{where} \quad w^\varepsilon = b^{-1}\left(\int_\xi f^\varepsilon \right). \quad \text{(hyp4)}$$

Both (26) and (28) verify such a hypothesis due to (27) and (30) (see Sec 3.2 for the model (28)). Under this framework it is a technical issue to show that the limiting $u = b(w)$ satisfies the scalar conservation law

$$\partial_t b(w) + \operatorname{div}_x \int_\xi a(\xi) \mathcal{M}(w, \xi) = 0 . \tag{31}$$

3.2   The Fluid-Dynamic Limit for a Kinetic BGK-Model

We provide the main technical details for the fluid dynamic limit of the BGK-model

$$\partial_t f^\varepsilon + a(\xi) \cdot \nabla_x f^\varepsilon = -\frac{1}{\varepsilon}(f^\varepsilon - \mathcal{M}(u^\varepsilon, \xi))$$
$$f^\varepsilon(0, x, \xi) = f_0^\varepsilon(x, \xi) \tag{32}$$

where $u^\varepsilon = \int f^\varepsilon$, $x \in \mathbb{R}^d$, $\xi \in \mathbb{R}$. It is assumed that $a(\xi)$ is uniformly bounded and that the Maxwellians are smooth functions that satisfy $\mathcal{M}(0, \xi) = 0$,

$$\mathcal{M}(u, \cdot) \in L^1_\xi, \quad \mathcal{M}(\cdot, \xi) \text{ is strictly increasing}$$
$$u = \int_\mathbb{R} \mathcal{M}(u, \xi)\, d\xi . \tag{a}$$

Then (hyp1) and (hyp2) and (hyp3) are fulfilled. Let $\omega(\tau)$ be a positive, increasing function denoting a modulus of continuity, $\limsup_{\tau \to 0+} \omega(\tau) = 0$.

Theorem 4. *Let $|a(\xi)| \le M$ and assume the initial data satisfy*

$$\mathcal{M}(a, \xi) \le f_0^\varepsilon(x, \xi) \le \mathcal{M}(b, \xi) \qquad \text{for some } a < b$$
$$\int_x \int_\xi |f_0^\varepsilon(x, \xi)|\, dx\, d\xi \le C \tag{33}$$
$$\int_x \int_\xi |f_0^\varepsilon(x + h, \xi) - f_0^\varepsilon(x, \xi)|\, dx\, d\xi \le \omega(|h|) \quad \text{for } h \in \mathbb{R}^d .$$

Then

$$u^\varepsilon = \int_\xi f^\varepsilon \to u \quad \textit{a.e. and in } L^p_{\text{loc}}((0,T) \times \mathbb{R}^d) \textit{ for } 1 \leq p < \infty \ . \qquad (34)$$

The limiting $u \in C([0,T]; L^1(\mathbb{R}^d)) \cap L^\infty((0,T) \times \mathbb{R}^d)$ is an entropy solution:

$$\partial_t |u - k| + \text{div}_x (F(u) - F(k)) \, \text{sgn}\,(u - k) \leq 0 \quad \textit{in } \mathcal{D}', \textit{ for } k \in \mathbb{R} \ , \qquad (35)$$

where $F(u) = \int_\xi a(\xi) \mathcal{M}(u, \xi) \, d\xi$.

Proof. The proof proceeds in three steps. From the L^1 contraction property, the invariance under translations, and the use of Maxwellians as comparison functions we have

$$\mathcal{M}(a, \xi) \leq f^\varepsilon(t, x, \xi) \leq \mathcal{M}(b, \xi) \qquad \text{for } a < b$$

$$\int_x |u^\varepsilon(t, x)| \leq \int_x \int_\xi |f^\varepsilon| \leq \int_x \int_\xi |f_0^\varepsilon| \leq C \qquad (36)$$

and

$$\int_x |u^\varepsilon(t, x + h) - u^\varepsilon(t, x)| \leq \int_x \int_\xi |f^\varepsilon(t, x + h, \xi) - f^\varepsilon(t, x, \xi)|$$

$$\leq \int_x \int_\xi |f_0^\varepsilon(x + h, \xi) - f_0^\varepsilon(x, \xi)| \leq \omega(|h|) \ . \qquad (37)$$

Using the lemma in the appendix and the bound $|a(\xi)| \leq M$ we conclude that for $k > 0$

$$\int_x |u^\varepsilon(t + k, x) - u^\varepsilon(t, x)| \, dx \leq C\omega(k) \ . \qquad (38)$$

From (36), (37) and (38) we obtain that u^ε is precompact in $L^1_{\text{loc}}((0,T) \times \mathbb{R}^d)$ and, along a subsequence, $u^\varepsilon \to u$ a.e. From (38) and Fatou's lemma $u \in C([0,T]; L^1(\mathbb{R}^d))$.

Note that $a \leq u^\varepsilon \leq b$ is uniformly bounded and that (30) implies $|f^\varepsilon - \mathcal{M}(u^\varepsilon, \xi)| \to 0$ a.e (t, x, ξ). Using (36) we conclude

$$\int_\xi |f^\varepsilon - \mathcal{M}(u^\varepsilon, \xi)| \to 0 \quad \text{a.e. } (t, x) \ .$$

Along a further subsequence, $f^\varepsilon \to \mathcal{M}(u, \xi)$ a.e., and passing to the limit in the kinetic entropies (24) we see that

$$\partial_t \int_\xi |\mathcal{M}(u, \xi) - \mathcal{M}(k, \xi)| + \text{div}_x \int_\xi a(\xi) |\mathcal{M}(u, \xi) - \mathcal{M}(k, \xi| \, d\xi \leq 0$$

in $\mathcal{D}'$. The latter inequality is recast in the form (35) by using (a) and the property $\text{sgn}\,(\mathcal{M}(u, \xi) - \mathcal{M}(k, \xi)) = \text{sgn}\,(u - k)$. Since u is an entropy solution, it is unique and the whole family $u^\varepsilon \to u$ in $L^p_{\text{loc}}, 1 \leq p < \infty$. $\qquad \square$

3.3   The Connection with the Kinetic Formulation

Consider the scalar conservation law

$$\begin{cases} \partial_t u + \operatorname{div} F(u) = 0 \\ \qquad\qquad u(x,0) = u_0(x) \end{cases} \tag{39}$$

with data $u_0 \in L^1 \cap L^\infty$. There are two equivalent notions of solution for this problem: The notion of Kruzhkov entropy solution stating that u is an entropy solution of the initial value problem (39) if

$$\operatorname{ess\,lim}_{t \to 0} \int |u(x,t) - u_0(x)| \, dx = 0 \tag{40}$$

and u satisfies the entropy conditions

$$\partial_t \eta(u) + \operatorname{div} q(u) \le 0 \tag{41}$$

in $\mathcal{D}'$ for all entropy pairs η, q with η convex. (Recall that entropy pairs η, q are required to satisfy $q_i'(u) = \eta'(u) F_i'(u)$ with $i = 1, \ldots, d$.)

The second notion is the kinetic formulation of Lions-Perthame-Tadmor [22] and is based on the Maxwellian

$$\mathbf{1}(u, \xi) = \begin{cases} \mathbf{1}_{0 < \xi < u} & \text{if } u > 0 \\ 0 & \text{if } u = 0 \\ -\mathbf{1}_{u < \xi < 0} & \text{if } u < 0 \end{cases} = \frac{1}{2} \big[\operatorname{sgn}(u - \xi) + \operatorname{sgn} \xi \big]$$

It is equivalent to the notion of Kruzhkov entropy solution and states that u is a solution of (39) if it takes the initial data as in (40) and there exist a positive bounded measure $m = m(t, x, \xi)$ on $\mathbf{R}_t^+ \times \mathbf{R}_x^d \times \mathbf{R}_\xi$ so that

$$\partial_t \mathbf{1}(u, \xi) + a(\xi) \cdot \nabla_x \mathbf{1}(u, \xi) = \partial_\xi m . \tag{42}$$

Moreover, the measure m is supported on the shocks.

This notion arises naturally as the $\varepsilon \to 0$ limit of the BGK model

$$\partial_t f^\varepsilon + a(\xi) \cdot \nabla_x f^\varepsilon = -\frac{1}{\varepsilon} \big(f^\varepsilon - \mathbf{1}(u^\varepsilon, \xi) \big) \tag{43}$$

$$f^\varepsilon(0, x, \xi) = \mathbf{1}(u_0(x), \xi) .$$

The variable $\xi \in \mathbb{R}$ and the model (43) is a special case of (21). The special form of the Maxwellian allows to calculate the kinetic equation that the limiting f satisfies. Following [30,22] we show:

Theorem 5. *As $\varepsilon \to 0$,*

$$u^\varepsilon \to u \quad \text{a.e,} \qquad f^\varepsilon \to F = \mathbf{1}(u, \xi) \quad \text{a.e.}$$

and F satisfies (42) for some positive bounded measure m.

Proof. As before the BGK-model (43) defines an L^1-contraction and one shows $u^\varepsilon \to u$ a.e. Comparisons with the Maxwellians $\mathbf{1}(V, \xi)$ and $\mathbf{1}(-V, \xi)$, $V = \sup |u_0(x)|$, give

$$-1 \le f^\varepsilon \le 1, \qquad \text{supp}_\xi f^\varepsilon \subset [-V, V]$$
$$f^\varepsilon \ge 0 \text{ for } \xi > 0, \quad f^\varepsilon \le 0 \text{ for } \xi < 0 .$$

Next introduce m^ε by

$$\partial_\xi m^\varepsilon = \frac{1}{\varepsilon}\left(\mathbf{1}(u^\varepsilon, \xi) - f^\varepsilon\right) = \begin{cases} > 0 & \text{for } \xi < u^\varepsilon \\ < 0 & \text{for } \xi > u^\varepsilon \end{cases}$$

The function

$$m^\varepsilon = \int_{-\infty}^\xi \frac{1}{\varepsilon}\left(\mathbf{1}(u^\varepsilon, \xi) - f^\varepsilon(\xi)\right) d\xi$$

satisfies $m^\varepsilon(-\infty) = 0$, $m^\varepsilon(+\infty) = 0$ by conservation and $m^\varepsilon > 0$ for $\xi \in \mathbb{R}$. We write the BGK-model in the form

$$\partial_t f^\varepsilon + a(\xi) \cdot \nabla_x f^\varepsilon = \partial_\xi m_\varepsilon .$$

We multiply by ξ and integrate over $[0, T] \times \mathbb{R}^d \times \mathbb{R}$. Taking account of the compact support in ξ we obtain

$$\int_0^T \int_x \int_\xi m^\varepsilon = -\int_x \int_\xi \xi f^\varepsilon(t, x, \xi) d\xi \, dx + \int_x \int_\xi \xi f_0(x, \xi) d\xi \, dx$$
$$\le V \int_x \int_\xi |f^\varepsilon| + V \int_x \int_\xi |f_0| \le C .$$

Using the relations

$$u^\varepsilon \to u \text{ a.e.}, \quad f^\varepsilon - \mathbf{1}(u^\varepsilon, \xi) \to 0 \text{ in } \mathcal{D}', \quad \mathbf{1}(u^\varepsilon, \xi) \to \mathbf{1}(u, \xi) \text{ a.e.}$$

and the property (along subsequences)

$$m^\varepsilon \rightharpoonup m \quad \text{weak-}\star \text{ in measures}$$

we pass to the limit $\varepsilon \to 0$ in $\mathcal{D}'$ to obtain (42).   □

Bibliographic remarks. The development of the model (43) is due to Perthame-Tadmor [30], the model (26) is produced in Katsoulakis-Tzavaras [18] as a discrete velocity approximation of the scalar multi-d conservation law. The discrete kinetic version of the BGK-model (28) is proposed in Natalini [26,27] while the continuous variant is developed in Bouchut [4]; see also Bouchut-Guarguaglini-Natalini [5] for a discussion of the diffusive limit. The convergence in the hyperbolic limit is based on the Kruzhkov theory [20]. Early kinetic theory motivated schemes for scalar conservation laws appear in [13,2].

Related issues appear in relaxation limits to scalar one-dimensional conservation laws [26,40]. Applications of fluid-limits can be found in derivations of hydrodynamic limits for stochastic interacting particle systems [31,19].

The kinetic formulation of Lions-Perthame-Tadmor [22] provides a notion of solution for scalar conservation laws equivalent to the Kruzhkov entropy solution [20]. It provides a description of the regularizing effect [22] through the use of the averaging lemma, proofs of uniqueness and error estimates [33], and a perspective to issues of propagation and cancellation of oscillations [34]. It is a rapidly developing subject both in the context of scalar equations, *e.g.* [3,29], but also for systems of two conservation laws, *e.g.* [23,24,16,34].

4   Fluid Limits to Systems of Two Conservation Laws

In this section we discuss certain discrete kinetic models whose fluid limits are systems of two conservation laws in one space dimension. These results are motivated from the theory of relaxation approximations for hyperbolic systems. We begin with a system for the evolution of the kinetic vector variable $f^\varepsilon = (f_1^\varepsilon, f_2^\varepsilon, f_3^\varepsilon)$

$$
\begin{aligned}
\partial_t f_1 - c\partial_x f_1 &= -\frac{1}{\varepsilon}\big(f_1 - \mathcal{M}_1(u,v)\big) \\
\partial_t f_2 + c\partial_x f_2 &= -\frac{1}{\varepsilon}\big(f_2 - \mathcal{M}_2(u,v)\big) \qquad (44) \\
\partial_t f_3 &= -\frac{1}{\varepsilon}\big(f_3 - \mathcal{M}_3(u,v)\big)
\end{aligned}
$$

where

$$
u = f_1 + f_2 + f_3, \quad v = cf_1 - cf_2, \quad \sigma = c^2 f_1 + c^2 f_2
$$

are the first three moments. The system is a discrete kinetic model of BGK type. Under the hypotheses

$$
\begin{aligned}
\mathcal{M}_1(u,v) + \mathcal{M}_2(u,v) + \mathcal{M}_3(u,v) &= u \\
c\mathcal{M}_1(u,v) - c\mathcal{M}_2(u,v) &= v
\end{aligned}
\qquad (45)
$$

it is equipped with two conservation laws

$$
u_t - v_x = 0, \quad v_t - \sigma_x = 0 .
$$

The Maxwellians are now selected as

$$
\mathcal{M}_1 = \frac{g(u)}{2c^2} + \frac{v}{2c}, \quad \mathcal{M}_2 = \frac{g(u)}{2c^2} - \frac{v}{2c}, \quad \mathcal{M}_3 = u - \frac{g(u)}{c^2} . \qquad (46)
$$

where $g(u)$ is a strictly increasing function with $g(0) = 0$. This choice is consistent with (45) and $g(u) = \mathcal{M}_1(u,v)c^2 + \mathcal{M}_2(u,v)c^2$. The system governing the evolution of the moments is closed and reads

$$\partial_t u - \partial_x v = 0$$
$$\partial_t v - \partial_x \sigma = 0$$
$$\partial_t (\sigma - c^2 u) = -\frac{1}{\varepsilon}(\sigma - g(u)) \ . \tag{47}$$

The kinetic model (44) can be viewed as the system for the evolution of the Riemann invariants

$$f_1 = \frac{\sigma}{2c^2} + \frac{v}{2c}, \quad f_2 = \frac{\sigma}{2c^2} - \frac{v}{2c}, \quad f_3 = u - \frac{\sigma}{c^2}, \tag{48}$$

associated with the hyperbolic operator in (47).

The formal limit of a solution $(u^\varepsilon, v^\varepsilon, \sigma^\varepsilon)$ of the equation (47) is the system of isothermal elasticity

$$\partial_t u - \partial_x v = 0$$
$$\partial_t v - \partial_x g(u) = 0 \ . \tag{49}$$

Insight on the nature of the approximation process is obtained via an adaptation of the Chapman-Enskog expansion familiar from the theory of the Boltzmann equation. One seeks to identify the effective response of the relaxation process as it approaches the surface of local equilibria. It is postulated that the relaxing variable σ^ε can be described in an asymptotic expansion that involves *only* the local macroscopic values u^ε, v^ε and their derivatives, *i.e.*

$$\sigma^\varepsilon = g(u^\varepsilon) + \varepsilon S(u^\varepsilon, v^\varepsilon, u_x^\varepsilon, v_x^\varepsilon, \dots) + O(\varepsilon^2) \ .$$

To calculate the form of S, we substitute the expansion in (47),

$$\partial_t u^\varepsilon - \partial_x v^\varepsilon = 0$$
$$\partial_t v^\varepsilon - \partial_x g(u^\varepsilon) = \varepsilon S_x + O(\varepsilon^2)$$
$$\partial_t (g(u^\varepsilon) - c^2 u^\varepsilon) + O(\varepsilon) = -S + O(\varepsilon),$$

whence we obtain

$$S = [c^2 - g_u(u^\varepsilon)]v_x^\varepsilon + O(\varepsilon),$$

and conclude that the effective equations describing the process are

$$\partial_t u^\varepsilon - \partial_x v^\varepsilon = 0$$
$$\partial_t v^\varepsilon - \partial_x g(u^\varepsilon) = \varepsilon \partial_x \left([c^2 - g_u(u^\varepsilon)]v_x^\varepsilon\right) + O(\varepsilon^2) \ . \tag{50}$$

This is a stable parabolic system when $g_u < c^2$ is satisfied. The formal expansion suggests that the limit of (47) will be the equations (49) provided $0 < g_u < c^2$, a condition stating that the characteristic speeds $\pm\sqrt{g_u}$ of the hyperbolic system (49) lie between (and not in resonance to) the characteristic speed $\pm c, 0$ of the hyperbolic system (47).

We work under the standing hypotheses

$$g(0) = 0, \quad 0 < \gamma \leq g_u \leq \Gamma < c^2, \tag{h}$$

for some constants γ, Γ. The system (47) may be viewed as a model in viscoelasticity. Motivated from deliberations of consistency of constitutive theories of materials with internal variables with the second law of thermodynamics [12,42], one can check that smooth solutions (u, v, σ) of this viscoelasticity model satisfy the energy dissipation identity

$$\partial_t \left(\frac{1}{2} v^2 + \Psi(u, \sigma - c^2 u) \right) - \partial_x (\sigma v) + \frac{1}{\varepsilon} (u - h^{-1}(\alpha))(\alpha - h(u)) \Big|_{\alpha = \sigma - c^2 u} = 0$$

where

$$\Psi(u, \alpha) = - \int_0^\alpha h^{-1}(\zeta) \, d\zeta + \alpha u + \int_0^u E\xi \, d\xi$$

$\alpha = \sigma - c^2 u$, $h(u) = g(u) - c^2 u$ and h^{-1} is the inverse function of h. We remark that the associated constitutive theory is compatible with the second law of thermodynamics iff $g_u < c^2$, and that the function $\frac{1}{2}v^2 + \Psi$ provides an "entropy" function for the associated relaxation process, which is convex in (v, u, α) iff $g_u > 0$.

It follows that under (h) the viscoelasticity system (47) admits global smooth solutions (for smooth data) which satisfy the ε-independent bounds

$$\int_{\mathbf{R}} (u^2 + v^2 + \sigma^2) \, dx + \frac{1}{\varepsilon C} \int_0^t \int_{\mathbf{R}} (\sigma - g(u))^2 \, dx \, dt \leq C \int_{\mathbf{R}} (u_0^2 + v_0^2 + \sigma_0^2) \, dx \tag{51}$$

for some C independent of ε and t. This estimate indicates that the natural stability estimate of the problem is in L^2. It is also clear that while this estimate is in a sense the analog of the H-theorem in this simplified context, and while it provides a control of the distance from equilibrium as $\varepsilon \to 0$, it is not sufficient to guarantee strong convergence.

However, as it turns out (47) is endowed with a stronger dissipative structure, analogous to the one associated with viscosity approximations of the equations (49). To this end, note that (47) can be put in the form,

$$\partial_t u - \partial_x v = 0$$
$$\partial_t v - \partial_x g(u) = \partial_x (\sigma - g(u)) = \varepsilon (c^2 v_{xx} - v_{tt}), \tag{52}$$

of an approximation of (49) via the wave equation.

Lemma 6. *For initial data satisfying*

$$\int_{\mathbf{R}} v_0^2 + u_0^2 + \sigma_0^2 \, dx + \varepsilon^2 \int_{\mathbf{R}} u_{0x}^2 + v_{0x}^2 + \sigma_{0x}^2 \, dx \leq O(1) \tag{d}$$

and under hypothesis (h) *solutions of* (47) *satisfy the* ε *independent estimates*

$$\varepsilon \int_0^t \int_{\mathbf{R}} u_x^2 + v_x^2 + \sigma_x^2 \, dx \, dt \leq O(1) . \tag{53}$$

Proof. From (52) we obtain the natural energy identity

$$\partial_t\left(\frac{1}{2}v^2 + W(u) + \varepsilon v v_t\right) - \partial_x\left(vg(u)\right) + \varepsilon(c^2 v_x^2 - v_t^2) = \varepsilon\partial_x(c^2 v v_x)$$

where the stored energy function $W(u) = \int_0^u g(\xi)\,d\xi$. The term $c^2 v_x^2 - v_t^2$ is not positive definite. To compensate, we multiply the second equation in (52) by the natural multiplier of the wave equation v_t to obtain

$$\varepsilon^2\partial_t\left(c^2 v_x^2 + v_t^2\right) + \varepsilon(2v_t^2 - 2g_u u_x v_t) = 2\varepsilon^2\partial_x(c^2 v_t v_x)\ .$$

Using the identity $a_x b_t - a_t b_x = \partial_t(a_x b) - \partial_x(a_t b)$, we have

$$g_u u_x^2 = u_x\partial_t(v + \varepsilon v_t) - \varepsilon c^2 u_x v_{xx}$$
$$= \left[u_t\partial_x(v + \varepsilon v_t) + \partial_t\left(u_x(v + \varepsilon v_t)\right) - \partial_x\left(u_t(v + \varepsilon v_t)\right)\right] - \varepsilon\partial_t\left(\frac{1}{2}c^2 u_x^2\right),$$

and, in turn,

$$\varepsilon^2\partial_t\left(\frac{1}{2}c^4 u_x^2 - \frac{1}{2}c^2 v_x^2\right) - \varepsilon\partial_t\left(c^2 u_x(v + \varepsilon v_t)\right) + \varepsilon(c^2 g_u u_x^2 - c^2 v_x^2)$$
$$= -\varepsilon\partial_x\left(c^2 u_t(v + \varepsilon v_t)\right)\ .$$

Adding the identities, we arrive at the strengthened dissipation estimate

$$\partial_t\left(\frac{1}{2}(v + \varepsilon v_t - \varepsilon c^2 u_x)^2 + \frac{1}{2}\varepsilon^2(v_t^2 + c^2 v_x^2) + W(u)\right) - \partial_x(vg(u))$$
$$+ \varepsilon\left[v_t^2 - 2g_u u_x v_t + c^2 g_u u_x^2\right] = \varepsilon^2(c^2 v_t v_x)_x\ .$$

Because of (h) the second term is positive definite:

$$\varepsilon\left[v_t^2 - 2g_u u_x v_t + c^2 g_u u_x^2\right] \geq \varepsilon g_u(c^2 - g_u)u_x^2 \geq 0\ .$$

We conclude

$$\int_{\mathbf{R}} \frac{1}{2}(v + \varepsilon v_t - \varepsilon c^2 u_x)^2 + \frac{1}{2}\varepsilon^2(v_t^2 + c^2 v_x^2) + W(u)\,dx$$
$$+ \varepsilon\int_0^t\int_{\mathbf{R}} g_u(c^2 - g_u)u_x^2\,dx\,dt$$
$$\leq \int_{\mathbf{R}} \frac{1}{2}(v_0 + \varepsilon\sigma_{0x} - \varepsilon c^2 u_{0x})^2 + \frac{1}{2}\varepsilon^2(\sigma_{0x}^2 + c^2 v_{0x}^2) + W(u_0)\,dx$$
$$\leq O(1)$$

and, due to (h) and (d),

$$\varepsilon\int_0^t\int_{\mathbf{R}} g_u(c^2 - g_u)u_x^2\,dx\,dt \leq O(1)$$

and

$$\varepsilon \int_0^t \int_{\mathbb{R}} \sigma_x^2 \, dx \, dt \le O(1)$$

$$\varepsilon \int_0^t \int_{\mathbb{R}} v_x^2 \, dx \, dt \le O(1)$$

which completes the proof of (53).   □

We come next to the convergence Theorem.

Theorem 7. *Let $g \in C^3$ satisfy the subcharacteristic condition* (h) *and*

$$g''(u_0) = 0 \text{ and } g''(u) \ne 0 \text{ for } u \ne u_0 \ ,$$
$$g'', g''' \in L^2 \cap L^\infty \ .$$

Let $(u^\varepsilon, v^\varepsilon, \sigma^\varepsilon)$ be a family of smooth solutions of (47) *on $\mathbb{R} \times [0, T]$ emanating from smooth initial data subject to the bounds* (d). *Then, along a subsequence if necessary,*

$$u^\varepsilon \to u, \quad v^\varepsilon \to v, \quad a.e. \ (x, t) \text{ and in } L^p_{\text{loc}}(\mathbb{R} \times (0, T)), \text{ for } p < 2 \ ,$$

and (u, v) is a weak solution of (49).

Sketch of Proof. Let $\eta(u, v)$, $q(u, v)$ be an entropy pair for the equations of isothermal elasticity. Using (52) we obtain

$$\begin{aligned}
\partial_t \eta(u^\varepsilon, v^\varepsilon) &+ \partial_x q(u^\varepsilon, v^\varepsilon) \\
&= \eta_v \partial_x(\sigma^\varepsilon - g(u^\varepsilon)) \\
&= \partial_x(\eta_v(\sigma^\varepsilon - g(u^\varepsilon))) - (\eta_{vu}\varepsilon^{\frac{1}{2}} u_x^\varepsilon + \eta_{vv}\varepsilon^{\frac{1}{2}} v_x^\varepsilon) \frac{\sigma^\varepsilon - g(u^\varepsilon)}{\varepsilon^{\frac{1}{2}}} \\
&= I_1^\varepsilon + I_2^\varepsilon \ .
\end{aligned}$$

The dissipation measure $\partial_t \eta^\varepsilon + \partial_x q^\varepsilon$ is tested for entropy pairs $\eta - q$ that are uniformly bounded up to second order derivatives. Due to (51) and (53) the term I_1^ε lies in a compact of H^{-1}, the term I_2^ε is uniformly bounded in L^1, and the sum $I_1^\varepsilon + I_2^\varepsilon$ is uniformly bounded in $W^{-1,\infty}$. One concludes from Murat's lemma

$$\partial_t \eta(u^\varepsilon, v^\varepsilon) + \partial_x q(u^\varepsilon, v^\varepsilon) \quad \text{lies in a compact of } H^{-1}_{\text{loc}} \ .$$

Using the L^p theory of compensated compactness developed in [38,36] for the equations of elasticity, one obtains strong convergence $u^\varepsilon \to u$ and $v^\varepsilon \to v$ a.e. (x, t) along a subsequence (see [42] for the details).   □

Next we present a second discrete kinetic model that may be treated with techniques of similar flavor. For κ, $\lambda > 0$ consider

$$\partial_t f_1 - \kappa \partial_x f_1 = -\frac{1}{\varepsilon}(f_1 - \mathcal{M}_1)$$

$$\partial_t f_2 + \kappa \partial_x f_2 = -\frac{1}{\varepsilon}(f_2 - \mathcal{M}_2)$$

$$\partial_t f_3 - \lambda \partial_x f_3 = -\frac{1}{\varepsilon}(f_3 - \mathcal{M}_3) \tag{54}$$

$$\partial_t f_4 + \lambda \partial_x f_4 = -\frac{1}{\varepsilon}(f_4 - \mathcal{M}_4)$$

where $\mathcal{M}_i = \mathcal{M}_i(u, v)$, $i = 1, \ldots, 4$, are Maxwellians depending on the "moments"

$$u = \kappa f_1 - \kappa f_2, \quad v = \lambda f_3 - \lambda f_4 , \tag{55}$$

that are assumed to satisfy

$$u = \kappa \mathcal{M}_1(u, v) - \kappa \mathcal{M}_2(u, v), \quad v = \kappa \mathcal{M}_3(u, v) - \kappa \mathcal{M}_4(u, v) . \tag{56}$$

Then (54) is equipped with two conservation laws for u, v. In fact, if we introduce also the moments

$$\kappa^2(f_1 + f_2) = a, \quad \lambda^2(f_3 + f_4) = b, \tag{57}$$

the moments (u, v, a, b) evolve according to the closed system

$$u_t - a_x = 0$$

$$a_t - \kappa^2 u_x = -\frac{1}{\varepsilon}\left(a - \kappa^2(\mathcal{M}_1 + \mathcal{M}_2)\right)$$

$$v_t - b_x = 0 \tag{58}$$

$$b_t - \lambda^2 v_x = -\frac{1}{\varepsilon}\left(b - \lambda^2(\mathcal{M}_3 + \mathcal{M}_4)\right) .$$

If the Maxwellians are selected

$$\mathcal{M}_1 = \frac{u}{2\kappa} + \frac{v}{2\kappa^2} \qquad \mathcal{M}_3 = \frac{v}{2\lambda} + \frac{g(u)}{2\lambda^2}$$

$$\mathcal{M}_2 = -\frac{u}{2\kappa} + \frac{v}{2\kappa^2} \qquad \mathcal{M}_4 = -\frac{v}{2\lambda} + \frac{g(u)}{2\lambda^2}$$

a choice consistent with (56), the system (58) takes the form

$$u_t - a_x = 0$$

$$v_t - b_x = 0$$

$$a_t - \kappa^2 u_x = -\frac{1}{\varepsilon}(a - v) \tag{59}$$

$$b_t - \lambda^2 v_x = -\frac{1}{\varepsilon}(b - g(u))$$

of a relaxation approximation for the equation of elasticity (49) of the type proposed in Jin-Xin [17]. The convergence of (59) to (49) when $\kappa = \lambda$ is carried out in Serre [37]. The proof utilizes the L^∞ compensated compactness framework of DiPerna [11], but differs in the methodology for controlling the dissipation measure from the one presented above. It is based on the idea of extending entropy pairs of (49) by viewing them as "equilibrium" entropy pairs for the hyperbolic operator in (59), an idea first developed for scalar conservation laws by Chen, Levermore and Liu [7,8]. It is proved in [37] that, remarkably, the system (59) is endowed with invariant regions and its solutions converge to the entropy solutions of (49).

For the model (59) one can also give a convergence proof based on strengthened dissipation estimates (see Gosse-Tzavaras [15]). One proceeds by writing (59) in the form

$$
\begin{aligned}
u_t - v_x &= \varepsilon\left(\kappa^2 u_{xx} - u_{tt}\right) \\
v_t - g(u)_x &= \varepsilon\left(\lambda^2 v_{xx} - v_{tt}\right)
\end{aligned}
\tag{60}
$$

of an approximation of (49) by two wave equations. It turns out that under the subcharacteristic condition $\min\{\kappa^2, \lambda^2\} > g_u$ the system (60) is endowed with a stronger dissipative structure, similar to the one characteristic of viscosity approximations. This is the key observation allowing to carry out the $\varepsilon \to 0$ limit to the equations of elasticity (see [15]).

Such strong dissipation estimates are known in more general contexts of approximations by wave operators [42]. Consider the hyperbolic system

$$
\partial_t u + \partial_x F(u) = 0, \quad x \in \mathbb{R}, \, t > 0
\tag{61}
$$

where $u(x,t)$ takes values in $\mathbb{R}^n$ and assume that it is endowed with a strictly convex entropy $\eta(u)$. Consider the approximation of (61) by a wave operator,

$$
\partial_t u + \partial_x F(u) = \varepsilon(A u_{xx} - u_{tt})
\tag{62}
$$

where A is a positive definite symmetric $n \times n$ matrix. This can be written in the form of the relaxation approximation

$$
\begin{aligned}
\partial_t u - \partial_x v &= 0 \\
\partial_t v - A \partial_x u &= -\frac{1}{\varepsilon}\left(v - F(u)\right) .
\end{aligned}
\tag{63}
$$

Using the notations $\eta_u := \nabla \eta$, η_{uu} for the Hessian of η, and I for the $n \times n$ identity matrix, we prove:

Proposition 8. *Assume that (61) is equipped with a strictly convex entropy* $\eta(u)$ *that satisfies, for some* $\alpha > 0$,

$$
\eta_{uu}(u) \le \alpha I ,
$$

and suppose that the positive definite, symmetric matrix A satisfies

$$\frac{1}{2}\left(A^T \eta_{uu}(u) + \eta_{uu}(u)A\right) - \alpha F'^T(u)F'(u) \geq \nu I \ .$$

Then smooth solutions of (62), that decay fast at infinity satisfy the dissipation estimate

$$\int_R \eta(u + \varepsilon u_t) + \frac{1}{2}\varepsilon^2 \alpha |u_t|^2 + \varepsilon^2 \alpha u_x \cdot A u_x \, dx$$

$$+ \int_0^t \int_R \varepsilon^3 \alpha |u_{tt} - A u_{xx}|^2 + \varepsilon \nu |u_x|^2 \, dx \, d\tau$$

$$\leq \int_R \eta(u_0 + \varepsilon u_t(0)) + c\varepsilon^2 |u_t(0)|^2 + \varepsilon^2 u_{0x} \cdot A u_{0x} \, dx$$

where c is a constant independent of ε.

Proof. Taking the inner product of (62) with u_t, we obtain

$$\partial_t \left(\frac{1}{2}\varepsilon |u_t|^2 + \frac{1}{2}\varepsilon u_x \cdot A u_x\right) + \left(|u_t|^2 + u_t \cdot F'(u)u_x\right) = \partial_x(\varepsilon u_t \cdot A u_x)$$

Next, taking the inner product with η_u, we arrive at

$$\partial_t \left(\eta(u) + \varepsilon \eta_u \cdot u_t\right) + \partial_x q(u) + \varepsilon \left(\eta_{uu} u_x \cdot A u_x - u_t \cdot \eta_{uu} u_t\right) = \varepsilon \partial_x(\eta_u \cdot A u_x) \ . \quad (64)$$

We multiply the first identity by $2\alpha\varepsilon$, add the second identity, and use that

$$\eta(u + \varepsilon u_t) = \eta(u) + \varepsilon \eta_u(u) \cdot u_t + \varepsilon u_t \cdot \left(\int_0^1 \int_0^s \eta_{uu}(u + \varepsilon \tau u_t) \, d\tau \, ds\right)\varepsilon u_t$$

to obtain, after some rearrangements of terms,

$$\partial_t \left(\eta(u + \varepsilon u_t) + \varepsilon^2 u_t \cdot \left[\frac{1}{2}\alpha I - \int_0^1 \int_0^s \eta_{uu}(u + \varepsilon \tau u_t) \, d\tau \, ds\right]u_t\right.$$

$$+ \frac{1}{2}\varepsilon^2 \alpha |u_t|^2 + \varepsilon^2 \alpha u_x \cdot A u_x\right) + \partial_x q(u) + \varepsilon u_t \cdot (\alpha I - \eta_{uu})u_t$$

$$+ \varepsilon \alpha |u_t + F'(u)u_x|^2 + \varepsilon u_x \cdot (\eta_{uu} A - \alpha F'^T F')u_x$$

$$= \partial_x \left(\varepsilon \eta_u \cdot A u_x + 2\varepsilon^2 \alpha u_t \cdot A u_x\right) \ .$$

In view of the hypotheses

$$u_t \cdot \left[\frac{1}{2}\alpha I - \int_0^1 \int_0^s \eta_{uu}(u + \varepsilon \tau u_t) \, d\tau \, ds\right]u_t \geq 0 \ ,$$

and (64) follows.   □

Bibliographic remarks. Dissipation induced by damping appears in a variety of subjects from kinetic theory and continuum physics, prime examples being the theory of viscoelasticity and many models in the kinetic theory of gases. There is a prolific literature in the domain of viscoelastiicity (*c.f.* Renardy, Hrusa and Nohel [35]) and in particular related issues appear in studies of weak solutions for conservation laws with memory [10,28,9]. On the realm of relaxation, the importance of the subcharacteristic condition was recognized from the early studies of Liu [25] and Whitham [43]. A general framework for investigating relaxation to processes containing shocks is proposed in Chen-Levermore-Liu [7,8], and the mechanism is exploited in Jin-Xin [17] to introduce a class of nonoscillatory numerical schemes. We refer to Yong [44] for a discussion of stability conditions for general relaxation systems.

The connection between discrete kinetic models and relaxation approximations is exploited in Aregba-Driollet and Natalini [27,1] in order to develop numerical schemes. The problem of constructing entropies for relaxation and kinetic systems can be systematically addressed by considerations motivated by either continuum physics [12,41] as well as by kinetic theory [4]. The existence of strong dissipative estimates for certain relaxation models is noted in [42]. Convergence results to weak solutions of systems of two conservation laws can be found in Tzavaras [42], Serre [37] and Gosse-Tzavaras [15], and in Slemrod-Tzavaras [39] for self-similar limits of the Broadwell system.

5   Appendix: An Estimate of S.N. Kruzhkov

In this Appendix we consider the conservation law

$$\partial_t u + \operatorname{div} f = \mu \Delta g \tag{65}$$

and we prove that an L^1-modulus of continuity for the functions u, f and g in the variable x induces a modulus of continuity in t of the function u. This idea was introduced in one of the central lemmas of the celebrated work of Kruzhkov [20]. The presented version indicates that there is no loss of modulus of continuity in t.

In what follows we use the notation

$$\omega_w(h) = \sup_{|y|<h} \int_{\mathbf{R}^d} |w(x+y) - w(x)| \, dx \tag{66}$$

for the L^1-modulus of continuity of the function w. Also, for $k, h > 0$ and $t > 0$ we let M_f denote the quantity

$$M_f(k,h) = \int_t^{t+k} \sup_{|y|<h} \int_{\mathbf{R}^d} |f(x+y,\tau) - f(x,\tau)| \, dx \, d\tau$$
$$= \int_t^{t+k} \omega_f(h,\tau) \, d\tau \tag{67}$$

associated to the function f. We prove the following lemma.

Lemma 9. *Let u, g and f^j, $j = 1, \ldots, d$, in $L^1((0, T) \times \mathbb{R}^d)$ satisfy (65) in the sense of distributions. If*

$$\omega_u(\varepsilon) = \sup_{|z| \leq 1} \int_{\mathbb{R}^d} |u(x + \varepsilon z, t) - u(x, t)| \, dx \qquad (68)$$

is the L^1-modulus of continuity of the function u then for $t > 0$, $k > 0$ (with $t + k < T$) and any $\varepsilon > 0$ we have

$$\int_{\mathbb{R}^d} |u(x, t + k) - u(x, t)| \, dx \leq C\left(\omega_u(\varepsilon) + M_f(k, \varepsilon)\frac{1}{\varepsilon} + \mu M_g(k, \varepsilon)\frac{1}{\varepsilon^2}\right) \quad (69)$$

and

$$\int_{\mathbb{R}^d} |u(x, t + k) - u(x, t)| \, dx \leq C \min_{\varepsilon, k > 0} \left(\omega_u(\varepsilon, t) + k\frac{1}{\varepsilon} \sup_{t \leq \tau \leq t+k} \omega_f(\varepsilon, \tau)\right.$$

$$\left. + \mu k \frac{1}{\varepsilon^2} \sup_{t \leq \tau \leq t+k} \omega_g(\varepsilon, \tau)\right) \quad (70)$$

Proof. From the weak formulation of (65)

$$-\int \int u\varphi_t + f \cdot \nabla\varphi + \mu g \Delta\varphi \, dx \, dt = 0, \qquad \varphi \in C_c^\infty((0, T) \times \mathbb{R}^d),$$

we readily obtain using Lebesgue's theorem that

$$\int (u(x, t + k) - u(x, t))\psi(x) \, dx = \int_t^{t+k} \int f(x, \tau) \cdot \nabla\psi(x)$$

$$+ \mu g(x, \tau) \Delta\psi(x) \, dx \, d\tau \qquad (71)$$

for any $0 < k < T - t$ and for $\psi = \psi(x) \in C^2(\mathbb{R}^d)$.

Step 1. We proceed to estimate the right hand side of (71). In what follows we use extensively mollifiers. Let ρ_1 be a positive symmetric kernel supported on $(0, 1)$ with $\int_{\mathbb{R}} \rho_1 = 1$ and consider the positive symmetric kernel on $[0, 1]^d$ defined by $\rho(x) = \rho_1(x_1) \ldots \rho_1(x_d)$. The kernel ρ generates a sequence of mollifiers $\rho_\varepsilon(x) = \frac{1}{\varepsilon^d}\rho(\frac{x}{\varepsilon})$ which satisfy

$$\rho_\varepsilon \in C_c^\infty(\mathbb{R}^d), \quad \text{supp}\rho_\varepsilon \subset B_\varepsilon, \quad \int \rho_\varepsilon \, dx = 1, \quad \rho_\varepsilon \geq 0.$$

Let $\rho_\varepsilon * \psi$ denote the convolution of ρ_ε and ψ.

Consider the splitting of the integral

$$\int_t^{t+k} \int f(x, \tau) \cdot \nabla\psi(x) \, dx \, d\tau$$

$$= \int_t^{t+k} \int f(x, \tau) \cdot \left(\nabla\psi(x) - \nabla(\rho_\varepsilon * \psi)(x)\right) dx \, d\tau$$

$$+ \int_t^{t+k} \int f(x, \tau) \cdot \nabla(\rho_\varepsilon * \psi)(x) \, dx \, d\tau$$

$$= I_1 + I_2 \, .$$

The term I_1 is rewritten as

$$I_1 = \int_t^{t+k} \int f(x,\tau) \cdot \left(\nabla \psi(x) - \nabla \int_z \rho(z) \psi(x - \varepsilon z) \, dz \right) dx \, d\tau$$

$$= \int_t^{t+k} \int_x \int_z \Big(f(x,\tau) - f(x + \varepsilon z, \tau) \Big) \cdot \nabla \psi(x) \rho(z) \, dx \, dz \, d\tau$$

and is estimated by

$$|I_1| \leq \left(\int_t^{t+k} \sup_{|z| \leq 1} \int_x |f(x,\tau) - f(x + \varepsilon z, \tau)| \, dx \, d\tau \right) \sup_x |\nabla \psi(x)| \int_z \rho \, dz$$

$$= M_f(k, \varepsilon) \sup_x |\nabla \psi(x)|$$

$$(72)$$

Using the property $\int_z \nabla_z \rho(z) \, dz = 0$ the term I_2 is rewritten as

$$I_2 = \int_t^{t+k} \int_x f(x,\tau) \cdot \frac{1}{\varepsilon} \int_z \nabla_z \rho(z) \psi(x - \varepsilon z) \, dz \, dx \, d\tau$$

$$= \int_t^{t+k} \int_x f(x,\tau) \cdot \frac{1}{\varepsilon} \left[\int_z \nabla_z \rho(z) \psi(x - \varepsilon z) \, dz - \int_z \nabla_z \rho(z) \psi(x) \, dz \right] dx \, d\tau$$

$$= \int_t^{t+k} \int_x \int_z \Big(f(x + \varepsilon z, \tau) - f(x,\tau) \Big) \cdot \frac{1}{\varepsilon} \nabla_z \rho(z) \psi(x) \, dz \, dx \, d\tau$$

which in turn yields

$$|I_2| \leq \left(\int_t^{t+k} \sup_{|z| \leq 1} \int_x |f(x,\tau) - f(x + \varepsilon z, \tau)| \, dx \, d\tau \right) \frac{1}{\varepsilon} \sup_x |\psi(x)| \left(\int_z |\nabla \rho| \, dz \right)$$

$$\leq C M_f(k, \varepsilon) \frac{1}{\varepsilon} \sup_x |\psi(x)| \ .$$

$$(73)$$

In a similar fashion the last integral in (71) is split in two terms

$$\int_t^{t+k} \int g(x,\tau) \Delta \psi(x) \, dx \, d\tau = \int_t^{t+k} \int g(x,\tau) \Big(\Delta \psi(x) - \Delta(\rho_\varepsilon * \psi)(x) \Big) \, dx \, d\tau$$

$$+ \int_t^{t+k} \int g(x,\tau) \Delta(\rho_\varepsilon * \psi)(x) \, dx \, d\tau$$

$$= J_1 + J_2 \ .$$

The terms J_1 and J_2 are estimated by similar arguments as the terms I_1 and I_2 using the property $\int_z \Delta \rho(z) \, dz = 0$. This leads to the bounds

$$|J_1| \leq \left(\int_t^{t+k} \sup_{|z| \leq 1} \int_x |g(x,\tau) - g(x + \varepsilon z, \tau)| \, dx \, d\tau \right) \sup_x |\Delta \psi(x)| \qquad (74)$$

$$|J_2| \leq C \left(\int_t^{t+k} \sup_{|z| \leq 1} \int_x |g(x,\tau) - g(x + \varepsilon z, \tau)| \, dx \, d\tau \right) \frac{1}{\varepsilon^2} \sup_x |\psi(x)| \qquad (75)$$

Finally, combining (71) with (72), (73), (74) and (75) we have

$$
\left| \int \left(u(x, t + k) - u(x, t) \right) \psi(x) \, dx \right|
$$

$$
\leq M_f(k, \varepsilon) \left(\frac{C}{\varepsilon} \sup_x |\psi(x)| + \sup_x |\nabla \psi(x)| \right) \tag{76}
$$

$$
+ \mu M_g(k, \varepsilon) \left(\frac{C}{\varepsilon^2} \sup_x |\psi(x)| + \sup_x |\Delta \psi(x)| \right) .
$$

Step 2. Set now

$$
w(x) = u(x, t + k) - u(x, t), \quad v(x) = \operatorname{sgn} w(x) = \operatorname{sgn} \left(u(x, t + k) - u(x, t) \right),
$$

and consider the choice $\psi = \rho_\varepsilon * v$. Since v is bounded we have

$$
\sup_x |\psi(x)| \leq 1, \quad \sup_x |\nabla \psi(x)| \leq \frac{C}{\varepsilon}, \quad \sup_x |\Delta \psi(x)| \leq \frac{C}{\varepsilon^2} .
$$

For $\psi = \rho_\varepsilon * v$ the estimate (76) gives

$$
\left| \int w(x) \left(\rho_\varepsilon * \operatorname{sgn} w \right)(x) \, dx \right| \leq \frac{C}{\varepsilon} M_f(k, \varepsilon) + \mu \frac{C}{\varepsilon^2} M_g(k, \varepsilon) . \tag{77}
$$

On the other hand the identity

$$
\int |w(x)| - w(x)(\rho_\varepsilon * \operatorname{sgn} w)(x) \, dx
$$

$$
= \int_x w(x) \operatorname{sgn} w(x) - w(x) \left(\int_z \rho(z) \operatorname{sgn} w(x - \varepsilon z) \, dz \right) dx
$$

$$
= \int_x \int_z \left(w(x - \varepsilon z) - w(x) \right) \operatorname{sgn} w(x - \varepsilon z) \rho(z) \, dz \, dx
$$

yields the estimate

$$
\left| \int |w(x)| - w(x)(\rho_\varepsilon * \operatorname{sgn} w)(x) \, dx \right|
$$

$$
\leq \sup_{|z| \leq 1} \int_x |w(x - \varepsilon z) - w(x)| \, dx \leq 2\omega_u(\varepsilon, t) . \tag{78}
$$

Combining (77) and (78) we obtain the bound

$$
\int_{\mathbf{R}^d} |w(x)| \, dx \leq C \left(\omega_u(\varepsilon) + M_f(k, \varepsilon) \frac{1}{\varepsilon} + \mu M_g(k, \varepsilon) \frac{1}{\varepsilon^2} \right)
$$

which is precisely (69). Then (70) follows from the estimation

$$
M_f(k, \varepsilon; t) \leq k \sup_{t \leq \tau \leq t + k} \omega_f(\varepsilon, \tau) .
$$

$\square$

References

1. D. Aregba-Driollet and R. Natalini: Discrete kinetic schemes for multidimensional conservation laws, (1998), (preprint).
2. Y. Brenier: Résolution d' equations d' évolution quasilinéaires en dimension N d'espace à l'aide déquations linéaires en dimension $N + 1$, J. Differential Equations 50 (1983), 375-390.
3. Y. Brenier, L. Corrias: A kinetic formulation for multi-branch entropy solutions of scalar conservation laws, Ann. Inst. H. Poincaré Anal. Non Linéaire 15 (1998), 1450-1461.
4. F. Bouchut: Construction of BGK models with a family of kinetic entropies for a given system of conservation laws, J. Statist. Phys. 95 (1999), 113-170.
5. F. Bouchut, F.R. Guarguaglini and R. Natalini: Diffusive BGK approximations for nonlinear multidimensional parabolic equations, to appear in Indiana Univ. Math. J.
6. C. Cercignani, R. Illner and M. Pulvirenti: The mathematical theory of dilute gases, Applied Mathematical Sciences, 106. Springer, New-York, 1994.
7. G.-Q. Chen and T.-P. Liu: Zero relaxation and dissipation limits for hyperbolic conservation laws. Comm. Pure Appl. Math. 46 (1993), 755-781.
8. G.-Q. Chen, C.D. Levermore and T.-P. Liu: Hyperbolic conservation laws with stiff relaxation terms and entropy, Comm. Pure Appl. Math. 47 (1994), 787-830.
9. G.-Q. Chen and C.M. Dafermos: Global solutions in L^∞ for a system of conservation laws of viscoelastic materials with memory. J. Partial Differential Equations 10 (1997), 369–383.
10. C.M. Dafermos: Solutions in L^∞ for a conservation law with memory. Analyse mathematique et applications, 117-128, Gauthier-Villars, Paris, 1988.
11. R. DiPerna: Convergence of approximate solutions to conservation laws, Arch. Rational Mech. Analysis 82 (1983), 27-70.
12. C. Faciu and M. Mihailescu-Suliciu: The energy in one-dimensional rate-type semilinear viscoelasticity, Int. J. Solids Structures 23 (1987), 1505-1520.
13. Y. Giga and T. Miyakawa: A kinetic construction of global solutions of first-order quasilinear equations, Duke Math. J 50 (1983), 505-515.
14. F. Golse: From kinetic to macroscopic models, (1998), (preprint).
15. L. Gosse and A. Tzavaras: Convergence of relaxation schemes to the equations of elastodynamics, to appear in Math. Comp.
16. F. James, Y.-J. Peng and B. Perthame: Kinetic formulation for chromatography and some other hyperbolic systems. J. Math. Pures et Appl. 74 (1995), 367-385.
17. S. Jin and Z. Xin: The relaxing schemes for systems of conservation laws in arbitrary space dimensions, Comm. Pure Appl. Math. 48 (1995), 235-277.
18. M. Katsoulakis and A.E. Tzavaras: Contractive relaxation systems and the scalar multidimensional conservation law, Comm. Partial Differential Equations 22 (1997), 195-233.
19. M. Katsoulakis and A.E. Tzavaras: Multiscale analysis for interacting particles: relaxation systems and scalar conservation laws, J. Statist. Phys. 96 (1999), 715-763.
20. S.N. Kruzhkov: First order quasilinear equations with severla independent variables, Math. USSR Sbornik 10 (1970), 217-243.
21. E.M. Lifshitz and L.P. Pitaevskii: Physical Kinetics, Landau and Lifshitz: Course of Theoretical Physics, Vol. 10, Pergamon, 1981.

22. P.L. Lions, B. Perthame and E. Tadmor: A kinetic formulation of scalar multidimensional conservation laws, *J. AMS* **7** (1994), 169-191.
23. P.L. Lions, B. Perthame and E. Tadmor: Kinetic formulation of the isentropic gas dynamics and p-systems, *Comm. Math. Physics* **163** (1994), 415-431.
24. P.L. Lions, B. Perthame and P.E. Souganidis: Existence and stability of entropy solutions for the hyperbolic systems of isentropic gas dynamics in Eulerian and Lagrangian coordinates, *Comm. Pure Appl. Math.* **49** (1996), 599-638.
25. T.-P. Liu: Hyperbolic conservation laws with relaxation, *Comm. Math. Phys.* **108** (1987), 153-175.
26. R. Natalini: Convergence to equilibrium for the relaxation approximations of conservation laws, *Comm. Pure Appl. Math.* **49** (1996), 795-823.
27. R. Natalini: A discrete kinetic approximation of entropy solutions to multidimensional scalar conservation laws. *J. Differential Equations* **148** (1998), 292-317.
28. J.A. Nohel, R.C. Rogers and A.E. Tzavaras: Weak solutions for a nonlinear system in viscoelasticity. *Comm. Partial Differential Equations* **13** (1988), 97-127.
29. A. Nouri, A. Omrane and J.P. Vila: Boundary conditions for scalar conservation laws from a kinetic point of view, to appear in *J. Stat. Phys.*.
30. B. Perthame and E. Tadmor: A kinetic equation with kinetic entropy functions for scalar conservation laws, *Comm. Math. Phys.* **136** (1991), 501-517.
31. B. Perthame and M. Pulvirenti: On some large systems of random particles which approximate scalar conservation laws, *Asymptotic Anal.* **10** (1995), 263-278.
32. B. Perthame: Introduction to the collision models in Boltzmann's theory, (1995) (preprint).
33. B. Perthame: Uniqueness and error estimates in first order quasilinear conservation laws via the kinetic entropy defect measure, *J. Math. Pures et Appl.* (to appear).
34. B. Perthame and A.E. Tzavaras: Kinetic formulation for systems of two conservation laws and elastodynamics, (preprint).
35. M. Renardy, W.J. Hrusa and J.A. Nohel: *Mathematical problems in viscoelasticity.* Pitman Monographs and Surveys in Pure and Applied Mathematics, 35, Longman Sci. and Techn., New York, 1987.
36. D. Serre and J. Shearer: Convergence with physical viscosity for nonlinear elasticity, (1993) (unpublished manuscript).
37. D. Serre: Relaxation semi-linéaire et cinétique des systèmes de lois de conservation, to appear in *Ann. Inst. H. Poincaré, Anal. Non Linèaire.*
38. J. Shearer: Global existence and compactness in L^p for the quasi-linear wave equation. *Comm. Partial Differential Equations* **19** (1994), 1829-1877.
39. M. Slemrod and A.E. Tzavaras: Self-similar fluid dynamic limits for the Broadwell system. *Arch. Rational Mech. Anal.* **122** (1993), 353-392.
40. A. Tveito and R. Winther: On the rate of convergence to equilibrium for a system of conservation laws including a relaxation term. *SIAM J. Math. Anal.* **28** (1997), 136-161.
41. A.E. Tzavaras: Viscosity and relaxation approximations for hyperbolic systems of conservation laws. In *An Introduction to Recent Developments in Theory and Numerics for Conservation Laws*, D. Kroener, M. Ohlberger and C. Rohde, eds, Lecture Notes in Comp. Science and Engin., Vol. 5, Springer, Berlin, 1998, pp. 73-122.

42. A.E. Tzavaras: Materials with internal variables and relaxation to conservation laws, *Arch. Rational Mech. Anal.* **146** (1999), 129-155.
43. G. B. Whitham: *Linear and nonlinear waves*. Pure and Applied Mathematics, Wiley-Interscience, New York, 1974.
44. W.-A. Yong: Singular perturbations of first-order hyperbolic systems with stiff source terms. *J. Diff. Equations* **155** (1999), 89-132.

Printing: Weihert-Druck GmbH, Darmstadt
Binding: Buchbinderei Schäffer, Grünstadt